T0181663

Democratization of Artificial Intelligence for the Future of Humanity

Chandrasekar Vuppalapati
Senior Vice President – Products & Programs
Hanumayamma Innovations
and Technologies, Inc.
Fremont, California, USA

CRC Press
Taylor & Francis Group
Boca Raton   London   New York

CRC Press is an imprint of the
Taylor & Francis Group, an **informa** business

A SCIENCE PUBLISHERS BOOK

First edition published 2021
by CRC Press
6000 Broken Sound Parkway NW, Suite 300, Boca Raton, FL 33487-2742

and by CRC Press
4 Park Square, Milton Park, Abingdon, Oxon OX14 4RN

© 2021 Taylor & Francis Group, LLC

CRC Press is an imprint of Taylor & Francis Group, an Informa business

Reasonable efforts have been made to publish reliable data and information, but the author and publisher cannot assume responsibility for the validity of all materials or the consequences of their use. The authors and publishers have attempted to trace the copyright holders of all material reproduced in this publication and apologize to copyright holders if permission to publish in this form has not been obtained. If any copyright material has not been acknowledged please write and let us know so we may rectify in any future reprint.

Except as permitted under U.S. Copyright Law, no part of this book may be reprinted, reproduced, transmitted, or utilized in any form by any electronic, mechanical, or other means, now known or hereafter invented, including photocopying, microfilming, and recording, or in any information storage or retrieval system, without written permission from the publishers.

For permission to photocopy or use material electronically from this work, access www.copyright.com or contact the Copyright Clearance Center, Inc. (CCC), 222 Rosewood Drive, Danvers, MA 01923, 978-750-8400. For works that are not available on CCC please contact mpkbookspermissions@tandf.co.uk

Trademark notice: Product or corporate names may be trademarks or registered trademarks and are used only for identification and explanation without intent to infringe.

Library of Congress Cataloging-in-Publication Data
Names: Vuppalapati, Chandrasekar, 1972- author.
Title: Democratization of artificial intelligence (AI) for the future of
   humanity / Chandrasekar Vuppalapati.
Description: First edition. | Boca Raton : CRC Press ; Taylor & Francis
Group, 2021. | Includes bibliographical references and index.
Subjects: LCSH: Artificial intelligence--Social aspects. | Artificial
   intelligence--Industrial applications. | Artificial   intelligence--Government policy.
Classification: LCC Q335 .V88 2021 | DDC 006.3--dc23
LC record available at https://lccn.loc.gov/2020036006

ISBN: 978-0-367-52409-8 (hbk)
ISBN: 978-0-367-52412-8 (pbk)
ISBN: 978-1-003-05778-9 (ebk)

Typeset in Times New Roman by
Radiant Productions

Dedication

The author dedicates his works, book, and efforts of this book to his late father Murahari Rao Vuppalapati, who had inspired him to write the book and guided the author's pursuits of life, and to his loving mother Hanumayamma Vuppalapati, who was the inspiration to develop the company and dairy products for the betterment of humanity and future generations.

Preface

Artificial Intelligence (AI) refers to the ability of machines to perform cognitive tasks like thinking, perceiving, learning, problem solving and decision making. Initially conceived as a technology that could mimic human intelligence, AI has evolved in ways that far exceed its original conception. With incredible advances made in data collection, processing and computation power, intelligent systems can now be deployed to take over a variety of tasks, enable connectivity and enhance productivity.

As AI's capabilities have dramatically expanded, so have its utility in a growing number of fields. AI holds tremendous promise to benefit nearly all aspects of society, including the economy, healthcare, security, the law, transportation, even technology itself. Increased use of Artificial Intelligence (AI) can bring major social and economic benefits to the World. AI offers massive gains in efficiency and performance to most or all industry sectors, from drug discovery to logistics. It is no wonder that the Artificial intelligence (AI) stands out as a transformational technology of our digital age—and its practical application throughout the economy is growing apace.

One of the key factors that has contributed to the success of AI is data revolution and availability of large data. To continue developing and applying AI, the world and humanity will need to increase ease of access to data in a wider range of sectors. These sectors are outside traditional business forms and are at large spread throughout the world operated as social businesses, small corporations, individual businesses, and other industrial sectors. One common character these sectors endure, importantly, is lack of Information Communication Technology (ICT) to reap benefits of AI. For these AI-enabled interventions to be effectively applied, several barriers must be overcome. These include the challenges of data, computing, and talent availability, as well as more basic challenges of access, infrastructure, and financial resources that are particularly acute in remote or economically challenged places and communities. To successfully disseminate AI to masses and enable successful democratization, we need to bring these sectors and rural communities to digital revolution (people, technology and data together).

In order to harvest the benefits of AI revolution to humanity, the traditional AI software development paradigms must be fully upgraded to function successfully in environments that have resource constraints; small form factor compute devices with limited crunch power, intermittent or no connectivity and/or powered by non-perpetual source or battery-powered. The mission of our book, *Democratization of Artificial Intelligence (AI) for the Future of Humanity*, is to prepare current and future software engineering teams with the skills and tools to fully utilize AI capabilities in extremely resource-constrained devices. The book introduces essential AI concepts from the perspectives of full-scale software development with emphasis on creating niche Blue Ocean small form factored compute environment products. It also

- Outlines AI software architecture & Cloud architecture with emphasis on Edge Computing
- Provides comprehensive comparison and applicability of AI algorithms in constrained environments: Supervised and unsupervised.
- Emphasizes real-time embedded storages for AI applications, specifically operating in constrained environments.
- Develops AI driver software code with real-time deep learning on small footprint frameworks such as tensor flow, python, C and Android

- Provides exclusive examples of AI field deployments that operate in remote & non-connected environments
- Explains AI solution development from a product management perspective.

The strategy should strive to leverage AI for economic growth, social development and inclusive growth, and finally as an "enabler" for emerging and developing economies and humanity at large. We strongly believe that "technologies and innovations for helping humanity" to achieve such we need to democratize AI to the people who need the most.

Acknowledgements

The author is sincerely thankful and indebted to the emergency staff first responders, medical professionals, doctors, nurses, grocery shop personnel, farmers, teachers, individuals and all the wonderful leaders of the world for actively taking care lovingly and honourably of citizens across the world from Corona virus. As of completion of the book, 20 March, 2020, the Coronavirus disease (COVID-19) is affecting people all over the world and many cities across the world are in lockdown. Without the support of these true heroes, it would be an impossible task to focus on the book and complete the book. I sincerely salute to all the people who are working hard to maintain normalcy in this exceptional and unbelievable hard times.

Second, the author is deeply indebted to the love, support, and encouragement of his wife Anitha Ilapakurti, and the active support of his daughters, Sriya Vuppalapati and Shruti Vuppalapati who helped to complete the manuscript.

The author is deeply thankful to the support of his elder brother, Rajasekar Vuppalapati, and the joyful support of his younger brother, Jaya Shankar Vuppalapati. Additionally, the author is also very thankful to his elder sisters Padmavati Vuppalapati and Sridevi Vuppalapati.

The author sincerely thanks Santosh Kedari and his team in India for conducting field studies in India for Dairy and Sanjeevani Healthcare analytics products and bringing the market feedback for the betterment of services.

The author is deeply indebted to Sharat Kedari, who helped in setting Cloud and ML algorithms. In addition, the author is thankful for the support, hard work and dedication of Vanaja Mamidi who helped to solve embedded systems and build of iDispenser for running AI & ML algorithms on embedded device architectures. Finally, the author is sincerely and deeply thankful to Sandhya Vissapragada for her contributions in Data Science and in creating Machine Learning models on Microsoft Azure and Google Cloud.

The author is sincerely thankful to the support of fellow researchers and engineers including Sudha Rama Chandran for creating Android Mobile Applications and reviewing Chapters 1 through 4 from engineering point of view, Sneha Iyer for reviewing the engineering details of Chapter 3 and Chapter 4, Surbhi Rautji for developing ML validations scripts and Tessie Powers for her review of the first three chapters of the book from business and product management point of view and providing valuable feedback to improve the overall quality of the book.

Contents

SECTION III—Model Development and Deployment

SECTION IV—Democratization & Future of AI

SECTION-I

Introduction to Artificial Intelligence & Frameworks

CHAPTER 1

Introduction

"Imagination is more important than knowledge."

Albert Einstein

This chapter introduces Artificial Intelligence and covers techniques that constitute AI and maps AI to classical Machine Learning (ML) techniques. From computing history point of view, this chapter covers "waves of compute" and presents market adoptions of AI through 2017–2019 Gartner Hype cycles. The chapter concludes with AI for the social good.

Artificial Intelligence (AI) stands out as a transformational technology of our digital age—and its practical application throughout the economy is growing apace [1]. One of the chief reasons why AI applications are getting prominence and industry acceptance is in its software ability to learn, albeit continuously, from real-world use and experience, and its capability to improve its performance. It is no wonder that the applications of AI span from complex high-technology equipment manufacturing to personalized exclusive recommendations.

Scientists around the world are harnessing AI's data-mining ability in the fight against poverty.[1] Poverty, of course, is a multifaceted phenomenon.[2] However, the condition of poverty entails one of these three: a lack of income (joblessness); a lack of preparedness (education); and a dependency on government (welfare). AI can address all of these. AI with its ability to identify and locate jobs to the right candidates and tailor education for the needs to new job markets and enabling employers to hire the right candidates for the specialty fields. Additionally, AI can also improve performance of the government in delivering cost-effective and higher quality public services. It can assist governments in controlling crime, improving law and order, expedite hearing in courts, and improve the criminal justice system and bring overall peace and prosperity to society [2].

---

[1] AI is a game changer in the fight against hunger and poverty - https://www.nbcnews.com/mach/tech/ai-game-changer-fight-against-hunger-poverty-here-s-why-ncna774696

[2] A.I. and Big Data Could Power a New War on Poverty - https://www.nytimes.com/2018/01/01/opinion/ai-and-big-data-could-power-a-new-war-on-poverty.html

Finally, AI plays an important role in addressing some of the human civilization[3,4] threats that we are facing collectively [3][4]: Climate change, global warming and food security. Climate change is one of the greatest challenges that society has ever faced, with increasingly severe consequences for humanity as natural disasters multiply, sea levels rise, and ecosystems falter [5]. While no silver bullet, ML can be an invaluable tool in fighting climate change via a wide array of applications and techniques. Climate Change AI aims to facilitate work at the nexus of climate change and machine learning.[5] To tackle[6] climate change, researchers and data scientists across the industries, businesses, academia, and governmental and non-governmental organizations are developing new and revolutionary AI processes and techniques that identify and decipher changing global climate patterns and apply data-infused business practices to combat the ever-increasing threats from the climate change [5].

> *"AI is probably the most important thing humanity has ever worked on. I think of it as something more profound than electricity or fire."*
>
> *Sundar Pichai[7] (CNN Tech)[8]*

What is AI?

Artificial Intelligence (AI) has been defined and understood in many ways, as it encompasses a vast field within its realm [2]. Perspectives of AI also vary based upon many factors and actors involved. Therefore, no definition of AI can be so comprehensive that it will cover all the dimensions of AI.

The US Association for the Advancement of Artificial Intelligence (AAAI)[9] defines AI as "the scientific understanding of the mechanisms underlying thought and intelligent behavior and their embodiment in machines".

Stuart Russel and Peter Norvig[10] in defining AI have used the following taxonomy with regard to systems that [6]

- Think like humans (e.g., cognitive architectures or neural network);
- Think rationally like humans (e.g., inferences optimization);
- Act like humans (e.g., language processing, knowledge representation, automated reasoning and learning); and
- Act rationally like humans (e.g., intelligent software agents, embodied robots that achieve goals via perception, planning, reasoning, learning, etc.).

In Computer Science, AI, sometimes called machine intelligence, is intelligence demonstrated by machines, in contrast to the natural intelligence displayed by the humans. According to Gartner Market Research, Artificial Intelligence[11] applies advanced analyses and logic-based techniques, including ML, to interpret events, support and automate decisions, and take actions. The typical definition of AI looks something like the composite[12] design of AI, Machine Learning and Deep Learning (source KDnuggets); please refer Figure 1.

AI embodies techniques that enable computers to mimic human behavior. Simply put, AI is the ability of a machine to perform cognitive functions we associate with human mind, such as perceiving, reasoning,

---

[3] How artificial Intelligence can tackle the climate change - https://www.nationalgeographic.com/environment/2019/07/artificial-intelligence-climate-change/#close

[4] Here are 10 ways AI could help fight climate change -https://www.technologyreview.com/s/613838/ai-climate-change-machine-learning/

[5] Climate Change - https://www.climatechange.ai/

[6] Tackling Climate Change with Machine Learning - https://arxiv.org/pdf/1906.05433.pdf

[7] Google CEO: AI is more profound than electricity or fire - https://money.cnn.com/2018/01/24/technology/sundar-pichai-google-ai-artificial-intelligence/index.html

[8] https://blog.datarobot.com/get-ai-or-die-tryin-part-1

[9] AAAI - http://www.aaai.org/

[10] Russell Stuart and Peter Norvig , Artificial Intelligence: A Modern Approach (3rd edition) Essex, England; Pearson, 2009

[11] Gartner Glossary – Artificial Intelligence (AI) - https://www.gartner.com/en/information-technology/glossary/artificial-intelligence

[12] What is AI? https://www.kdnuggets.com/2017/07/rapidminer-ai-machine-learning-deep-learning.html

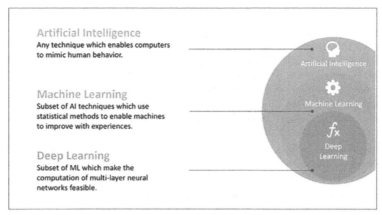

Figure 1: AI Definition

learning, and problem solving.[13] A full end-to-end AI solution—what we sometimes call a system of intelligence or a Smart system—is able to ingest human-level knowledge (e.g., via machine reading and computer vision) and use this information to automate and accelerate tasks that were previously performed only by humans [7].

Examples of technologies that enable AI to solve problems are computer vision, language, virtual agents, and machine learning. Like humans, AI algorithms are very valuable, albeit complex. Unlike "purely algorithmic tasks", as discussed in the common computational complexity literature, the input size is clearly defined; the algorithm gets an input instance, say, a list to be sorted, or an arithmetic operation to be calculated, which has a well-defined size. For machine learning tasks, the notion of an input size is not so clear [8]. An algorithm aims to detect some pattern in a data set and can only access random samples of that data. Another asymptotic behavior that the data scientists and practitioners of AI generally tend to observe is computational, i.e., runtime, complexity: time and space [9]. This is particularly important when designing AI for constrained and Tiny Edge devices subject to scarce or constrained resource capacities.

Machine Learning

Machine Learning (ML) is a subset of AI that uses statistical techniques to enable machines to improve with experience. ML algorithms detect patterns and learn how to make predictions and recommendations by processing data and experiences, rather than by receiving explicit programming instructions [10]. The algorithms also adapt in response to new data and experiences to improve efficacy over time.

Types of Analytics

Machine Learning typically outputs three types of analytics (please see Figure 2):
- Descriptive
- Predictive
- Prescriptive

Descriptive Analytics

Descriptive Analytics[14] helps to answer the question "What happened?" (or What is happening?) in all its forms:[15] What was our sales figure last quarter or last month or yesterday? Which customers required the

---

[13] An Executive's Guide to AI – McKinsey - https://www.mckinsey.com/~/media/McKinsey/Business%20Functions/McKinsey%20Analytics/Our%20Insights/An%20executives%20guide%20to%20AI/An-executives-guide-to-AI.ashx

[14] Descriptive Analytics - https://www.gartner.com/en/information-technology/glossary/descriptive-analytics

[15] Descriptive Analytics 101 - https://www.ibm.com/blogs/business-analytics/descriptive-analytics-101-what-happened/

Figure 2: Analytics Types

most customer service help? Which product had the most defects? What new accounts were opened in the last quarter? Questions like these form the foundation of entire analytics strategy. Descriptive Analytics [11] is characterized by traditional business intelligence (BI) and visualizations such as pie charts, bar charts, dendograms, line graphs, tables, or generated narratives. As an example of Descriptive Analytics (please see Figure 3), consider the following dashboard table that displays the calendar year along with counts of patent applications and grants, by document category (updated 4/2019).[16]

U.S. Patent Statistics Chart
Calendar Years 1963 - 2018

The following table displays the calendar year along with counts of patent applications and grants, by document category (updated 4/2019):

Year of Application or Grant	Utility Patent Applications, U.S. Origin	Utility Patent Applications, Foreign Origin	Utility Patent Applications, Foreign Origin Percent Share	Utility Patent Application, All Origin Total	Design Patent Applications	Plant Patent Applications	Total Patent Applications	Utility Patent Grants, U.S. Origin	Utility Patent Grants, Foreign Origin	Utility Patent Grants, Foreign Origin Percent Share	Utility Patent Grants, All Origin Total	Design Patent Grants	Plant Patent Grants	Reissue Patent Grants	Total Patent Grants	Total Patent Grants, Foreign Origin Percent Share	Year of Application or Grant
2018	285,095	312,646	52.3	597,141	45,083	1,079	643,303	144,413	163,346	53.1	307,759	30,497	1,208	528	339,992	52	2018
2017	293,904	313,052	51.6	606,956	43,340	1,059	651,355	150,952	167,876	52.7	318,828	30,870	1,311	394	351,403	52	2017
2016	295,327	310,244	51.2	605,571	42,571	1,177	649,319	143,725	159,324	52.6	303,049	28,873	1,235	426	333,583	52	2016
2015	288,335	301,075	51.1	589,410	39,097	1,140	629,647	140,969	157,439	52.8	298,408	25,986	1,074	512	325,080	52	2015
2014	285,096	293,706	50.7	578,802	35,378	1,063	615,243	144,621	156,057	51.9	300,677	23,657	1,072	626	326,032	51	2014
2013	287,831	283,781	49.6	571,612	36,034	1,406	609,052	133,593	144,242	51.9	277,835	23,468	847	798	302,948	51	2013
2012	268,782	274,033	50.5	542,815	32,799	1,149	576,763	121,026	133,129	52.2	253,155	21,951	860	822	276,788	52	2012
2011	247,750	255,832	50.8	503,582	30,467	1,139	535,188	108,622	115,883	51.6	224,505	21,356	823	1,029	247,713	51	2011
2010	241,977	248,249	50.6	490,226	29,059	992	520,277	107,791	111,823	50.9	219,614	22,799	981	947	244,341	50	2010
2009	224,912	231,194	50.7	456,106	25,806	959	482,871	82,382	84,967	50.8	167,349	23,116	1,009	453	191,927	50	2009
2008	231,588	224,733	49.2	456,321	27,782	1,209	485,312	77,502	80,270	50.9	157,772	25,565	1,240	647	185,224	50	2008
2007	241,347	214,807	47.1	456,154	27,752	1,049	484,955	79,526	77,756	49.4	157,282	24,062	1,047	508	182,899	49	2007
2006	221,784	204,183	47.9	425,967	25,515	1,151	452,633	89,823	83,949	48.3	173,772	20,965	1,149	519	196,405	48	2006
2005	207,867	182,866	46.8	390,733	25,553	1,222	417,508	74,637	69,169	48.1	143,806	12,951	716	245	157,718	48	2005
2004	189,536	167,407	46.9	356,943	25,975	1,221	392,139	84,270	80,020	48.7	164,290	15,695	1,016	296	181,299	48	2004
2003	188,941	153,500	44.8	342,441	22,602	1,000	356,043	87,893	81,130	48.0	169,023	16,574	994	421	187,012	47	2003
2002	184,245	150,200	44.9	334,445	20,904	1,344	356,493	86,971	80,360	48.0	167,331	15,451	1,133	460	184,375	47	2002
2001	177,511	148,997	45.6	326,508	18,280	944	345,732	87,600	78,435	47.2	166,035	16,871	584	480	183,970	46	2001

Figure 3: US Patent Statistics
DATA SOURCES: Data sources for counts of patent applications:[17]

Predictive Analytics

Predictive Analytics answers the question, "What will happen?" or "What is likely to happen next?"

Predictive Analytics is inherently probabilistic and it is the domain of data scientists. Predictive Analytics [11] is the process of extracting information from the data and predicting future outcomes and trends. Being able to effectively apply predictive analytics creates an opportunity for targeted resource allocation that is

---

[16] U.S. PATENT AND TRADEMARK OFFICE Patent Technology Monitoring Team (PTMT) - https://www.uspto.gov/web/offices/ac/ido/oeip/taf/us_stat.htm

[17] Patents - https://www.uspto.gov/web/offices/ac/ido/oeip/taf/us_stat.htm

not driven simply by experience or a "gut feeling". Time series forecasting,[18] one of the important examples of Predictive Analytics, is one of the most important topics in data science. Almost every business needs to predict the future in order to make better decisions and allocate resources more effectively [12]. Examples of time series forecasting use cases include financial forecasting, product sales forecasting, web traffic forecasting, energy demand forecasting for buildings and data centers, and many more.

Human vs. BOT Web Traffic Prediction ML Use Case

Web Application Firewall (WAF) is designed to protect web applications by filtering and monitoring HTTP traffic between a web application and the internet. During my data science consulting at a Web Application Firewall[19] (WAF) Cybersecurity company, I was charted to develop ML models to protect e-commerce websites by predicting traffic patterns of Human vs. BOT agents. The use case was of a high value and business-critical as it helped to protect e-commerce websites from intruders' automated attacks during the peak web traffic seasons—Christmas holidays, promotional events, and black-Friday sales. The need to develop ML model stems from the fact that the WAF was manual and pre-configured with rule-based architectures with extreme low adaptability and built-in intelligence. Adding ML to the rules processing enables automatic learning of the new patterns from website browsing patterns. The use case for ML was first applied on Human vs. BOT agents web pages navigation pattern.

The application navigation process of Humans and BOT agents are quite different. There is, generally, a signature to human browsing behavior. For example, human actors first go to a product page of an e-commerce site; second, they select a product; third, checks them in the shopping cart, and finally checks out by triggering the payment process. There could be one to many product page navigations but overall, metaphorically, a product selection concludes with a financial credit card transaction. Of course, given the human idiosyncrasies, unpredictable product abandonment, assuming the web searching damping factor[20] is minimal, or over-buying than prevailing averages, and emotional buying is quite common.

On the other hand, BOT navigation is quite different. BOT, as we have seen in the web traffic data, may issue successive transactions in a short span of time; you seldom see idiosyncrasies in buying pattern. Additionally, the time to request (TTR), a signature of BOT, is aggressively wild, in sub-milliseconds. The page navigation sequence is random or a non-human pattern.

WAF Architecture:

WAF operates through a set of rules often called 'policies'. These policies aim to protect against vulnerabilities in the application by filtering out malicious traffic. The value of a WAF comes in part from the speed and ease with which policy modification can be implemented, allowing for faster response to varying attack vectors; during a Distributed Denial of Service (DDoS) attack, rate limiting can be quickly implemented by modifying the WAF policies.

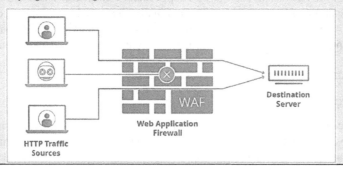

---

[18] Recurrent neural networks for time series forecasting - https://conferences.oreilly.com/artificial-intelligence/ai-eu-2018/public/schedule/detail/70226

[19] What is a WAF? | Web Application Firewall explained - https://www.cloudflare.com/learning/ddos/glossary/web-application-firewall-waf/

[20] The Anatomy of a Large-Scale Hypertextual Web Search Engine - http://infolab.stanford.edu/~backrub/google.html

Data:

The HTTP request and response headers[21] data is required to analyze traffic pattern and TTRs.

The header data includes: Charset, Content-Location, Content-Type, Date, From, Expect, To, MIME-Versions, Server, and User-agent.

Example of data capture:

Request URL: http://www.hanuinnotech.com/
Request Method: GET
Status Code: 304 Not Modified
Remote Address: 50.63.197.141:80
Referrer Policy: no-referrer-when-downgrade
```
GET/HTTP/1.1 Host: www.hanuinnotech.com Connection: keep-alive Cache-
Control: max-age=0 Upgrade-Insecure-Requests: 1 User-Agent: Mozilla/5.0
(iPad; CPU OS 11_0 like Mac OS X) AppleWebKit/604.1.34 (KHTML, like
Gecko) Version/11.0 Mobile/15A5341f Safari/604.1 Accept: text/
html,application/xhtml+xml,application/xml;q=0.9,image/webp,image/
apng,*/*;q=0.8,application/signed-exchange;v=b3;q=0.9 Accept-Encoding: gzip,
deflate Accept-Language: en-US,en;q=0.9 If-None-Match: "6b17adbebcdfd41:0"
If-Modified-Since: Thu, 21 Mar 2019 08:04:44 GMT
```

Response Headers:
HTTP/1.1 200 OK
Last-Modified: Thu, 21 Mar 2019 08:04:44 GMT
Accept-Ranges: bytes
Server: Microsoft-IIS/7.5
X-Powered-By: ASP.NET
Date: Tue, 24 Dec 2019 23:09:44 GMT
Content-Type: text/html
Content-Encoding: gzip
ETag: "6b17adbebcdfd41:0"
Vary: Accept-Encoding
Content-Length: 36679

ML Algorithms:

For predicting human vs. BOT agent behavior, we have taken historical Web navigations for selected users and BOT agents; created a transaction list and applied Association Rule Mining (ARM). Next, we extracted rules from ARM with configurable support and confidence. These rules helped to predict the erratic BOT behavior or human near real-time, Finally, the model refresh rate, i.e., the new model getting deployed with adaptive data, is cut down from quarterly to weekly. The model has become adaptive, importantly, as the new data sets progressively enables more data driven intelligence and pattern insights.

Prescriptive Analytics

Prescriptive Analytics[22] takes inputs from prediction and—combined with rules and constraint-based optimization—enables better decisions about what to do. The decision might be to send an automated task to a human decision maker along with a set of next action recommendations, or to send a precise next action command to another system. Prescriptive Analytics is therefore best suited for situations where the constraints are precise [13].

---

[21] RFC 7231 Hypertext Transfer Protocol - https://tools.ietf.org/html/rfc7231
[22] Prescriptive Analytics - https://www.ibm.com/blogs/business-analytics/prescriptive-analytics-done/

Construction Management and Prescriptive Analytics Use Case

At the time of writing this book, December 24, 2019, the construction industry is embracing AI to address some of the industry use cases such as financial optimization, workplace safety improvement, skilled labor shortage, and equipment optimization.

- **Financial Optimization:** Goal of the use case was to optimize financial returns on the construction job. Given the high rate of underperforming projects—"over 50% of engineering and construction professionals report one or more underperforming projects"[23]—the construction industry is primed to be invaded by AI insights to improve financial optimizations [13]. Following are some of the common financial optimizations:
  o Optimized estimates for accurate job bids
  o Reliably forecast cash flow
  o Ability to predict competitor's bid for public work projects
- **Workplace Safety:** Occupational Safety and Health Administration (OSHA) data reveals that 60% of construction workplace accidents occur within the employee's first year of the job.[24] The purpose of use case was to apply predictive techniques to enable safer workplaces.[25] The metrics include:
  o Be aware of the likelihood of an incident occurring on-site
  o Know when and where injuries are likely to happen
  o Flag/bring attention to violations and prevent incidents

ML and Types of data

Recent advancement in compute (i.e., in-memory compute - Apache Spark,[26] Batch Processing- Apache Hadoop,[27] Columnar compute - Big Query)[28] and storage architectures (i.e., Query Columnar based Storage -Google Big, Multi-modal Document Database - Microsoft Azure Cosmos DB,[29] Massive Scale Object storage - Amazon S3[30] & Microsoft BLOB Storage[31]) enabled ML to work with several data types to identify patterns in the data. The several data types include:

- Structured data
- Unstructured data and
- Semi-structured data

---

[23] KPMG Global Construction Survey 2016 - https://assets.kpmg/content/dam/kpmg/xx/pdf/2016/09/global-construction-survey-2016.pdf
[24] Construction safety, by the numbers - https://naspweb.com/construction-safety-by-the-numbers/
[25] Vancouver Tile and Granite Company Fined more than $250,000 for Workplace Violations - https://www.ehstoday.com/ppe/respirators/article/21918115/vancouver-tile-and-granite-company-fined-more-than-250000-for-workplace-violations
[26] Apache Spark Lightning-fast unified analytics engine - https://spark.apache.org/
[27] Apache Hadoop - https://hadoop.apache.org/
[28] Big Query - https://cloud.google.com/bigquery/
[29] Microsoft Azure Cosmos DB - https://azure.microsoft.com/en-us/services/cosmos-db/
[30] Amazon S3 - https://aws.amazon.com/s3/
[31] Microsoft BLOB Storage - https://azure.microsoft.com/en-us/services/storage/blobs/

Structured data

Structured data is data that adheres to a schema, so all of the data has the same fields or properties. Structured data can be stored in a database table with rows and columns. Structured data relies on keys to indicate how one row in a table relates to data in another row of another table. Structured data is also referred to as relational data, as the data's schema defines the table of data, the fields in the table, and the clear relationship between the two. Examples of structured data include Sensor Data and Financial Data. For example, in the following figure (please see Figure 4), the Sensor Data is collected—as can be seen, the data has a rigid structure with Record ID, Date, Time, Sensor ID, X, Y, Z, Sensor Body Temperature, Ambient Temperature, Sensor Body Humidity and Ambient Humidity.

Record ID	Date	Time	Sensor ID	X	Y	Z	SensorBody_Temperature	Ambient_Temperature	SensorBody_Humidity	Ambient_I
[103]	11/30/2019	15:00:01	HANU_NECKLACE_0001	-931	-188	-16185	11.4	11.7	69.5	66.7
[103]	11/30/2019	15:00:01	HANU_NECKLACE_0001	-931	-188	-16185	11.4	11.7	69.5	66.7
[103]	11/30/2019	15:00:01	HANU_NECKLACE_0001	-931	-188	-16185	11.4	11.7	69.5	66.7
[104]	11/30/2019	16:00:01	HANU_NECKLACE_0001	-893	-239	-16153	11.4	11.7	68.6	67
[105]	11/30/2019	17:00:01	HANU_NECKLACE_0001	-914	-198	-16153	11.4	11.7	68.4	67
[106]	11/30/2019	18:00:01	HANU_NECKLACE_0001	-961	-210	-16184	11.3	11.6	68.4	67
[107]	11/30/2019	19:00:01	HANU_NECKLACE_0001	-884	-199	-16194	11.3	11.6	68.6	67
[108]	11/30/2019	20:00:01	HANU_NECKLACE_0001	-908	-216	-16149	11.2	11.5	69	67.2
[109]	11/30/2019	21:00:01	HANU_NECKLACE_0001	-933	-210	-16169	11.3	11.5	69	67.3
[110]	11/30/2019	22:00:01	HANU_NECKLACE_0001	-935	-238	-16155	11.3	11.6	72.7	67.4
[110]	11/30/2019	22:00:01	HANU_NECKLACE_0001	-935	-238	-16155	11.3	11.6	72.7	67.4
[110]	11/30/2019	22:00:01	HANU_NECKLACE_0001	-935	-238	-16155	11.3	11.6	72.7	67.4
[110]	11/30/2019	22:00:01	HANU_NECKLACE_0001	-935	-238	-16155	11.3	11.6	72.7	67.4
[110]	11/30/2019	22:00:01	HANU_NECKLACE_0001	-935	-238	-16155	11.3	11.6	72.7	67.4
[112]	12/1/2019	0:00:01	HANU_NECKLACE_0001	-921	-240	-16195	11.8	12	70.3	68.1
[112]	12/1/2019	0:00:01	HANU_NECKLACE_0001	-921	-240	-16195	11.8	12	70.3	68.1
[112]	12/1/2019	0:00:01	HANU_NECKLACE_0001	-921	-240	-16195	11.8	12	70.3	68.1
[112]	12/1/2019	0:00:01	HANU_NECKLACE_0001	-921	-240	-16195	11.8	12	70.3	68.1

Figure 4: Sensor Data

Unstructured data

The organization of unstructured data is generally ambiguous. Unstructured data is often delivered in files, such as photos, images or videos. For example, in the following diagram (please see Figure 5), Mel Spectrogram [14] of an audio file that was collected as part of Far Eastern Memorial Hospital, FEMH, voice order detection.[32] The audio/video file itself may have an overall structure and come with semi-structured metadata, but the data that comprises the audio/video itself is unstructured. Therefore, photos, videos, audios and other similar files are classified as unstructured data. Examples of unstructured data include media files such as movie files and audio files, Office files such as Word documents, text files, and log files.

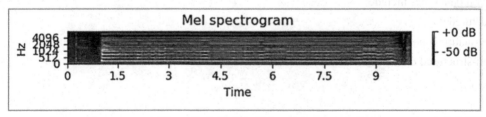

Figure 5: Audio Data Mel spectrogram

---

[32] 2018 FEMH Voice Data Challenge - https://femh-challenge2018.weebly.com/

The voice samples[33] were obtained from a voice clinic in a tertiary teaching hospital (Far Eastern Memorial Hospital, FEMH), which included 50 normal voice samples and 150 samples of common voice disorders, including vocal nodules, polyps, and cysts (collectively referred to as phono trauma), glottis neoplasm, and unilateral vocal paralysis. Voice samples of a 3-second sustained vowel sound /a:/ were recorded at a comfortable level of loudness, with a microphone-to-mouth distance of approximately 15–20 cm, using a high-quality microphone (Model: SM58, SHURE, IL), with a digital amplifier (Model: X2u, SHURE) under a background noise level between 40 and 45 dBA. The sampling rate was 44,100 Hz with a 16-bit resolution, and data were saved in an uncompressed wav format [14].

Semi-structured data

Semi-structured data is less organized than structured data, and is not stored in a relational format, as the fields do not neatly fit into tables, rows, and columns. Semi-structured data contains tags that make the organization and hierarchy of the data apparent. Semi-structured data is also referred to as non-relational or NoSQL data. Examples of Semi-structured data include Key/Value pairs, Network Graph data, JSON (JavaScript Object Notation), and XML (Extensible Markup Language) files. Please see (Figure 6) JSON meta data provided by the US Department of Agriculture Economic Research Service (ERS).[34]

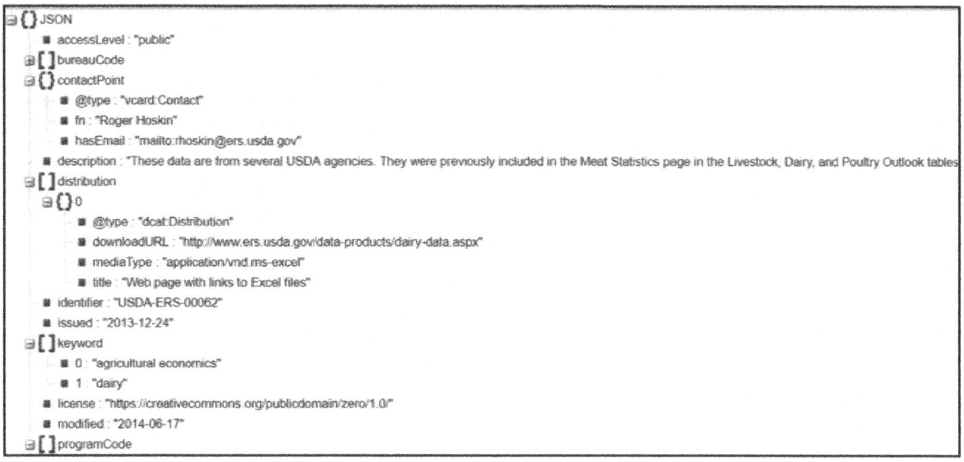

Figure 6: Semi-Structured Data
Source: https://catalog.data.gov/harvest/object/cea97321-07a7-4004-bd22-dc9fd1078c32

Machine Learning and Large-Scale Analytics

The latest Large-Scale Analytics which compute architectures such as Hadoop and Apache Spark, have enabled ML systems to work on large-scale data and this large-scale Big Data processing has led to advancement in the field of AI.

Big data

Big Data[35] has five important characteristics: Volume, Velocity, Variety, Veracity and Value.

---

[33] FEMH Data Samples - https://femh-challenge2018.weebly.com/faq.html
[34] USDA ERS Dairy Data - https://catalog.data.gov/dataset/dairy-data
[35] IBM Understanding Big Data - https://www.ibmbigdatahub.com/whitepaper/understanding-big-data-e-book

Volume is the scale of data. The sheer volume of available data—mainly driven by the IoT—has grown exponentially over the past five years and is expected to continue to do so[36] [15]. Ninety per cent of today's data has been created in the past two years![37] Now, every day, we create 2.5 quintillion of data—enough to fill 10 million blue ray discs.[38] Falling IoT prices related to sensors and wireless device will drive the growth of available data (please see Figure 7).

Velocity: It is the speed of data coming in and the speed in which we need to process the data. This is where streaming and real-time processing come to play. Please note: every 60 seconds, there are 72 hours of footage uploaded to YouTube, 216,000 Instagram posts, and 204,000,000 emails sent.[39]

Variety: It is the diversity of data. Data comes from various sources, from structured financial transaction data to unstructured logs, documents, medial (please note: 80% of data growth is video, images and documents), and IoT devices data, etc. Currently, 90% of the generated data is unstructured[40] [16].

Veracity: Data can be unreliable and flawed, from bad sensor data to human error. Veracity is certainty of data.

Value: Big Data is the ability to achieve greater value through insights from superior analytics.

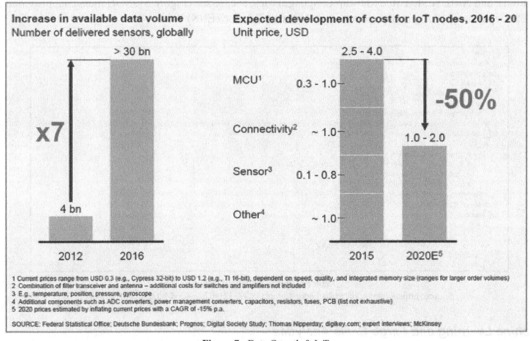

Figure 7: Data Growth & IoT

[36] Achieve Business Impact with the data - https://www.mckinsey.com/~/media/mckinsey/business%20functions/mckinsey%20analytics/our%20insights/achieving%20business%20impact%20with%20data/achieving-business-impact-with-data_final.ashx

[37] 2.5 quintillion bytes of data created every day. How does CPG & Retail manage it? - https://www.ibm.com/blogs/insights-on-business/consumer-products/2-5-quintillion-bytes-of-data-created-every-day-how-does-cpg-retail-manage-it/

[38] IBM Understanding Big Data - https://www.ibmbigdatahub.com/whitepaper/understanding-big-data-e-book

[39] Extracting Business Value from the 4V's of Big Data - https://www.ibmbigdatahub.com/sites/default/files/infographic_file/4Vs_Infographic_final.pdf

[40] 90 Percent of the Big Data We Generate Is an Unstructured Mess - https://www.pcmag.com/news/364954/90-percent-of-the-big-data-we-generate-is-an-unstructured-me

Example of Large-scale Analytical Systems

In order to process Big Data, the following Large-Scale Analytics are used:

Large-Scale Compute Architectures	Description
Google File System	Google File System,[41] a scalable distributed file system for large distributed data intensive applications. It provides fault tolerance while running on inexpensive commodity hardware, and it delivers high aggregate performance to a large number of clients [17].
Hadoop File System & Map Reduce	MapReduce[42] is a programming model and an associated implementation for processing and generating large data sets. Users specify a map function that processes a key/value pair to generate a set of intermediate key/value pairs, and a reduce function that merges all intermediate values associated with the same intermediate key. Many real world tasks are expressible in this model [18].
Spark in-memory System	Apache Spark [19] is a unified engine for large-scale data processing.[43]

Types of Learning

Machine Learning algorithms can be built with different styles of learning in order to model a problem[44] or address a business need or a pain-point [20]. The learning style is dictated by the interaction with the data environment expressed as the input to the model (for example, contiguous streams of data with time behavior, extrapolate historical patterns to predict re-occurrence or ad-hoc system with fuzziness to be modeled). The user must understand the role of the input data and the model's construction process. Finally, either from the users or from the business outcomes point of view, the goal is to select an ML model that can solve the problem with the best prediction result (value in 5Vs model). In this sense, ML sometimes overlaps with the goal of data mining.

 Eager Learner vs. Lazy Learner[45]

Eager Learners, when given a set of training tuples, will construct a generalization (i.e., classification) model before receiving new (e.g., test) tuples to classify. We can think of the learned model as being ready and eager to classify previously unseen tuples. Examples of Eager Learners include Decision Tree, Naive Bayes and Artificial Neural Networks.

Imagine a contrasting lazy approach, in which the learner instead waits until the last minute before doing any model construction in order to classify a given test tuple. That is, when given a training tuple, a lazy learner simply stores it (or does only a little minor processing) and waits until it is given a test tuple. Only when it sees the test tuple does it perform generalization in order to classify the tuple based on its similarity to the stored training tuples. Unlike eager learning methods, lazy learners do less work when a training tuple is presented and more work when making a classification or prediction. Because lazy learners store the training tuples or "instances", they are also referred to as instance-based learners, even though all learning is essentially based on instances. Examples of lazy learners include K-Nearest Neighbor and Case-Based Reasoning [10].

Supervised Learning

In supervised learning, an algorithm uses training data and feedback from humans to learn the relationship of given inputs to a given output [10]. That is, the humans label desired class or output so that the machine identifies a pattern between a given input and a labeled output. The labeled output is dependent-variable,

---

[41] The Google File System - https://static.googleusercontent.com/media/research.google.com/en//archive/gfs-sosp2003.pdf

[42] MapReduce – Simplified Data Processing on Large Clusters

[43] Apache Spark - https://spark.apache.org/

[44] Big-Data Analytics for Cloud, IoT and Cognitive Learning John Wiley & Sons © 2017

[45] Data Mining Concepts and Techniques, Jaiwei Han and Macheline Kambler – Publisher: Morgan Kaufmann; 3 edition (June 15, 2011)

i.e., depends upon the values of attributes, independent variables, in a given data set. For example, take USA Housing[46] data set (please see Table 1: Housing Data). The dataset contains independent variables: Average Area Income, Average Area House Age, Average Area Number of Rooms, Average Area Number of Bedrooms, Area Population, and Address. The class variable or dependent variable Price. In other words, the purpose of the model construction is to identify and predict Price of the house based on independent variable.

Table 1: Housing Data

Avg. Area Income	Avg. Area House Age	Avg. Area Number of Rooms	Avg. Area Number of Bedrooms	Area Population	Address	Price
79545.46	5.682861	7.009188	4.09	23086.8	208 Michael Ferry Apt. 674 Laurabury, NE 37010-5101	1059034
79248.64	6.0029	6.730821	3.09	40173.07	188 Johnson Views Suite 079 Lake Kathleen, CA 48958	1505891
61287.07	5.86589	8.512727	5.13	36882.16	9127 Elizabeth Stravenue Danieltown, WI 06482-3489	1058988
63345.24	7.188236	5.586729	3.26	34310.24	USS Barnett FPO AP 44820	1260617
59982.2	5.040555	7.839388	4.23	26354.11	USNS Raymond FPO AE 09386	630943.5
80175.75	4.988408	6.104512	4.04	26748.43	06039 Jennifer Islands Apt. 443 Tracyport, KS 16077	1068138
64698.46	6.025336	8.14776	3.41	60828.25	4759 Daniel Shoals Suite 442 Nguyenburgh, CO 20247	1502056

The learning process continues until the model achieves a desired level of accuracy on the training data. Future incoming data (without known labels) are tested on the constructed model. Examples of supervised learning is loan approval (Yes or No-binary) of an applicant based on income forecast, risk, loan guarantee, and education level. Another example is housing prices (dependent variable) dependent on Average Area Income, Average Area House Age, Average Area Number of Rooms, Average Area Number of Bedrooms, Area Population, and Address. The following table summarizes some of the supervised learning:

Algorithm	Definition
Linear Regression	Regression is concerned with modeling the relationship between variables that is iteratively refined using a measure of error in the predictions made by the model.[47] The most popular regression algorithms are [21]: • Ordinary Least Squares Regression (OLSR) • Linear Regression • Logistic Regression • Stepwise Regression • Multivariate Adaptive Regression Splines (MARS) • Locally Estimated Scatterplot Smoothing (LOESS) Highly interpretable,[48] linear regression models between input independent variables and output variable (class or dependent variable) to generate a model or classifier equation that helps to predict future values.

[46] USA Housing: Kaggle dataset - https://www.kaggle.com/gpandi007/usa-housing-dataset
[47] A Tour of Machine Learning Algorithms - https://machinelearningmastery.com/a-tour-of-machine-learning-algorithms/
[48] Model interpretability in Azure Machine Learning service - https://docs.microsoft.com/en-us/azure/machine-learning/service/how-to-machine-learning-interpretability

Logistic Regression	Logistic Regression[49] is an ML classification algorithm that is used to predict the probability of a categorical dependent variable. In Logistic Regression [22], the dependent variable is a binary variable that contains data coded as 1 (yes, success, etc.) or 0 (no, failure, etc.). In other words, the Logistic Regression model predicts $P(Y=1)$ as a function of X. Simply put, Logistic Regression is an extension of Linear Regression that is used for classification tasks, meaning the output variable binary rather than contiguous [21]. Example of Logistic Regression[50] includes identification of Digits.
Decision Tree	Decision Tree is a simple ML algorithm and does not require knowing a business domain to apply [10] [21]. Highly interpretable classification or regression model that splits data-feature values into branches at decision nodes (e.g., if a feature is a color, each possible color becomes a new branch) until a final decision output is made.[51] The most popular Decision Tree algorithms are: • Classification and Regression Tree (CART) • Iterative Dichotomiser 3(ID3) • C4.5 and C5.0 (different versions of a powerful approach) • Chi-squared Automatic Interaction Detection (CHAID) • Decision Stump • M5 • Conditional Decision Trees
Naive Bayesian	Bayesian algorithm applies conditional probabilities to predict the outcome. Classification technique that applies Bayes theorem, which allows the probability of an event to be calculated based on knowledge of factors that might affect that event. The most popular Bayesian algorithms are [10][21]: • Naive Bayes • Gaussian Naive Bayes • Multinomial Naive Bayes • Averaged One-Dependence Estimators (AODE) • Bayesian Belief Network (BBN) • Bayesian Network (BN)
Random Forest	The fundamental concept behind Random Forest is a simple but powerful one—the wisdom of crowds.[52] A large number of relatively uncorrelated models (trees) operating as a committee will outperform any of the individual constituent models [21]. In other words, Random Forest is a classification or regression model that improves the accuracy of a simple decision tree by generating multiple decision trees and taking a majority vote of them to predict the output, which is a continuous variable (e.g., age) for a regression problem and a discrete variable (e.g., either black, white, or red) for classification.
Gradient-Boosting Trees	Classification or regression technique that generates decision trees sequentially, where each tree focuses on correcting the errors coming from the previous tree model. The final output is a combination of the results from all the trees [21].
XGBoost	XGBoost[53] is a decision-tree-based ensemble ML algorithm that uses a gradient boosting framework. In prediction problems involving unstructured data (images, text, etc.), artificial neural networks tend to outperform all other algorithms or frameworks. However, when it comes to small-to-medium structured/tabular data, decision tree-based algorithms are considered best-in-class right now [23].

---

[49] Building A Logistic Regression in Python, Step by Step - https://towardsdatascience.com/building-a-logistic-regression-in-python-step-by-step-becd4d56c9c8

[50] Logistic Regression using Python (scikit-learn) - https://towardsdatascience.com/logistic-regression-using-python-sklearn-numpy-mnist-handwriting-recognition-matplotlib-a6b31e2b166a

[51] Tour of ML Algorithms - https://machinelearningmastery.com/a-tour-of-machine-learning-algorithms/

[52] Random Forest Classifier - https://towardsdatascience.com/understanding-random-forest-58381e0602d2

[53] XGBoost - https://towardsdatascience.com/https-medium-com-vishalmorde-xgboost-algorithm-long-she-may-rein-edd9f99be63d

---

⚬ Model Interpretability

Interpretability is critical for data scientists and business decision makers alike to ensure compliance with company policies, industry standards, and government regulations:

- Data scientists need the ability to explain their models to executives and stakeholders, so they can understand the value and accuracy of their findings

- Business decision makers need peace of mind of the ability to provide transparency for end users to gain and maintain their trust

Enabling the capability of explaining an ML model is important during two main phases of model development:

- During the training phase of the ML model development cycle. Model designers and evaluators can use interpretability output of a model to verify the hypotheses and build trust with the stakeholders. They also use the insights into the model for debugging, for validating that model behavior matches their objectives, and to check for bias or insignificant features.

- During the inferencing phase, as having transparency around deployed models empowers executives to understand "when deployed" how the model is working and how its decisions are treating and impacting people in real life.

Unsupervised Learning

An algorithm explores input data without being given an explicit output variable. A model is generated by exploring the structures presented in the input data. This may be achieved by extracting general rules, going through a mathematical process to reduce redundancy, or organizing data by similarity testing. Examples of Unsupervised algorithm include exploring customer demographic data to identify buying patterns, arrangement of similarity of customers based on similar movie choices and segmentation of retail chain stores based on sales categories [10].

Algorithm	Description
K-means Clustering	The k-means algorithm searches for a pre-determined number of clusters within an unlabeled multidimensional data set. It accomplishes this using a simple conception of what the optimal clustering looks like: • The "cluster center" is the arithmetic mean of all the points belonging to the cluster. • Each point is closer to its own cluster center than to other cluster centers Those two assumptions are the basis of the k-means model [10][21].
Gaussian mixture model	A generalization of k-means clustering that provides more flexibility in the size and shape of groups (clusters).
Hierarchical clustering	Hierarchical clustering is an agglomerative (top down) clustering method. As its name suggests, the idea of this method is to build a hierarchy of clusters, showing relations between the individual members and merging clusters of data based on similarity. Splits or aggregates clusters along a hierarchical tree to form a classification system [10][21].
Recommender System	Often uses cluster behavior prediction to identify the important data necessary for making a recommendation (Item based and Collaborative filter based)

Dimensionality Reduction	Seeks and exploits similarities in the structure of data in a manner similar to clustering algorithms, but using unsupervised methods [21]. The purpose is to summarize or describe data using less information so that the dataset becomes smaller and easier to manage In some cases, people use these algorithms for classification or regression problems Here is a list of common dimensionality reduction algorithms:[54] • Principal Component Analysis (PCA) • Principal Component Regression (PCR) • Partial Least Squares Regression (PLSR) • Sammon Mapping • Multidimensional Scaling (MDS) • Projection Pursuit • Linear Discriminant Analysis (LDA) • Mixture Discriminant Analysis (MDA) • Quadratic Discriminant Analysis (QDA) • Flexible Discriminant Analysis (FDA)
Association Rules	Association Rule Learning methods extract rules that best explain observed relationships between variables in data. The most popular association Rule Learning algorithms are[10]: • A priori algorithm • Eclat algorithm

Reinforcement Learning

An algorithm learns to perform a task simply by trying to maximize rewards it receives for its actions.[55] Examples of reinforcement learning includes maximizes points it receives for increasing returns of an investment portfolio [24].

Deep Learning

Deep Learning is a type of machine learning that can process a wider range of data resources, requires less data preprocessing by humans, and can often produce more accurate results than traditional ML approaches. This particular group of algorithms works well with semi-supervised learning problems in which the amount of labeled data is minimal. In deep learning, interconnected layers of software-based calculators known as "neurons" form a neural network. The network can ingest vast amounts of input data and process them through multiple layers that learn increasingly complex features of the data at each layer. The network can then make a determination about the data, learn if its determination is correct, and use what it has learned to make determinations about new data. For example, once it learns what an object looks like, it can recognize the object in a new image.

Data requirements for deep learning are substantially greater [25] than for other analytics.[56] Here are the most popular Deep Learning algorithms[57] [21]:

- Convolutional Neural Network (CNN)
- Recurrent Neural Network (RNN)
- Long Short-Term Memory Networks (LSTMs)
- Stacked Auto-Encoders
- Deep Boltzmann Machine (DBM)
- Deep Belief Networks (DBN)

---

[54] A Tour of Machine Learning Algorithms - https://machinelearningmastery.com/a-tour-of-machine-learning-algorithms/
[55] Fuzzy Logic Infused Intelligent Scent Dispenser For Creating Memorable Customer Experienceof Long-Tail Connected Venues - https://ieeexplore.ieee.org/document/8527046
[56] Notes from AI Frontier: Applications and value of deep learning - https://www.mckinsey.com/featured-insights/artificial-intelligence/notes-from-the-ai-frontier-applications-and-value-of-deep-learning
[57] A Tour of Machine Learning Algorithms - https://machinelearningmastery.com/a-tour-of-machine-learning-algorithms/

TensorFlow Lite

TensorFlow provides computational platform to run neural networks. The models, nonetheless, developed using TensorFlow needs huge computational space. So, the trained models using TensorFlow is not compatible to run on small devices. To overcome, TensorFlow Lite[58] has been introduced. TF Lite is ideally suited to run models on microcontrollers.

TensorFlow Lite is a set of tools to help developers run TensorFlow models on mobile, embedded, and IoT devices. It enables on-device machine learning inference with low latency and a small binary size.

Algorithm	Description
Convolutional Neural Network (CNN)	A multilayered neural network with a special architecture designed to extract increasingly complex features of the data at each layer to determine the output. Generally used when you have an unstructured data set (e.g., images) and you need to infer information from it [21].
Recurrent Neural Network (RNN)	A multilayered neural network that can store information in context nodes, allowing it to learn data sequences and output a number or another sequence. General use cases involve when you are working with time-series data or sequences (e.g., audio recordings or text)

Neural networks are a subset of ML techniques. Essentially, they are AI systems based on simulating connected "neural units", loosely modeling the way that neurons interact in the brain. Computational models inspired by neural connections have been studied since the 1940s and have returned to prominence as computer processing power has increased and large training data sets have been used to successfully analyze input data such as images, videos, and speeches. AI practitioners refer to these techniques as "deep learning", since neural networks have many ("deep") layers of simulated interconnected neurons[59] [25].

 What is a Heuristic?

A heuristic[60] is a method drawn from experience, common sense, or an educated guess that aims at providing or contributing to providing a practical solution to a problem that is usually very difficult to solve (NP-Hard), and consequently an optimal or good, feasible solution is too complicated to obtain. Heuristic methods can be used to speed up the process of finding a good, feasible solution by providing us with a shortcut. This speed up process is usually carried out via search algorithms where we traverse a tree representing the space of possible solutions. The application of certain problem-specific heuristics can significantly reduce tree search [9].

Heuristic methods possess some information about the proximity of every state to the goal state, which allows them to explore the most promising paths first. Summarizing, some of the most general features of heuristic methods are as follows:

• They do not guarantee that a solution will be found, even though it may exist.

• If it finds a solution, it does not guarantee that it will be optimal (minimal or maximal).

• Sometimes (not defined a priori) it will find a good solution in a reasonable time.

---

[58] TensorFlow Lite - https://www.tensorflow.org/lite/guide

[59] Notes from AI Frontier: Applications and value of deep learning - https://www.mckinsey.com/featured-insights/artificial-intelligence/notes-from-the-ai-frontier-applications-and-value-of-deep-learning

[60] Practical Artificial Intelligence: Machine Learning, Bots, and Agent Solutions Using C# - Chapter 14, Apress 2018

We usually work with a heuristic through a heuristic function. This function assigns a numeric value to every state of the problem and defines how promising that state is as far as attempting to reach a goal state from a given point (node); it's usually denoted as $H(e)$. The heuristic function can have two interpretations. It could indicate how close state e is to the goal state, meaning states with the lowest heuristic value are preferred, or it could indicate how far state e is from a goal state, meaning we prefer states with the highest heuristic values. In such cases, we are either minimizing or maximizing the heuristic function [9].

Heuristics underlie[61] the whole field of AI and the computer simulation of thinking, as they may be used in situations where there are no known algorithms [26].

Choosing the Right Estimator

Often the hardest part of solving a machine learning problem can be finding the right estimator for the job.[62] Different estimators are better suited for different types of data and different problems. The flowchart below is designed to give users a bit of a rough guide on how to approach problems with regard to which estimators to try on your data. The estimator enables to learn insights from the data or able to develop a model to predict (please see Figure 8).

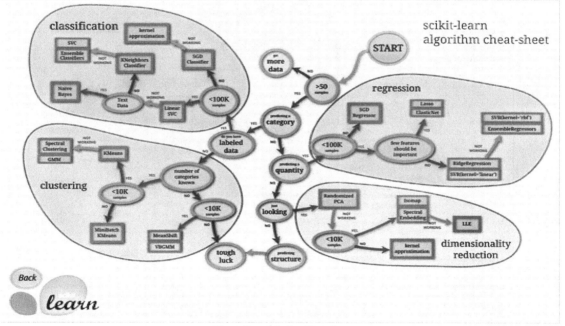

Figure 8: scikit-algorithm sheet
Source: https://scikit-learn.org/stable/tutorial/machine_learning_map/index.html#

When considering automated learning, computational resources also play a major role in determining the complexity of a task: that is, how much computation is involved in carrying out a learning task. Once a sufficient training sample is available to the learner, there is some computation to be done to extract a hypothesis or figure out the label of a given test instance.

---

[61] The Computer Simulation of Behaviour By Michael J Apter
[62] Choosing right estimator - https://scikit-learn.org/stable/tutorial/machine_learning_map/index.html#

These computational resources are crucial in any practical application of ML. There are two other considerations to be taken into account: sample complexity and computational complexity [8].

Mapping AI Technique to Classical ML

As AI technologies advance, so does the definition of which techniques constitute AI (please see Figure 9). The following heat-map maps the common ML techniques used in AI: classification, continuous estimation and clustering are most used [25].

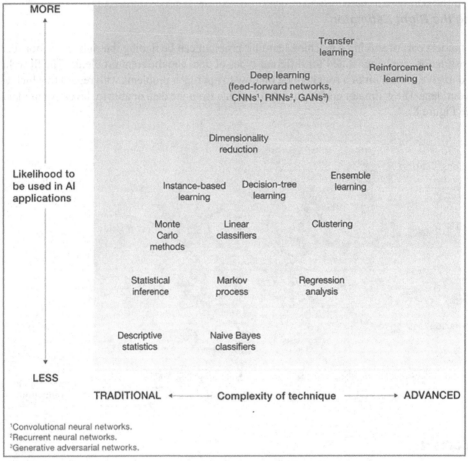

Figure 9: AI Mapping

AI Epochs: Waves of Compute

Artificial Intelligence was introduced to the compute landscape as part of the second wave of compute and has since been on the computing spectrum (please see Figure 10). The tipping point[63] for AI was the introduction of Cloud and large-scale availability of compute and storage [27].

---

[63] Own the A.I. Revolution: Unlock Your Artificial Intelligence Strategy to Disrupt Your Competition, McGraw-Hill © 2019

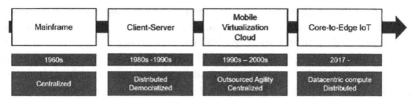

Figure 10: Waves of Computing

First Wave of Computing

The first wave began when companies started to manage their operations via mainframe computer systems over 50 years ago. Then computing got "personal" in the 1980s and '90s with the introduction of the Personal Computer (PC). For the most part, computing remained immobile and lacked contextual awareness.[64] The systems were more task-oriented and less human augmented with instructions [28].

Second Wave of Computing

In computing's second wave, the mainframe operated centralized monolithic computing diversified with the introduction of distributed tiered architecture and started the Client Server architecture era. Even though the compute decentralized and program execution tiered (Front-end, middleware and backend, the three tiers), the intelligence and context aware compute was still not present during the second wave [27]. Incapable of "thinking" in the sense of independent analysis and decision-making, computing systems simply executed the human user's instructions.

Third Wave of Computing

In computing's third wave, mobile computing and the smartphone took the center stage. Billions of people, some who might not have had access to clean water, electricity, or even housing, were connected. Developers created apps and provided consumers with access to just about everything through their phone at the cost of a monthly data plan.

The mobile platforms, importantly, enabled the computational platforms for big data processing and making knowledge automation more attractive due to rapid advances in computational technologies and reduction in costs and boost in performance [29]. Computing power continues to grow exponentially (approximately doubling every two years on a price/performance basis), and today (2013) a $400 iPhone4 offers roughly equal performance (in millions of floating point operations per second, or MFLOPS) to the CDC 7600 supercomputer, which was the fastest supercomputer in 1975 and cost $5 million at the time.[65]

The Cray 2 Computer System vs. iPhone XS

Computing performance and sophistication of smartphones continue to grow and with the introduction of advanced preprocessors for image and ML operations, now we can technically say smartphones have brought super-computing to the hands of the end users. A good example is performance comparison of iPhone XS and Cray2 computer systems. Please note both computing systems are from a different era but sheer magnitude of size and potentials that these systems have rendered are unprecedented.

Cray Research's mission is to lead in the development and marketing of high-performance systems that make a unique contribution to the markets they serve.[66] For close to a decade, Cray Research has been the

---

[64] The third wave of computing - https://fortune.com/2013/10/03/the-third-wave-of-computing/

[65] Disruptive technologies: advances that will transform life, business, and the global economy - https://www.mckinsey.com/~/media/McKinsey/Business%20Functions/McKinsey%20Digital/Our%20Insights/Disruptive%20technologies/MGI_Disruptive_technologies_Full_report_May2013.ashx

[66] The Cray-2 Computer System - http://archive.computerhistory.org/resources/text/Cray/Cray.Cray2.1985.102646185.pdf

industry leader in large-scale computer systems. During 1980s, most supercomputers installed worldwide were Cray systems. These systems are used in advanced research laboratories around the world and have gained strong acceptance in diverse industrial environments. No other manufacturer has Cray Research's breadth of success and experience in supercomputer development. The company's initial product, the CRAY-1 Computer System, was first installed in 1976. With its significant innovations in architecture and technology, the CRAY-2 Computer System sets the standard for the next generation of supercomputers. The CRAY-2 design allows many types of users to solve problems that cannot be solved with any other computers. CRAY-2 provides an order of magnitude increase in performance over CRAY-1 at an attractive price/performance ratio.

iPhone XS

iPhone XS and iPhone XS Max are smartphones designed and marketed by Apple Inc. They are 12th generation of iPhone. iPhone XS uses hexa-core (2.49 GHz Dual core Vortex, 1.52 GHz Quad core Tempest) CPU and Apple A12 Bionic chipset (64-bit RAM based system on Chip (SoC). The A12 brings advanced neural & AI capabilities.[67]

Comparison of CRAY-2 vs. iPhone XS

The performance in terms of FLOPS is comparable:

	Cray 2 (1985)	iPhone XS (2018)
Price (2017 USD)	> $30,000,000	~ $900
Main Processor	4	1 A12X
Memory (RAM)	256 Megaword	4 Gigabytes
	megaword –> a 'word' varies but the Cray-2 uses 64-bit words with 8 bit parity checks	
Storage (max)	~ 32 GB*	512 GB
	* Max storage requires 32 disk drives of 1.2 GB each	
Peak Power Consumption	195,000 W	< 1 W*
	*It's very hard to compare peak power, given iPhone has so many other functions, so let's just go with < 1 W	
Peak Performance	1.9 GFLOP *	~ 1 GFLOP*
	*GFLOP = Billions of Floating Point operations per second	
Weight	5,500 lb (2,494 kg)	128 g (0.3lb, 0.14 kg)
Volume	1.8 cubic meters	~ 0.007 cubic meters
Height	45 in (1.2 m)	5.8 in (0.16 m)
Width	54 in (1.4 m)*	3.05 in (0.075 m)

* Cray-2 was cylindrical in shape, so the 'width' is really its diameter
In addition, the iPhone also has 2 cameras, GPS, Cellular, Wi-Fi, Bluetooth, a display with over 2.7 MM pixels, a battery that lasts a day, compass, a3-axis gyroscope, speakers, an accelerometer, ambient ILight and proximity sensors, a barometer and a microphone. It can record 4K video. It can take 8MP photos while recording 4k video. It has NFC and IBeacon.

Fourth Wave of Computing

Edge computing ushered in the fourth wave of computing. The crucial principle that Edge compute delivers is data & algorithmic processing at the source of data. Edge compute use cases include pattern recognition, outlier detection, anomaly detection and real-time complex event processing (CEP). Today, in 2019, we are in the peak of Edge computing with many companies spearheading Edge deployments. Edge computing embodies AI and ML models at its core.[68]

---

[67] iPhone XS' industry-first A12 chip gives Apple big advantage over rivals - https://www.cnet.com/news/iphone-xs-industry-first-a12-chip-gives-apple-big-advantage-over-rivals/

[68] IoT Devices - https://www2.deloitte.com/us/en/insights/focus/internet-of-things/iot-primer-iot-technologies-applications.html

The following diagram (please see Figure 11) contains a high level view of Edge devices:

- **Communication module:** This gives the Edge device its communication capabilities. It is typically either a radio transceiver (Bluetooth Low-Energy, CoAP) with an antenna or a wired connection (Wi-Fi).

- **Microcontroller:** It is a small microprocessor that runs the software of the Edge device. The brain of the device is its software or firmware that is burned during the device construction. Many of ML & AI algorithms are part of firmware of the device.

- **Sensors or actuators:** These give the Edge devices a way to sense and interact with the physical world.

- **Power source:** This is needed because Edge devices contain electrical circuits. The most common power source is a battery, but there are other examples as well, such as piezoelectric power sources that provide power when a physical force is applied, or small solar cells that provide power when light shines on them [30]. The following table (please see Table 2) contains power source for Edge devices [30]:

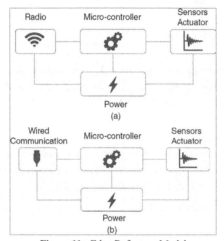

Figure 11: Edge Reference Model

Table 2: Power Source for Edge Devices

Power source	Typical maximum current (mA)	Typical charge (mAh)
CR2032 button cell	20	200
AA alkaline battery	20	3000
Solar cell	40	Limitless
RF power	25	Limitless

Microcontrollers have two types of memory: Read-only memory (ROM) and random-access memory (RAM). ROM is used to store the program code that encodes the behavior of the device and RAM is used for temporary data the software needs to do its task. For example, temporary data includes storage for program variables and buffer memory for handling radio traffic.

For constrained devices, the content of the ROM is typically burned into the device when it is manufactured and is not altered after deployment. Modern microcontrollers provide a mechanism for rewriting the ROM, which is useful for in-field updates of software after the devices have been deployed [30].

One level deep dive into the Edge device's brain and software architecture is as follows (please see Figure 12): as you can see, the reference architecture is split into three levels: System Level, Software Level and AI Levels.

System Level consists of hardware components and libraries. Software Level consists of ML models, Custom ML models, Tiny ML models, TensorFlow Lite and Software/Firmware libraries. Finally, AI Level consists of Edge Analytics modules that drive ML models to exploit the system capabilities.

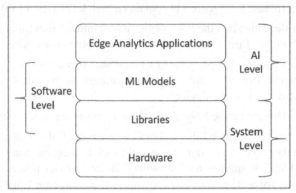

Figure 12: Edge Device - Component Levels

-☼- AI on Microcontrollers or Tiny ML

IoT devices, constrained and small footprint, are powered by microcontrollers. Microcontrollers are typically small, low-powered computing devices that are often embedded within the hardware that requires basic computation, including household appliances and IoT devices. No wonder, billions of microcontrollers are manufactured every year.[69]

Microcontrollers are often optimized for low energy consumption and small size, at the cost of reduced processing power, memory, and storage. By running ML models on microcontrollers, AI can be deployed to a vast range of hardware devices without relying on network connectivity, which is often subject to bandwidth and power constraints and results in high latency. Running inference on-device can also help preserve privacy, since no data has to leave the device.

Fifth Wave of Computing

AI, 5G and Secure IoT are the drivers of fifth wave of computing. On device, AI will have a transformative effect on technology. The 5G gives an AI enabled IoT applications the speed it needs, which was a limitation in fourth wave of compute [31]. The fifth wave of computing will be fundamental in making the world we live in more efficient and more sustainable.[70]

Finally, from technological point of view, the combinatorial impact of technology[71] (horizontal view of Wave of computing) is as below [32].

The dominant technology curves (please see Figure 13) from 1950 to today (2019), i.e., from 1. Mainframe, 2. Client Server and PCs, 3. Web 1.0, 4. Web 2.0 with Mobile, 5. Big Data, 6. IoT Smart Machine, 7. AI and 8. Quantum computing [32].

"AI is the ultimate breakthrough technology."[72]

Satya Nadella, CEO, Microsoft

---

[69] Why Microcontrollers are important? - https://www.tensorflow.org/lite/microcontrollers

[70] Fifth wave of compute - https://www.arm.com/blogs/blueprint/the-fifth-wave-of-computing-ai-5g-iot

[71] Applied Intelligence - https://www.accenture.com/_acnmedia/pdf-84/accenture-ai-explained-overview.pdf

[72] Technology revolution like no other - https://www.accenture.com/_acnmedia/pdf-84/accenture-ai-explained-overview.pdf

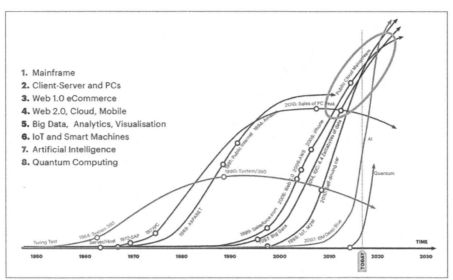

Figure 13: The Combinatorial of Technology

AI Hype Cycle—Current and Emerging Technologies

Gartner Hype cycle for AI provides technology expectations with respect to time period that a particular AI technology is either in one of five phases (years to mainstream adoption): innovation trigger, peak of inflated expectation, trough of disillusionment, slope of enlightenment, and plateau of productivity. In other words, the Hype curve tracks the progress of emerging technology trends from early innovation to becoming a household must-have.[73] For Chief Information Officer (CIO) teams, the Hype cycle provides a guidance of new technology landscape that would be in the market in the immediate future so that proper technology scoping and preparation could be planned. The Hype cycle can serve as an early warning system for executives, and at least jumping on technologies that serve business objectives in two to five years could eventually pay big dividends [33, 34]. For Data Scientists and Data Engineers the review of Gartner Hype curve would provide an informative direction on marketability of technology that is immediately in peak of expectation and with a short-term (five years) visibility on a newer technological transition to be taking place.

Hype Cycle for AI, 2017

Gartner Hype cycle for AI 2017 (please see Figure 14), highlights deep learning and machine learning are at what is called the "peak of inflated expectation", but are just two to five years away from mainstream adoption [33]. Artificial General Intelligence is in innovation trigger phase and Deep Reinforcement Learning and Edge Computing will get mainstream adoption in two years. Technologies pegged for the five- to ten-year range include Deep Reinforcement, IoT platforms and Cognitive computing.

Hype Cycle for AI, 2018

Technologies expected to reach the plateau of productivity (please see Figure 15) in two to five years include Virtual Assistants, Deep Neural Nets (DNN), Application Specific Integrated Circuits (ASICs) and 5G [34]. Artificial General Intelligence is in innovation trigger phase and Deep Reinforcement Learning & Edge Computing gets mainstream adoption in two years. Technologies pegged for the five to ten-year range include

---

[73] Gartner Hype Cycle 2017: Artificial intelligence at peak hype, blockchain heads for disillusionment, but say hello to 5G - https://www. cityam.com/gartner-hype-cycle-2017-artificial-intelligence-peak-hype/

Figure 14: Gartner Hype Cycle for AI, 2017

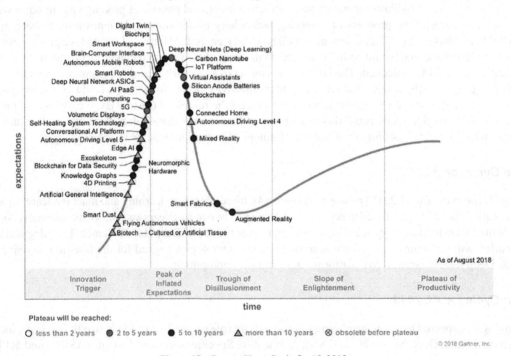

Figure 15: Gartner Hype Cycle for AI, 2018

augmented and mixed reality, blockchain, IoT platforms, digital twins, AI platform as a service, quantum computing and conversational AI.

Hype Cycle for AI, 2019

Between 2018 and 2019, organizations that have deployed AI grew from 4% to 14%, according to Gartner's 2019 CIO Agenda survey.[74] AI is reaching organizations in many ways compared with a few years ago [35] when there was no alternative to building your own solutions with ML. AutoML and intelligent applications have the greatest momentum, while other approaches are also popular—namely, AI platform as a service or AI Cloud services (please see Figure 16). Artificial Intelligence and Machine Learning are at the peak[75] of hype in organizations today [36].

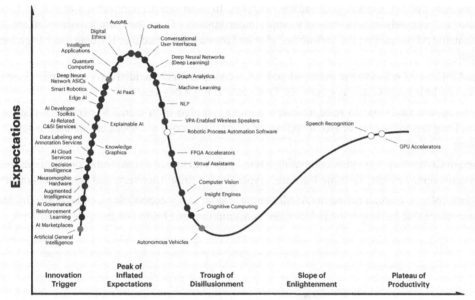

Figure 16: Gartner Hype Cycle for AI, 2019

> "Instead of trying to produce a program to simulate the adult mind, why not rather try to produce one which simulates the child's mind? If this were then subjected to an appropriate course of education one would obtain the adult brain."
>
> Alan Turing, 1950

This year, 2019, enterprises have introduced eight[76] new AI technologies to serve their customers [35].

1. AI Cloud Services: AI Cloud services are hosted services that enable development teams to incorporate the advantages inherent in AI and ML. Cognitive services from Microsoft Azure[77] and Rekognition[78] from Amazon Web services are examples of AI Cloud services.

2. AutoML: Automated Machine Learning is the capability of automating buildings, hyper-tuning, deploying, and managing ML models. With AutoML capabilities, data scientists could compare the accuracies of

[74] Gartner Survey of More Than 3,000 CIOs Reveals That Enterprises Are Entering the Third Era of IT - https://www.gartner.com/en/newsroom/press-releases/2018-10-16-gartner-survey-of-more-than-3000-cios-reveals-that-enterprises-are-entering-the-third-era-of-it

[75] Laying the foundations for AI and ML: Gartner Trend Insight report - https://emtemp.gcom.cloud/ngw/globalassets/en/doc/documents/3890363-laying-the-foundation-for-artificial-intelligence-and-machine-learning-a-gartner-trend-insight-report.pdf

[76] What's new in Gartner's Hype Cycle for AI, 2019 - https://www.forbes.com/sites/louiscolumbus/2019/09/25/whats-new-in-gartners-hype-cycle-for-ai-2019/#24254f8f547b

[77] Microsoft Azure Cognitive Services - https://azure.microsoft.com/en-us/services/cognitive-services/

[78] AWS Rekoginition - https://aws.amazon.com/rekognition/

various ML models and pivot the model selection to the most accurate technique in an automated and time efficient manner. Microsoft Auto ML,[79] Cloud AutoML[80] from Google, Amazon Sagemaker,[81] DataRobot[82] Automated Machine Learning and H_2O[83]—Automated ML.

3. Augmented Intelligence: Augmented Intelligence is a human-centered partnership model of people and AI working together to enhance cognitive performance. The goal of AI should be to empower humans to be better, smarter and happier, and not to create a 'machine world' for its own sake. Augmented Intelligence is a design approach to winning with AI, and it assists machines and people alike to perform at their best.

   According to Gartner Technology Market Research (please see Figure 17), AI Augmentation will create $2.9 trillion[84] business value in 2021 and both Decision support & AI augmentation will surpass all other types of AI [37].

4. Explainable AI: Explainable AI focuses on "black-box" AI model and explains how the various weights and measures of ML models generate their outputs. In other words, applications of A I, M L, and deep learning are relatively useless without a simple understanding of how their predictive models are derived.[85] Explainable AI focuses on the various weights and measures of machine learning models generate their outputs.

5. Edge AI: Edge AI refers to the use of AI and ML techniques in embedded devices in IoT. The embedded devices include Edge Gateways, Sensors, Devices and Edge devices.

6. Reinforcement Learning: In reinforcement learning, an algorithm learns to perform a task by simply maximizing rewards it receives for its action.[86] Generally, Fuzzy logic operators are used in reinforcement learning.

7. Quantum Computing: Quantum computing has the potential to make significant contributions to the areas of systems optimization, machine learning, cryptography, drug discovery, and organic chemistry.

8. AI Marketplaces: Gartner defines an AI Marketplace as an easily accessible place supported by a technical infrastructure that facilitates the publication, consumption, and billing of reusable algorithms.

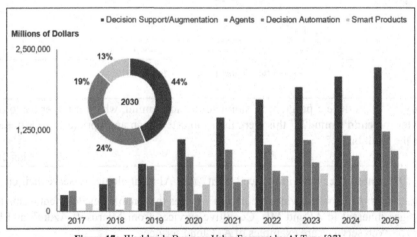

Figure 17: Worldwide Business Value Forecast by AI Type [37]

[79] Microsoft Auto ML - https://www.microsoft.com/en-us/research/project/automl/
[80] Google Cloud AutoML - https://cloud.google.com/automl/
[81] Amazon Sagemaker - https://aws.amazon.com/sagemaker/
[82] DataRobot Automated Machine Learning - https://www.datarobot.com/
[83] H₂O Automated ML - http://docs.h2o.ai/h2o/latest-stable/h2o-docs/automl.html
[84] Gartner Says AI Augmentation Will Create $2.9 Trillion of Business Value in 2021 - https://www.gartner.com/en/newsroom/press-releases/2019-08-05-gartner-says-ai-augmentation-will-create-2point9-trillion-of-business-value-in-2021
[85] The Essence of Explainable AI - https://aibusiness.com/explainable-ai-interpretability/
[86] Reinforcement learning - https://www.mckinsey.com/business-functions/mckinsey-analytics/our-insights/an-executives-guide-to-ai

Digital Strategy

To maximize return on investment (ROI) on AI technologies, organizations should focus on digital strategy. The same applies to nations that embark on AI and make it as their central theme for considerable future (more in the next section). To take advantage of AI, technology professionals must prepare the right foundational steps, such as developing AI strategy, devising data management and data quality processes, leveraging AI vendor offering to jump start AI/ML efforts in the enterprise. In addition to Development Operations, technology professionals must look into operationalize AI and ML components.

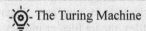 The Turing Machine

A mathematical model of a hypothetical computing machine which can use a predefined set of rules to determine a result from a set of input variables.

AI End-To-End (E2E) Process—Turning Data into Actionable Insights

Gartner Hype cycles [33][34][35] clearly indicates various AI services that are in peak of the curve. These AI services include: Machine Learning, AI Services, and Deep Neural Networks (DNN). The purpose of these services is to turn data into actionable insights, both for business purposes and personal recommendation services.

Data Science Lifecycle

Many of AI compute platform follow a typical End-To-End (E2E) process that generally encompasses Data Science Lifecycle for building AL models (please see Figure 18). The *data science lifecycle, namely data ingestion and preparation, model development, and deployment.* The End-to-End Data Platform [38] is vital for handling data workloads and transform data into actionable insights. The components of AI data platform generally include Data Sources, Data Ingestion, Data Preparation, Data Storage, Data Modeling, Data Serving and Visualization.[87]

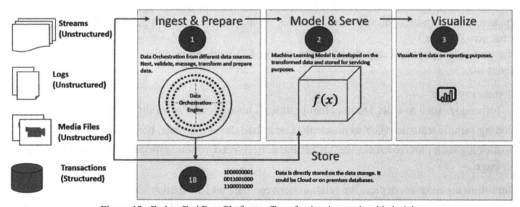

Figure 18: End-to-End Data Platform—Transfer data into actionable insights

---

Data Sources

Data Sources include structured, unstructured and semi-structured data sources. A general example includes Sensor data, Real-time social media data, Log data and Image/Video data.

Prepare & Transform

Data from data sources need to be ingested into the platform using an orchestration or data movement tools. Technically, there are two paths to the data. The structured data is directly ingested and stored in the data platform. For example, data transaction records are ingested into data storage with transforming as the data confined to strict structure or schema. The unstructured data, for example, log files, media files or movies/images, is first transformed to ingest into data storage and prepared to ready for input to modeling. One of the techniques to understand data and prepare for ML purposes is Exploratory Data Analysis (EDA).

Exploratory Data Analysis (EDA)

Exploratory Data Analysis[88] (EDA) is an approach/philosophy for data analysis that employs a variety of techniques (mostly graphical) to

1. Maximize insight into a data set
2. Uncover underlying structure
3. Extract important variables
4. Detect outliers and anomalies
5. Test underlying assumptions
6. Develop parsimonious models and
7. Determine optimal factor settings.

Most EDA techniques are graphical in nature with a few quantitative techniques. The reason for the heavy reliance on graphics is that by its very nature the main role of EDA is to open-mindedly explore, and graphics gives the analysts unparalleled power to do so, enticing the data to reveal its structural secrets (please see Figure 19), and being always ready to gain some new, often unsuspected, insight into the data. In combination with the natural pattern-recognition capabilities that we all possess, graphics provide, of course, unparalleled power to carry this out.

The particular graphical techniques employed in EDA are often quite simple, consisting of various techniques of:

- Plotting raw data
  o Techniques such as Data Traces, Histograms, Bi histograms, Probability Plot and Youden Plots
- Plotting sample statistics such as mean plots, standard deviation plots, box plots
- Positioning such plots to maximize our natural pattern-recognition abilities, such as using multiple plots per page.

The following diagram depicts the relation between data and attributes:
Descriptive analytics evaluates data through a process known as exploratory data analysis (EDA).

Model

Model contains ML algorithms (Supervised, Unsupervised, Reinforcement) that use transform data to generate desired outcomes. The output of the model is stored in the database or data store.

---

[88] What is EDA? - https://www.itl.nist.gov/div898/handbook/eda/section1/eda11.htm

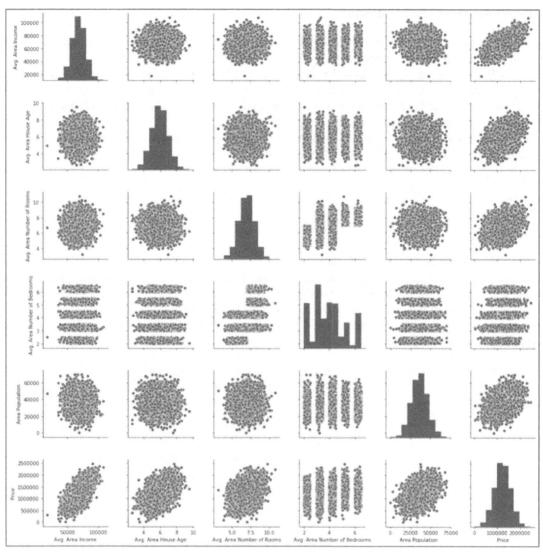

Figure 19: EDA

Visualization

Reporting and Visualization enable to view the transformed data and output of ML.

Some of the platforms generate recommendations in terms of actionable recommendations (real-time or near real-time)

> 4 out of 5 executives (81%) agree that within the next two years, AI will work next to humans in their organizations, as a co-worker, collaborator and trusted advisor.[89]
>
> Accenture

[89] Accenture Technology Vision. (2018). Redefine Your Company Based on the Company You Keep: Intelligent Enterprise Unleashed. https://www.accenture.com/t20180208T172438Z__w__/us-en/_acnmedia/Accenture/next-   gen-7/tech-vision-2018/pdf/Accenture-TechVision-2018-Tech-Trends-Report.pdf

Microsoft Azure—AI E2E Platform

Microsoft Azure is a Cloud-based big data compute, storage and infrastructure service. Azure provides various AI platform-based services, one of them being Azure Databricks.

Azure Databricks is an Apache Spark-based analytics platform optimized for the Microsoft Azure Cloud services platform. Designed with the founders of Apache Spark, Databricks is integrated with Azure to provide a one-click setup, streamlined workflows, and an interactive workspace that enables collaboration between data scientists, data engineers, and business analysts.

Azure Databricks[90] is a fast, easy, and collaborative Apache Spark-based analytics service. The E2E view of Azure Databricks contain the following major components:

- Ingestion
- Store
- Prepare & Train
- Mode & Serve

For a big data pipeline (please see Figure 20), the data (raw or structured) is ingested into Azure through Azure Data Factory in batches, or streamed near real-time using Kafka, Event Hub, or IoT Hub. This data lands in a data lake for long-term persisted storage, in Azure Blob Storage or Azure Data Lake Storage. As part of your analytics workflow, use Azure Databricks to read data from multiple data sources such as Azure Blob Storage, Azure Data Lake Storage, Azure Cosmos DB, or Azure SQL Data Warehouse and turn it into breakthrough insights using Spark.

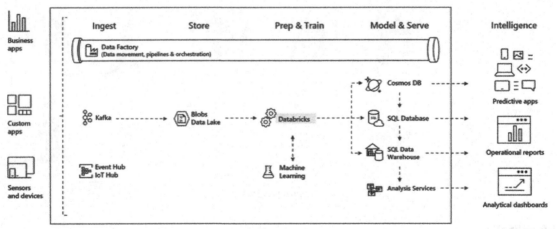

Figure 20: Data pipeline

AI Development Operations (DevOps) Loop for Data Science

The AI End-to-End platform described in the previous section is generally for product deployments. That is, running business critical customer facing applications.

To deploy such production platforms, the data science and development team would utilize developer tools with IT operations tools, the combination of both called DevOps (please see Figure 21). The DevOps

---

[90] What is Azure Databricks? https://docs.microsoft.com/en-us/azure/azure-databricks/what-is-azure-databricks

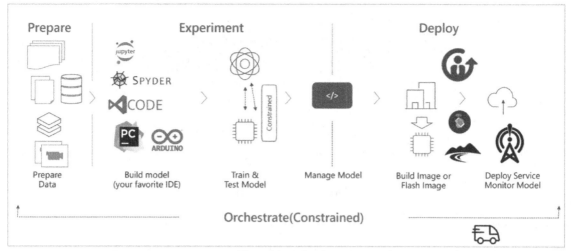

Figure 21: AI DevOps

platform would accelerate the development of AI applications with infrastructure developer services such as Source code repository, Bug tracking system, Version & Labeling systems and Project management systems.

In the following diagram a typical DevOps loop for Data Science teams is provided. In DevOps system the following are the major components:

- Data Preparation
- Experiment
- Deploy
- Orchestrate.

Data Preparation

Data Preparation is provision of data for the developer services. This process entails collating of different data sources, transforming the data sources, and ingesting data into a central repository.

Experiment

The basic constituent of every data science project or process is creation of an experiment. Experiment contains two major sequential steps: 1. Build Model and 2. Train & Test Model.

Build Model

Model building is the process of developing mathematical data science code that uses AI techniques (Machine Learning, Deep Learning) to model a business requirement, develop a hypothesis and test/validate the hypothesis to confirm the creations of the model.

Train & Test Model

The model was developed during the training phase of AI project and validated and tested for measuring the accuracy of the model. In the training phase, generally, the data is tested against seen data (please see Figure 22). And during the testing phase, the data is not seen or unknown.

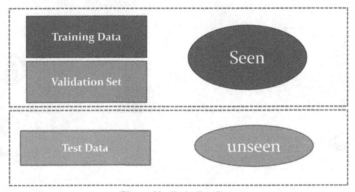

Figure 22: Train-Test Data

Deployment

AI Deployment phase contains building image of a model and deploying the model. Generally, this phase is closer to customer or services accessed by customers. In classical AI deployments, the targeted deployment would generally be a Cloud application or Software As A Service (SaaS) applications.

Edge deployments have become norm for many novel and advanced use cases.

Edge Devices

The detailed view of an Edge device consists of the following (please see Figure 23): Hardware is the base of the Edge platform from operation stack point of view. Components at the hardware include Random Access Memory (RAM (SRAM & DRAM)), Connectivity modules (Bluetooth Low Energy or Wi Fi), Power management (Battery or Electricity) and Storage component (External Storage, Flash Storage or Flat C File-based Storage).

Next, libraries are linked modules that provide abstraction layer on the hardware. Generally, the libraries provide system level services such as connectivity (Bluetooth Low Energy, Wi Fi), security (SSL) and run-time (LIB C). The ML modules process the data collected via Libraries to execute Edge analytics applications. The

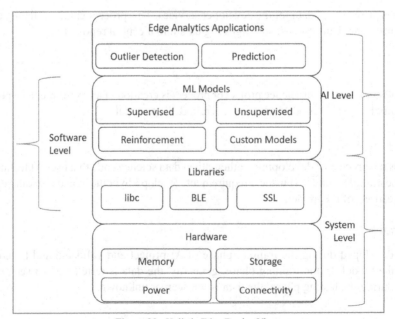

Figure 23: Holistic Edge Device View

ML modules can be broadly classified as Supervised Learning, Unsupervised Learning, Reinforcement and Custom Learning models. The Edge Analytics Applications is based on purpose built hardware. General Edge Analytics, in a broad view, include Outlier detection, Predictive applications and Decision-based applications.

For instance, following is the architecture of an Android smartphone—a classical Universal Edge.[91] The architecture consists of a hardware abstraction layer that consists of I/O and System Drivers such as Display Drivers, Camera Drivers, Flash Memory Drivers, Wi Fi Driver, Audio Drivers and Power Management. Hardware Abstraction Layer (HAL) consists of Libraries and Run-Time. Libraries include Surface Manager, SQLite, LIBC, SSL, SGL and other components. Applications and Application Framework layers use the Libraries and HAL to create applications on the Android platform (please see Figure 24).

Cost, limited size and minimal energy consumption are a few of the reasons that Edge devices have limited computational capabilities[92] [30]. Because of these functional and economic requirements, smart objects, especially those that are battery-powered, cannot afford to have heavy processing loads and use expensive communication protocols. On the one hand, limited processing capabilities means that it is hard to process large messages. On the other hand, less processing means lower energy consumption. As a result, Edge devices typically need to minimize the amount of transmitted data and processing capabilities [30].

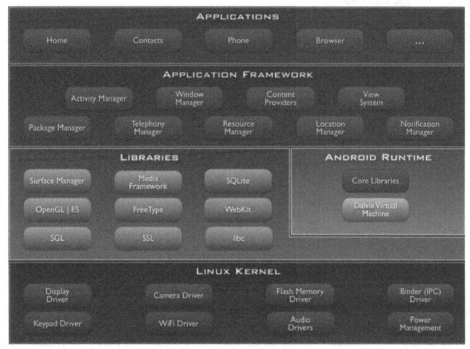

Figure 24: Android Kernel

ML Constrained Modeling

Constrained Modeling involves two phases. The first phase involves engineering design process of assessing device, infrastructure, environmental and data constraints. The second phase involves assessing ML modes from computational, operational and deployment point of view so that the models could perform with the utmost accuracy and energy efficiencies for meeting business objectives, technical requirements, and Service Level Agreements (SLA). Constrained modeling is very similar to that of Threat modeling in developing cyber-safe applications [8].

---

[91] Prototype: A Smartphone Universal Remote for the Internet of Things - https://thenewstack.io/smartphone-prototype-acts-like-universal-remote-internet-things/

[92] Internet of Things: Architectures, Protocols and Standards - Wiley; 1 edition (November 5, 2018)

-☼- Cyber Threat Modeling

In a typical Cyber Threat Modeling[93] of an application, the process of identification of threats, vulnerabilities and communication of mitigation strategies are the main theme. Similarly, in constrained modeling the process of identification of constraints on the ML model is the desired goal.

In the following diagram,[94] the process of instrumentation of attacks are listed. That is, the attack in Cyber event[95] involves the target "Unauthorized Result" and the means to achieve the result. This is called "attack enumeration". For instance, a design vulnerability can be exploited by using tools such as User Command, Script or Program, and Autonomous Agent to perform an action such as Scan, Flood, Spoof, Read on a target such as Process, Data, Computer, and Network to achieve the result Increased access, Disclosure of Information, Corruption of Information or Denial of Service.

In constrained modeling (please see Figure 25), the ML model needs to be evaluated based on environmental, hardware constraints, Connectivity and Business SLAs and Non-Functional aspects.

Constrained IoT Edge Devices

IoT devices are resource-constrained in terms of hardware resources (Memory, Connectivity, Power Consumption, Compute Processing) and usually with limited battery capacity.[96] Energy efficiency is the

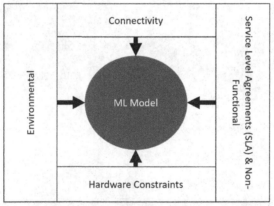

Figure 25: Tiny ML or Constrained Edge

crucial aspect for IoT. Majority of the IoT devices are resource constrained in nature. Therefore, battery or other constrained energy sources are used to operate it. IoT deployment scenarios are diverse, challenging and sometimes in very remote areas. Not only do device specific constraints play an important role but also external to device factors such as environmental, reachability, Terrain & Temperature Sensitivity (swings in temperature) play an integral part in the design and deployment of IoT devices. In turn, these internal and external constraints will have ML model design and AI inferences. There are several constraints applied to IoT devices:

- Hardware constraints
  o Device characteristics

---

[93] Application Threat Modeling - https://www.owasp.org/index.php/Application_Threat_Modeling
[94] A Common Language for Computer For Security Incidents - https://www.osti.gov/biblio/751004/
[95] Category: Threat Modeling - https://www.owasp.org/index.php/Category:Threat_Modeling
[96] IoT OS Management - https://www.ncbi.nlm.nih.gov/pmc/articles/PMC6514957/

- Operating environment constraints
- Connectivity
- SLAs & Non-Functional

The important factor to be noted is that these constraints will scope-in as part of ML model design and operating rubric.

The constrained space of Edge Computing-based [39] IoT is shown in Grid form (please see Figure 26).[97]

Infrastructure Constraints

IoT devices operate on geo-dispersed locations under various infrastructure limitations (please see Figure 27). For instance,

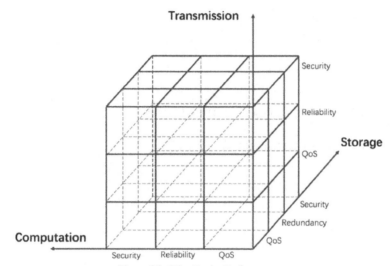

Figure 26: Edge Device Performance

- Intermittent or no Network Connectivity
- Electricity (lack of)
- Socio-economic factors.

Operating Environment

Operating environment poses a huge constraint on the IoT device price vs. performance. Operating environment constraints include:

- Geo-dispersed locations with limited reachability
- Hard terrain & rugged locations with high altitudes
- High humidity & temperature.

Device Characteristics

The form factor, price vs. performance and other physical limitations pose considerable constraints on the device

- Power (Electricity vs. Battery)

---

[97] A Survey on the Edge Computing for the Internet of Things - https://ieeexplore.ieee.org/document/8123913?denied=

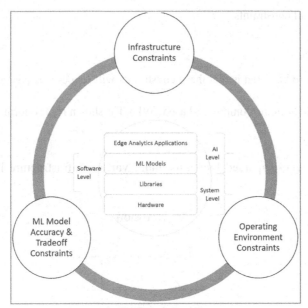

Figure 27: Edge Device Operational View

- Storage (on-device storage vs. remote storage)
- Memory
- Computational power
- Security
- Firmware footprint
- AI models accuracies vs. Performance Trade-offs.

In summary,[98] as depicted in the three-tier Clod-Fog-IoT devices model[99] [39][40]:

- **Memory constraints:** The memory of IoT devices is small, and most of the device's memory is used to store the embedded operating system. As a result, the system that uses IoT computing devices has limited memory to perform complex security protocols.

- **Speed of computation:** Almost all IoT computing devices have low-power processors; the processor needs to perform multiple tasks including managing, sensing, analyzing, saving,and communicating with a limited power source. As shown in Figure 27, the computational power and latency goes low on the Edge tier and this should be considered as part of the IoT device design. Therefore, forcing the processor to do the security procedure is a challenging issue.

- **Power consumption:** Most IoT devices have low battery capacity and computational power goes down at the IoT Edge devices due to form-factor and performance vs. cost benefits.

- **Scalability:** There is a sharp rise in the number of computing devices in the IoT network. Thus it is challenging to find the most suitable security algorithm for the growing number of devices in the IoT in the healthcare network.

- **Communication:** Channel IoT computing devices mostly participated in the network through multiple wireless communication protocols. As a result, it is challenging to find a standard security protocol that is suitable for various wireless communication protocols.

---

98 Survey of IoT - https://www.mdpi.com/2079-9292/8/7/768/pdf-vor
99 All one needs to know about fog computing and related edge computing paradigms: A complete survey - https://www.sciencedirect.com/science/article/pii/S1383762118306349

- **Security updates:** The security framework needs to be updated frequently to minimize potential security breaches. However, automatic updates also consume enormous power.

In the following diagram (Figure 28), the holistic view of device design and constraints view is presented.

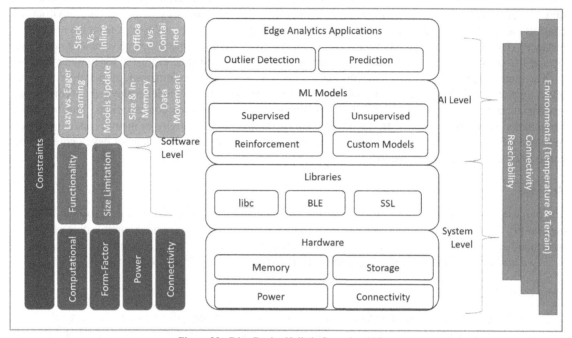

Figure 28: Edge Device Holistic Operational View

AI Performance and Computational Notations

The successful deployment of the AI model is predicated upon not only the model accuracy but also the fitness of the model in run-time environment. In order to evaluate a model fit, it is imperative to analyze the model algorithm performance with respect to traditional algorithm evaluation, check asymptotic behavior and impact of algorithm computation on the deployed infrastructure. The last component is the impact assessment of algorithm computation on the deployed infrastructure and it is vital when dealing with Tiny Edge devices, most of these devices being constrained, which are geographically distributed. Algorithmic complexity evaluation gives better footprint of an AI [9]. Algorithmic complexity[100] is concerned about how fast or slow particular algorithm performs. We define complexity as a numerical function $T(n)$—time versus the input size n. We want to define time taken by an algorithm without depending on the implementation details. However, you agree that $T(n)$ does depend on the implementation! A given algorithm will take different amounts of time on the same inputs depending on such factors as processor speed, instruction set, disk speed, brand of compiler, etc. The way around is to estimate efficiency of each algorithm asymptotically. We will measure time $T(n)$ as the number of elementary "steps" (defined in any way), provided each such step takes constant time [41].

Sample Complexity

Sample Complexity relates to data sampling [8]; the amount of data required to prepare the model. Other items to pay attention to include data availability, data holdout or stratification.

---

[100] Algorithmic complexity - https://www.cs.cmu.edu/~adamchik/15-121/lectures/Algorithmic%20Complexity/complexity.html

AI Algorithm and Computational Complexity

The domain of Computational Complexity applies to the model runtime. That is, when a model is developed with acceptable accuracy, the computational run-time needed to deploy the model to satisfy Service Level Agreements (SLA) as well as non-functional requirements such as concurrent model invocation, response time of model, and resiliency of the model play an important role. The runtime or computational environment controls the model execution and experiencing of user in getting inferences or insights from the model. For instance, AI model deployments serving in Large-scale Analytics environments, i.e., high velocity e-commerce sites or life-critical medical application or electronic health records, must be evaluated for runtime behavior such as model memory footprint, data workloads (streaming vs. batch) performance, model algorithm type (classification, prediction, time-series) and resilience. On the other hand, AI model deployment in constrained environments must be validated for model performance against resource consumptions. The impact of computation on power source (battery), or data storage needs no connectivity or memory data movement (between CPU and SRAM) on power source, or data collection for continuous model development under constrained geolocation deployments.

Analysis

The purpose of algorithm analysis is to check and then conclude how well a particular algorithm works in general [9]. Additionally, to document the algorithm in a performance agnostic manner with respect to a particular hardware or in other words how the evaluation on a Turing Hardware benchmarks.

Two measures help to analyze an algorithm: Time Complexity and Space Complexity.

Time Complexity

In classical algorithm analysis, Time Complexity is a function describing the amount of time an algorithm takes regarding the number of inputs to the algorithm. "Time" measures the number of memory accesses performed, the number of comparisons between integers, and the number of iterations the inner loop executes, since there are many factors unrelated to the algorithm that can affect the real time (like the language used, type of computing hardware, proficiency of the programmer, optimization in the compiler, etc.) [9]. Coming to AI algorithms, the time complexity varies with respect to data size, data process iterations and data depth level.

Syntax: $T(A, n)$ where T is the number of elementary instructions executed, A is algorithm and n denotes the size of the data input [9].

Space Complexity

In classical algorithm point of view, Space Complexity is a function describing the amount of memory (space) an algorithm takes regarding the number of inputs given to the algorithm. In AI point of view, the Space Complexity varies from Active vs. Lazy Learners; or varies from streaming vs. batch process.

Syntax: $S(n)$ where S is the space required by the set of functions computable in space, at most $c*S(n)$ for some constant $c > 0$, where n is the input size [9].

Algorithm Performance Metrics—The Asymptotic Notions

Algorithm performance metrics can be derived on how long an algorithm takes to run, and importantly, the order of growth of running time; this measure de-emphasizes on running time (scalar) and provides more emphasis on growth of running the algorithm.

The Asymptotic notation is a shorthand way to write down and talk about 'the fastest possible' and 'the slowest possible' running times for an algorithm, utilizing highest and lowest bounds on speed [9].

Big-O Notation (O-Notation)

For any monotonic functions $f(n)$ and $g(n)$ from the positive integers to the positive integers, we say that $f(n) = O(g(n))$ when there exist constants $c > 0$ and $n_0 > 0$ such that

$$f(n) \leq c * g(n), \text{ for all } n \geq n_0$$

Intuitively, this means that function $f(n)$ does not grow faster than $g(n)$, or that function $\mathbf{g}(n)$ is an upper bound for $f(n)$, for all sufficiently large $n \rightarrow \infty$

Read it as "f of n is big O of g of n" or "f is big O of g") [9]

⚙ Monotonic Functions [41]

A function $f(n)$ is monotonically increasing, if for $m \leq n \Rightarrow f(m) \leq f(n)$ and function $f(n)$ is monotonically decreasing, if for $m < n \Rightarrow f(m) \geq f(n)$

Similarly, we say that a function $f(n)$ is strictly increasing, if for $m < n \Rightarrow f(m) < f(n)$ and function $f(n)$ is strictly decreasing, if for $m < n \Rightarrow f(m) \geq f(n)$.

Here is a graphic representation of $f(n) = O(g(n))$ relation [9, 41]—please see Figures 29 and 30:

O-notation is a "tight upper bound notation" and it provides maximum value of running time for any given function [9].

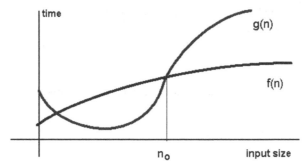

Figure 29: Big O $f(n)$ and $g(n)$

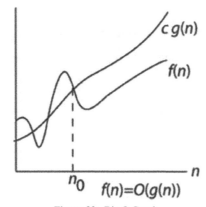

Figure 30: Big O Graph

Constant Time: O (1)

An algorithm is said to run in constant time if it requires the same amount of time regardless of the input size [41]. Examples:

- array: accessing any element
- fixed-size stack: push and pop methods
- fixed-size queue: enqueue and dequeue methods

Linear Time: O (n)

An algorithm is said to run in linear time if its time execution is directly proportional to the input size, i.e., time grows linearly as input size increases. Examples:

- array: linear search, traversing, find minimum
- ArrayList: contains method
- queue: contains method

Logarithmic Time: O (log n)

An algorithm is said to run in logarithmic time if its time execution is proportional to the logarithm of the input size. Example: This is an example of the general problem-solving method known as binary search:

Locate the element A in a sorted (in ascending order) array by first comparing A with the middle element and then (if they are not equal) dividing the array into two subarrays; if A is less than the middle element you repeat the whole procedure in the left subarray, otherwise—in the right subarray. The procedure repeats until A is found or subarray is a zero dimension.

Note, $\log(n) < n$, when $n \rightarrow \infty$. Algorithms that run in O(log n) does not use the whole input.

Quadratic Time: O (n^2)

An algorithm is said to run in logarithmic time if its time execution is proportional to the square of the input size. Examples:

- bubble sort, selection sort, insertion sort

Omega Notation (Ω Notation)

We need the notation for the lower bound [41]. A capital omega Ω notation (Figure 31) is used in this case. We say that $f(n) = \Omega(g(n))$ when there exists constant c that

$$f(n) \geq c * g(n) \text{ for all sufficiently large } n.$$

Ω notation is a "tight lower bound notation", and it provides the non-negative minimum value of running time for any given function.

Examples of Omega functions include [41]:

- $n = \Omega(1)$
- $n2 = \Omega(n)$
- $n2 = \Omega(n \log(n))$
- $2 n + 1 = O(n)$

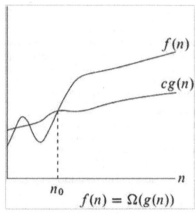

Figure 31: Omega Notation

Theta Notation (Θ Notation)

To measure the complexity of a particular algorithm, means to find the upper and lower bounds, Θ notation is used (Figure 32).

We say that $f(n) = \Theta(g(n))$ if and only $f(n) = O(g(n))$ and $f(n) = \Omega(g(n))$ [41]

Another way of defining:

If, $0 \leq C1.g(n) \leq f(n) \leq C2.g(n)$, then $f(n) = \Theta(g(n))$ for all $n \geq n_0$

Here, $f(n)$ and $g(n)$ are two functions, n is the set of positive integers, and n_0 is an input until which the above-given condition may or may not follow. However, after it, the relation is always valid. As shown in the graph below:

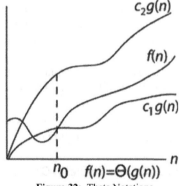

Figure 32: Theta Notations

Θ-Notation is "asymptotically tight bound" because of its equivalence from both upper and lower bound [41]. Examples:

- $2n = \Theta(n)$
- $n2 + 2n + 1 = \Theta(n^2)$

Algorithm performance metric is a scalar quantity with respect to time and provides guidance as to how fast an algorithm is, slow and average speeds of algorithm.

Performance Metrics	
Worst-case complexity	For a given input size, the algorithm takes the longest time; Complexity (number of times the basic operation executed) for the worst case is found for input of size n, hence the algorithm runs for longest time as compared to all possible inputs of size n. If complexity is indicated by function $f(n)$ then in worst case, it is indicated by the maximum value of $f(n)$ for any possible input [9]. **Worst-case runtime complexity is $O(n)$ [41]**
Best-case complexity	Algorithm execution time is the fastest; Complexity (number of times the basic operation executed) for the best case is calculated for input of size n, when the algorithm runs the fastest as compared with all possible inputs of size n. Here, the value of $f(n)$ is minimum for any possible input [9]. ***Best-case runtime complexity is $O(1)$ [41]***
Average-case complexity	Average time was taken (number of times the basic operation executed) to solve all the possible instances (random) of the input. The value of $f(n)$ lies in between maximum and minimum for any possible input. Note: Here the average does not mean by the average of worst and best case [9]. Average case runtime complexity is $O(n/2) = O(n)$ [41]

Space and Time Constraint Benchmarks

One of the criteria to evaluate computational complicity of an algorithm is to publish generic run time performance, i.e., performance agnostic to a particular machine. The actual runtime (in seconds) of an algorithm depends on the specific machine the algorithm is being implemented on (e.g., what the clock rate of the machine's CPU is or what is the RAM memory; or what is the data transfer rate between CPU and memory). To avoid dependence on the specific machine, it is common to analyze the runtime of algorithms in an asymptotic sense [8]. For example, we say that the computational complexity of the merge-sort algorithm, which sorts a list of n items, is $O(n \log (n))$. This implies that we can implement the algorithm on any machine that satisfies the requirements of some accepted abstract model of computation, and the actual runtime in seconds will satisfy the following: there exist constants c and n_0, which can depend on the actual machine, such that, for any value of $n > n_0$, the runtime in seconds of sorting any n items will be at most $c \, n \log (n)$. It is common to use the term feasible or efficiently computable for tasks that can be performed by an algorithm whose running time is $O(p(n))$ for some polynomial function p [8]. Specifically, for constrained environments, we need to access the complexity of an algorithm with respect to device resources that are scarce such as memory, battery power, computational power and need to access with operating environmental factors such as connectivity of the device, reachability and serviceability.

In this book, we focus on computational complexity, as it is very vital to evaluate the model performance to make sure model performs with acceptable SLAs in constrained and limited compute environments.

AI for Greater Good—Solving Humanity and Societal Challenges

The world population is expected to hit 9.8 billion by 2050,[101] according to the United Nations [42][43], and experts warn that if scientists do not find more efficient ways to use and protect limited agricultural resources such as land, water, and energy, there could be a global food crisis. Moreover, manifestation of current heavy resource dependence and intensive agricultural practices have resulted in unsustainable agricultural practices (Figure 33). The direct effects of unsustainable agricultural practices include degradation of land, reduction in soil fertility, increased dependence on inorganic fertilizers for higher production, rapidly dropping water tables and emerging pest resistance.

Aggravating the delicate balance, the global warming and the climate change exerting unprecedented pressure on agriculture productivity and forced rapid decline of natural resources, causing food crisis and food

---

[101] World population projected to reach 9.8 billion in 2050, and 11.2 billion in 2100 - https://www.un.org/development/desa/en/news/population/world-population-prospects-2017.html

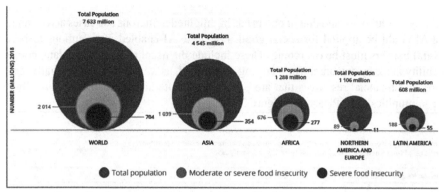

Figure 33: Prevalence of Moderate and Severe Food Insecurity in the Population

security,[102] major social order and governance issues across the world[103] [42]. It is for this reason that eliminating world hunger is a top priority among the UN's Sustainable Development Goals, which the supranational organization hopes to attain by 2030.[104]

 Food for Future!

One in nine people, or 795 million people, do not have enough food to lead a healthy active life. For the third year in a row, world hunger has increased — in 2017, around 821 million people faced undernourishment from chronic food deprivation, the Food and Agriculture Organization of the United Nations reported.[105]

Machine learning (ML) and artificial intelligence (AI) are among the new technologies' leaders are relying on to improve agriculture productivity and help alleviate a global food crisis. AI holds promise of driving the food revolution and meeting the increased demand for food[106] (global need to produce 50% more food and cater to an additional 2 billion people by 2050 as compared to today) [5].

 The Role of ML in Food Security[107]

Data can be used to monitor the risk of food insecurity in real time, to forecast near-term shortages, and to identify areas at risk in the long-term, all of which can guide interventions.

For real-time and near-term systems, it is possible to distill relevant signals from mobile phones, credit card transactions, and social media data. These have emerged as low-cost, high-reach alternatives to manual surveying. The idea is to train models that link these large, but decontextualized, data with ground truth consumption or survey information, collected on small representative samples. This process of *developing proxies to link small, rich datasets with large, coarse ones can be viewed as a type of semi-supervised learning and is fertile ground for research.*

Already, ML and AI technologies are predicting impoverished regions across the globe, using the data to find solutions to mitigate global hunger. AI has broad potential across range of human domains: Equity & Inclusion, Education, Health & Hunger, Security & Justice, Information Verification & Validation, Crisis Response, Economic Empowerment, Public & Social Sector, Environment and Infrastructure.[108] AI technologies make huge impacts in the social domains and could lead to less disparity between the haves and have-nots [44].

---

[102] Food and Agriculture Organization of the United Nations - http://www.fao.org/sustainable-development-goals/indicators/212/en/
[103] Food - https://www.un.org/en/sections/issues-depth/food/index.html
[104] How AI can help fight Poverty - https://www.delltechnologies.com/en-us/perspectives/how-ai-can-help-fight-poverty/
[105] The United Nations – Shaping our Future together - https://www.un.org/en/sections/issues-depth/food/index.html
[106] National Strategy for AI - https://niti.gov.in/writereaddata/files/document_publication/NationalStrategy-for-AI-Discussion-Paper.pdf
[107] Tackling Climate Change with Machine Learning - https://arxiv.org/pdf/1906.05433.pdf
[108] Applying AI for Social Good - https://www.mckinsey.com/featured-insights/artificial-intelligence/applying-artificial-intelligence-for-social-good

-☼- McKinsey:[109] Our work and that of others has highlighted numerous use cases across many domains where AI could be applied for social good. For these AI-enabled interventions to be effectively applied, several barriers must be overcome. These include the usual challenges of data, computing, and talent availability faced by any organization trying to apply AI, as well as more basic challenges of access, infrastructure, and financial resources that are particularly acute in remote or economically challenged places and communities [44]. Please see Figure 34.

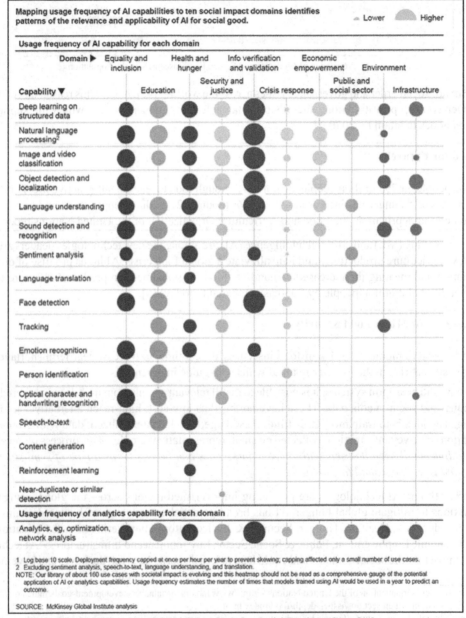

Figure 34: AI Capabilities Mapped to Social Good Use Cases [44]

---

[109] The challenge and promise of AI - https://www.mckinsey.com/featured-insights/artificial-intelligence/the-promise-and-challenge-of-the-age-of-artificial-intelligence

> **Chapter Summary:**
>
> After reading the chapter you should comfortably answer:
>
> • The definition of AI and the role of ML techniques in AI.
>
> • Deep Learning and its role in AI
>
> • Historical compute development of AI.
>
> • Current and future trends of AI.
>
> • End-to-End architecture of AI from production and development system's point of view.
>
> • The role of AI in social good.
>
> • AI algorithm benchmarks and performance metrics.

References

1. Michael Chui, James Manyika, Mehdi Miremadi, Nicolaus Henke, Rita Chung, Pieter Nel and Sankalp Malhotra. Notes from the AI frontier: Applications and value of deep learning. April 2018, https://www.mckinsey.com/featured-insights/artificial-intelligence/notes-from-the-ai-frontier-applications-and-value-of-deep-learning, Access Date: 09/18/2019.

2. Prabhat Kumar, Artificial Intelligence: Reshaping Life and Business, Chapter 2, BPB Publications; 1 edition (August 17, 2019), ISBN-10: 9388511077.

3. Jackie Snow. How artificial intelligence can tackle climate change. July 18, 2019, https://www.nationalgeographic.com/environment/2019/07/artificial-intelligence-climate-change/#close, Access date: 09/27/2019.

4. Karen Hao. Here are 10 ways AI could help fight climate change. June 20, 2019, https://www.technologyreview.com/s/613838/ai-climate-change-machine-learning/, Access Date: Nov. 08, 2019.

5. Rolnick, David, Donti, Priya L. Kaack, Lynn H. Kochanski, Kelly, Lacoste, Alexandre, Sankaran, Kris, Slavin Ross, Andrew, Milojevic-Dupont, Nikola, Jaques, Natasha, Waldman-Brown, Anna, Luccioni, Alexandra Maharaj, Tegan, Sherwin, Evan D., Karthik Mukkavilli, S., Kording, Konrad P., Gomes, Carla, Ng, Andrew Y., Hassabis, Demis, Platt, John C. Creutzig, Felix Chayes, Jennifer, Bengio, Yoshua. Tackling Climate Change with Machine Learning. Publication: eprint arXiv:1906.05433, "https://arxiv.org/pdf/1906.05433.pdf", June 2019 arXiv: arXiv:1906.05433 Bibcode: 2019arXiv190605433R, Access Date: Sep. 18, 2019.

6. Russell Stuart and Peter Norvig. Artificial Intelligence: A Modern Approach (3rd edition), Publisher: Pearson (December 11, 2009), ISBN-10: 0136042597.

7. Matt Taddy, Business Data Science: Combining Machine Learning and Economics to Optimize, Automate, and Accelerate Business Decisions 1st Edition, Publisher: McGraw-Hill Education; (August 21, 2019), ISBN-10: 1260452778.

8. Shai Shalev-Shwartz and Shai Ben-David, Understanding Machine Learning: Theory Cambridge University Press; 1 edition (May 19, 2014), ISBN-10: 1107057132.

9. Shefali Singhal and Neha Garg, Analysis and Design of Algorithms: A Beginner's Hope, Chapter 2—Complexity of Algorithms, Publisher: BPB PUBLICATIONS (May 31, 2018), ISBN-10: 9386551896.

10. Jiawei Han, Micheline Kamber and Jian Pei. Data Mining: Concepts and Techniques. Morgan Kaufmann; 3 edition (June 15, 2011), ISBN-10: 9780123814791.

11. Dan Vesset. Descriptive analytics 101: What happened? May 10, 2018, https://www.ibm.com/blogs/business-analytics/descriptive-analytics-101-what-happened/Access Date: April 30, 2019.

12. Yijing Chen, Dmitry Pechyoni, Angus Taylor, and Vanja Paunic. Recurrent neural networks for time series forecasting. 9 October 2018, https://conferences.oreilly.com/artificial-intelligence/ai-eu-2018/public/schedule/detail/70226, Access Date: April 30, 2019.

13. Dan Vesset. Prescriptive analytics 101: What should be done about it? May 15, 2018, https://www.ibm.com/blogs/business-analytics/prescriptive-analytics-done/, Access Date: Sep 18, 2019.

14. A. Ilapakurti, S. Kedari, J.S. Vuppalapati, S. Kedari and C. Vuppalapati. Artificial Intelligent (AI) Clinical Edge for Voice disorder Detection. IEEE Fifth International Conference on Big Data Computing Service and Applications (BigDataService), Newark, CA, USA, 2019, pp. 340–345, doi: 10.1109/BigDataService.2019.00060, URL: http://ieeexplore.ieee.org/stamp/stamp.jsp?tp=&arnumber=8848238&isnumber=8848193.

15. Niko Mohr and Holger Hürtgen. Achieving business impact with data. 04.2018, URL: https://www.mckinsey.com/~/media/mckinsey/business%20functions/mckinsey%20analytics/our%20insights/achieving%20business%20impact%20with%20data/achieving-business-impact-with-data_final.ashx, Access Date: Sep 18, 2019.

16. Eric Griffith. 90 Percent of the Big Data We Generate is an Unstructured Mess. November 15, 2018 5:00AM EST, URL: https://www.pcmag.com/news/364954/90-percent-of-the-big-data-we-generate-is-an-unstructured-me, Access Date: Nov. 22, 2019.

17. Sanjay Ghemawat, Howard Gobioff and Shun-Tak Leung. The Google File System. Proceedings of the 19th ACM Symposium on Operating Systems Principles, ACM, Bolton Landing, NY (2003), pp. 20–43, URL: https://research.google/pubs/pub51/, Access Date: Jan. 08, 2019.

18. Jeffrey Dean and Sanjay Ghemawat. MapReduce: simplified data processing on large clusters. Publication: Communications of the ACM, January 2008, URL: https://doi.org/10.1145/1327452.1327492, Access Date: Jan. 02, 2019.

19. Matei Zaharia, Mosharaf Chowdhury, Tathagata Das, Ankur Dave, Justin Ma, Murphy McCauly, Michael J. Franklin, Scott Shenker and Ion Stoica. Resilient Distributed Datasets: A Fault-Tolerant Abstraction for In-Memory Cluster Computing. Presented as part of the 9th {USENIX} Symposium on Networked Systems Design and Implementation ({NSDI}, Date: 2012, ISBN: 978-931971-92-8, URL: https://www.usenix.org/conference/nsdi12/technical-sessions/presentation/zaharia, Access Date: Nov. 14, 2019.

20. Kai Hwang and Min Chen, Big-Data Analytics for Cloud, IoT and Cognitive Computing, Wiley; 1 edition (August 14, 2017), ISBN-10: 9781119247029.

21. Jason Brownlee. A Tour of Machine Learning Algorithms. August 12, 2019, URL: https://machinelearningmastery.com/a-tour-of-machine-learning-algorithms/, Access Date: Sep. 18, 2019.

22. Susan Li. Building a Logistic Regression in Python, Step by Step. Sept. 28, 2017, URL: https://towardsdatascience.com/building-a-logistic-regression-in-python-step-by-step-becd4d56c9c8, Access Date: Nov. 22, 2019.

23. Vishal Morde. XGBoost Algorithm: Long May She Reign. Apr. 7, 2019, URL: https://towardsdatascience.com/https-medium-com-vishalmorde-xgboost-algorithm-long-she-may-rein-edd9f99be63d, Access Date: Sept. 18, 2019.

24. C. Vuppalapati, R. Vuppalapati, S. Kedari, A. Ilapakurti, J.S. Vuppalapati and S. Kedari. Fuzzy Logic Infused Intelligent Scent Dispenser for Creating Memorable Customer Experience of Long-Tail Connected Venues. 2018 International Conference on Machine Learning and Cybernetics (ICMLC), Chengdu, 2018, pp. 149–154. doi: 10.1109/ICMLC.2018.8527046,URL: http://ieeexplore.ieee.org/stamp/stamp.jsp?tp=&arnumber=8527046&isnumber=8526918, Access Date: April 30, 2019.

25. Michael Chui, James Manyika, Mehdi Miremadi, Nicolaus Henke, Rita Chung, Pieter Nel and Sankalp Malhotra. Notes from the AI frontier: Applications and value of Deep Learning. April 2018, URL: https://www.mckinsey.com/featured-insights/artificial-intelligence/notes-from-the-ai-frontier-applications-and-value-of-deep-learning, Access Date: Jan. 02, 2019.

26. Michael J., Apter, Artificial Intelligence: The Computer Simulation of Behaviour (Volume 1), Routledge; 1 edition (May 15, 2018), ISBN-10: 113849660X.

27. Neil Sahota, Own the A.I. Revolution: Unlock Your Artificial Intelligence Strategy to Disrupt Your Competition, McGraw-Hill Education; 1 edition (May 22, 2019), ISBN-10: 1260458377.

28. Theodore Forbath. The third wave of computing. October 3, 2013, URL: https://fortune.com/2013/10/03/the-third-wave-of-computing/, Access Date: June 24, 2019.

29. James Manyika, Michael Chui, Jacques Bughin, Richard Dobbs, Peter Bisson, and Alex Marrs. Disruptive technologies: Advances that will transform life, business, and the global economy. May 2013, URL: https://www.mckinsey.com/~/media/McKinsey/Business%20Functions/McKinsey%20Digital/Our%20Insights/Disruptive%20technologies/MGI_Disruptive_technologies_Full_report_May2013.ashx, Access Date: June 23, 2019.

30. Simone Cirani, Gianluigi Ferrari, Marco Picone, Luca Veltri, Internet of Things: Architectures, Protocols and Standards, Wiley; 1 edition (Nov. 5, 2018), ISBN-10: 1119359678.

31. Simon Segars, Chief Executive Officer, Arm. The Fifth Wave of Computing Is Built on AI, 5G and a Secure IoT. April 17, 2019, URL: https://www.arm.com/blogs/blueprint/the-fifth-wave-of-computing-ai-5g-iot, Access Date: Nov. 8, 2019.

32. Ray Eitel-Porter and Athina Kanioura. ExplAIned—A guide for executives. September 21, 2018, URL: https://www.accenture.com/us-en/insights/artificial-intelligence/artificial-intelligence-explained-executives, Access Date: Nov 22, 2019.

33. Lynsey Barber. Gartner Hype Cycle 2017: Artificial intelligence at peak hype, blockchain heads for disillusionment, but say hello to 5G. 17 August 2018, URL: "https://www.cityam.com/gartner-hype-cycle-2017-artificial-intelligence-peak-hype/", Access Date: Nov 22, 2019.

34. Alex Hickey. Gartner serves up 2018 Hype Cycle with a heavy side of AI", Aug. 20, 2018, URL: https://www.ciodive.com/news/gartner-serves-up-2018-hype-cycle-with-a-heavy-side-of-ai/530385/, Access Date: Sep. 18, 2019.

35. Laurence Goasduff. Top Trends on the Gartner Hype Cycle for Artificial Intelligence, 2019. Sept. 12, 2019, URL: https://www.gartner.com/smarterwithgartner/top-trends-on-the-gartner-hype-cycle-for-artificial-intelligence-2019/, Access Date: December 25, 2019.

36. Carlton Sapp. Laying the Foundation for Artificial Intelligence and Machine Learning: A Gartner Trend Insight Report. 20 Sept. 2018 ID: G00373110, URL: https://emtemp.gcom.cloud/ngw/globalassets/en/doc/documents/3890363-laying-the-foundation-for-artificial-intelligence-and-machine-learning-a-gartner-trend-insight-report.pdf, Access Date: Nov. 14, 2019.

37. Katie Costello. Gartner Says AI Augmentation Will Create $2.9 Trillion of Business Value in 2021. August 5, 2019, URL:https://www.gartner.com/en/newsroom/press-releases/2019-08-05-gartner-says-ai-augmentation-will-create-2point9-trillion-of-business-value-in-2021, Access Date: December 23, 2019.
38. Rob Thomas and Patrick McSharry, Big Data Revolution: What farmers, doctors and insurance agents teach us about discovering big data patterns, Wiley, 1st edition (March 2, 2015), ISBN-10: 9781118943717.
39. Dang, L.M., Piran, M.J., Han, D., Min, K. and H.A. Moon. Survey on Internet of Things and Cloud Computing for Healthcare", Electronics, Vol. 8, p.768, 9 July 2019, URL: https://www.mdpi.com/2079-9292/8/7/768/htm, Access Date: Nov. 22, 2019.
40. Ashkan Yousefpour. Caleb Fung, Tam Nguyen, Krishna Kadiyala, Fatemeh Jalali, Amirreza Niakanlahiji, Jian Kong, and Jason P. Jue. All one needs to know about fog computing and related edge computing paradigms: A complete survey. Journal of Systems Architecture, Vol. 98, September 2019, pp. 289–330, https://www.sciencedirect.com/science/article/pii/S1383762118306349, Access Date: December 23, 2019.
41. Victor S. and Adamchik, C.M.U. Algorithmic Complexity, 2009. https://www.cs.cmu.edu/~adamchik/15121/lectures/Algorithmic%20Complexity/complexity.html.
42. Lisa Rabasca Roepe. How AI Can Help Fight Poverty. Nov. 14, 2018, URL: https://www.delltechnologies.com/en-us/perspectives/how-ai-can-help-fight-poverty/, Access Date: December 07, 2019.
43. The United Nations Department of Economic and Social Affairs. World population projected to reach 9.8 billion in 2050, and 11.2 billion in 2100. 21 June 2017, URL: https://www.un.org/development/desa/en/news/population/world-population-prospects-2017.html, Access Date: Dec. 15, 2019.
44. James Manyika and Jacques Bughin. The promise and challenge of the age of Artificial Intelligence. Oct. 2018, URL: https://www.mckinsey.com/featured-insights/artificial-intelligence/the-promise-and-challenge-of-the-age-of-artificial-intelligence, Access Date: April 30, 2019.

## CHAPTER 2

Standard Processes and Frameworks

"Every digital transformation is going to begin and end with the customer, and I can see that in the minds of every CEO I talk to."

Marc Benioff

I was a Cloud & AI consultant at the Samsung US Headquarter in San Jose, CA. Many times, before approval of a new AI initiative, I heard from my manager, "What is the Business Value of AI?" Similarly, when I was consulting for a Construction Management Company in Portland, a product manager I worked with had a similar question from a different point of view: "Chandra, can you provide a C Level strategy & AI playbook that provides a standard operating procedure (SOP) to onboard AI initiatives from different line of businesses in our company?" In other words, "the game plan to onboard and execute AI project that deliver business value." During my work as a strategic business consultant at Cisco Systems, San Jose, my departmental charter was to interface with various Cisco Business groups to bring Analytics as a competitive differentiator to the Cisco enterprise. No wonder we were called "Connected Analytics Group". One of the vital responsibilities, as part of our business charter, was to deliver AI Long-Range Plan (LRP) to executives and business stakeholders. Two key metrics were required to be prepared for executive briefings: AI Total Addressable Market (TAM) and Competitive AI strategies. Needless to mention, successful LRP has a clear value-strategy outlined.

In this chapter, I would like to provide frameworks that answer key business imperatives to develop compelling AI strategy for the companies and organizations. This chapter introduces the cross industry standard process for Artificial Intelligence both from business and from software engineering point of view. The Digital Transformation, the Digital Feedback Loop, the Insights Value Chain, and the New Product Introduction are covered from Business Users point of view. The Knowledge Discovery in Databases (KDD), Lifecycle for Developing Learning System, Cross Industry Standard Process for Data Mining (CRISP-DM), Data Mining Compute Reference Architecture, Cloud to Edge Learning Systems are covered from data scientists, data engineers and data architects. Finally, the chapter concludes with explainable AI model—DARPA Explainable Artificial Intelligence (XAI).

Artificial Intelligence is everywhere and is shaping our personal, professional and business lives. AI is helping to improve the net value of businesses by developing improved consumer-friendly products, blue-ocean, latent or out-of-box products and monetization data services [1]. Many global organizations (84%) believe that investing in AI will lead to greater competitive advantage [2]. A recent report by Statista (Figure 1), one of the most successful statistical databases in the world, on its organizations' reasons for adopting AI has provided the following major decision factors [2][3]:

- AI will allow us to obtain or sustain a competitive advantage
- AI will allow us to move into new businesses
- New organizations using AI will enter our market
- Incumbent competitors will use AI
- Pressure to reduce costs will require us to use AI
- Suppliers will offer AI driven products and services, and
- Customer will ask for AI-driven offerings.

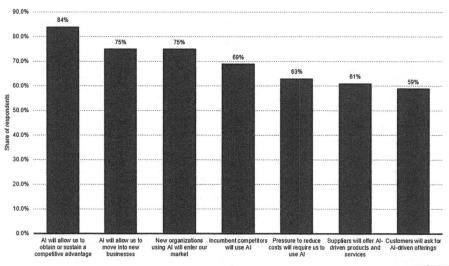

Figure 1: Business Organizations' Reasons for Adopting AI Worldwide, as of 2017 [3]

No wonder, the size of AI influence on businesses is phenomenal and continues to grow [1]. One measure of growth, albeit, can be witnessed from the global spending rate on AI projects. IDC, a global think tank on IT strategies and market research firm, confirms that the global spending on AI systems is expected to maintain its strong growth[110] trajectory as businesses continue to invest in projects that utilize the capabilities of AI software and platforms. Spending on AI systems will reach $97.9 billion in 2023, more than two- and one-half times the $37.5 billion that will be spent in 2019. The compound annual growth rate (CAGR) for the 2018–2023 forecast period will be 28.4% [4] (Figure 2). Global revenues from AI for enterprise applications is projected[111] to grow from $1.62B in 2018 to $31.2B in 2025, attaining a 52.59% CAGR in the forecast period [2].

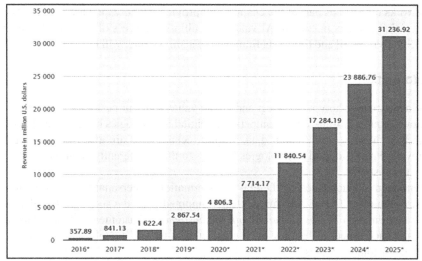

Figure 2: Revenues from the artificial intelligence for enterprise applications market worldwide, from 2016 to 2025 (in million U.S. dollars) [2][4]

---

[110] Worldwide Spending on Artificial Intelligence Systems Will Be Nearly $98 Billion in 2023, According to New IDC Spending Guide - https://www.idc.com/getdoc.jsp?containerId=prUS45481219

[111] Revenues from the artificial intelligence for enterprise applications market worldwide, from 2016 to 2025 (in million U.S. dollars) - https://www.statista.com/statistics/607612/worldwide-artificial-intelligence-for-enterprise-applications/

		Illustrative rates of technology improvement and diffusion	Illustrative groups, products, and resources that could be impacted[1]	Illustrative pools of economic value that could be impacted[1]
	Mobile Internet	**$5 million vs. $400**[2] Price of the fastest supercomputer in 1975 vs. that of an iPhone 4 today, equal in performance (MFLOPS) **6x** Growth in sales of smartphones and tablets since launch of iPhone in 2007	**4.3 billion** People remaining to be connected to the Internet, potentially through mobile Internet **1 billion** Transaction and interaction workers, nearly 40% of global workforce	**$1.7 trillion** GDP related to the Internet **$25 trillion** Interaction and transaction worker employment costs, 70% of global employment costs
	Automation of knowledge work	**100x** Increase in computing power from IBM's Deep Blue (chess champion in 1997) to Watson (Jeopardy winner in 2011) **400+ million** Increase in number of users of intelligent digital assistants like Siri and Google Now in past 5 years	**230+ million** Knowledge workers, 9% of global workforce **1.1 billion** Smartphone users, with potential to use automated digital assistance apps	**$9+ trillion** Knowledge worker employment costs, 27% of global employment costs
	The Internet of Things	**300%** Increase in connected machine-to-machine devices over past 5 years **80–90%** Price decline in MEMS (microelectromechanical systems) sensors in past 5 years	**1 trillion** Things that could be connected to the Internet across industries such as manufacturing, health care, and mining **100 million** Global machine to machine (M2M) device connections across sectors like transportation, security, health care, and utilities	**$36 trillion** Operating costs of key affected industries (manufacturing, health care, and mining)
	Cloud technology	**18 months** Time to double server performance per dollar **3x** Monthly cost of owning a server vs. renting in the cloud	**2 billion** Global users of cloud-based email services like Gmail, Yahoo, and Hotmail **80%** North American institutions hosting or planning to host critical applications on the cloud	**$1.7 trillion** GDP related to the Internet **$3 trillion** Enterprise IT spend
	Advanced robotics	**75–85%** Lower price for Baxter[3] than a typical industrial robot **170%** Growth in sales of industrial robots, 2009–11	**320 million** Manufacturing workers, 12% of global workforce **250 million** Annual major surgeries	**$6 trillion** Manufacturing worker employment costs, 19% of global employment costs **$2–3 trillion** Cost of major surgeries

Figure 3: McKinsey—Global Major Influencer [5]

It is no wonder that AI, Automation of Knowledge Work, is slated to have a potential impact of $5 to $7 trillion economic value by 2025 (Figure 3) [5].

Artificial intelligence and machine learning are top of mind for most organizations today, and IDC expects that AI will be the disrupting influence changing entire industries over the next decade [4]. Nonetheless, one of the challenges that many strategic decision makers across all industries are now grappling with is the question of how effectively they will proceed with their AI journey [4]. How to adopt the changing dynamics of AI markets and what is the success mantra for AI initiative?

The frameworks outlined as part of the chapter will provide a landscape view of identifying a potential business use case that delivers maximum AI value. The ultimate success of any AI initiative is its positive impact on the people's lives [6] and following frameworks in this chapter will insure the successful AI journey.

Digital Transformation

Digital Technologies change how we connect and create value with our customers [7]. Digital technologies transforming how we need to think about competition. Digital technologies have changed our world perhaps most significantly in how we think about data. Digital technologies are also transforming the ways that businesses innovate. Finally, digital technologies force us to think differently about how we understand and create value for the customer [7].

Having mentioned so many facets of "Digital Transformation", I personally feel there is no one definition that fits all. Marc Benioff, CEO of Salesforce, clearly summed[112] up the sole of "Digital Transformation"- "every digital transformation *begins and ends with the customer*". Customer success is the basis for digital transformation and each aspect of business well connected as part of digital transformation as indicated in Figure 4 to deliver customer success.

The Digital Transformation strategy [7] focuses on all the components of modern digital footprint graph: Customers, Competition, Data, Innovation and Value.

---

[112] Digital Transformation - https://www.salesforce.com/products/platform/what-is-digital-transformation/

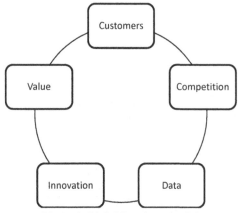

Figure 4: Digital Transformation [7]

Digital Transformation at Salesforce Definition

Digital transformation is the process of using digital technologies to create new or modify existing business processes, culture, and customer experiences to meet changing business and market requirements. This reimagining of business in the digital age is digital transformation. It transcends traditional roles like sales, marketing, and customer service. Instead, digital transformation begins and ends with how you think about, and engage with, customers. As we move from paper to spreadsheets to smart applications for managing our business, we have the chance to reimagine how we do business—how we engage our customers—with digital technology on our side.[113]

> 🔔 "Every digital transformation is going to begin and end with the customer, and I can see that in the minds of every CEO I talk to."
>
> Marc Benioff, Chairman and CEO, Salesforce[114]

Digital Feedback Loop

To disrupt and innovate, companies must take on strategies that connect their customers, employees, products and operations as effectively as possible. The Digital Feedback Loop[115] exemplifies such strategy and in its core (please see Figure 5), it uses data as a central engagement tool to connect with inside & outside customers and embodies entire business operations and value chain of a company to deliver value to all stakeholders and shareholders. The insights derived as part of feedback loop employs ML and AI [8].

There are five pillars[116] to Digital Feedback Loop [9].

1. Engage Customers
2. Empower Employees
3. Optimize Operations
4. Transform Products
5. Data & Intelligence

Digital Feedback Loop is part of Digital Transformation and aims to bring effective communications and product innovations to customers. When designing AI initiatives, it is imperative to bring the customer to the

[113] Salesforce Digital Transformation - https://www.salesforce.com/products/platform/what-is-digital-transformation/
[114] Digital Transformation - https://www.salesforce.com/products/platform/what-is-digital-transformation/
[115] AI Digital Feedback Loop - https://blogs.msdn.microsoft.com/stevengu/2018/07/18/ai-and-digital-feedback-loops/
[116] A Complete Guide to Microsoft's Digital Feedback Loop - https://www.aerieconsulting.com/blog/microsofts-digital-feedback-loop

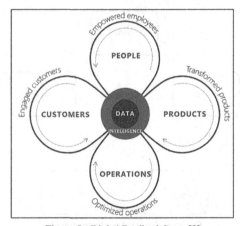

Figure 5: Digital Feedback Loop [8]

center of value addition and develop a compelling strategy to benefit the customer in its entirety of effective customer journey.

Insights Value Chain

The main emphasis of Insights Value Chain is "value capture" [10], the value from the customer point of view. As we clearly know, the data in its raw and most basic form is virtually worthless until we give it a voice by gleaning valuable insights from it.[117] The Insights Value Chain (Figure 6) outlines the process of converting the raw form of the data into a meaningful use and clearly call out each major business and technical foundations processes that should work in tandem to generate a value. The Insights Value Chain is a binary model—that is, it could generate a numeric value or zero value if there are disconnection in the process handshake. In order to deliver value for customers, all the components of Value Chain have to be contributing to the fullest

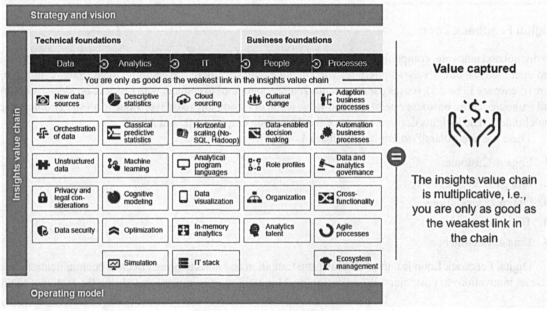

Figure 6: Insights Value Chain [10]

---

[117] Achieve Business Impact with the data - https://www.mckinsey.com/~/media/mckinsey/business%20functions/mckinsey%20analytics/our%20insights/achieving%20business%20impact%20with%20data/achieving-business-impact-with-data_final.ashx

or put it another way, any weak-link or non-contributing element could make overall value of the project to zero–multiplicative rule! [10]

The technical components include data, analytics and IT. The business components include people and process. The multiplicative rule kicks in when technical components of Insights Value Chain work in coordination to extract meaningful information from high-quality data by regressing/EPOCH smart algorithms constructed by data scientists and Business Components of Value Chain converts meaningful information into actions by applying people and business processes.

Data

Data must be the entire process of collecting, linking, cleaning, enriching, transforming—both with internal and external data sources. Please refer KDD section.

Analytics

Analytics process involves application of right ML techniques to derive insights from data. Analytics types could be supervised, unsupervised, semi-supervised and reinforcement. The type of analytics is based on business use case and type of data.

IT

IT is the provider compute, storage and connectivity. The IT role manifests based on the areas of value chain we are working on. For instance, the big data workloads or large-scale analytics data stores, the IT responsible to provide high performance compute (Cloud with storage). For Edge level workloads, IT provides embedded devices with connectivity.

People

People from the front lines of sales to deep within the business—not just "geeks"—are needed to run an analytics operation that turns data into insights and successfully implements those insights in the business. The crucial capability in today's Big Data world is being able to "translate" analytics- and data-driven insights into business implications and actions.

Process

Processes must be assessed for their ability to deliver at scale. Some old processes might need to be adapted, some might need to be fully automated, and others might need to be made more agile.

Strategy and Vision

Data Analytics should not be "done" just for the sake of it. Data Analytics process has to fit under corporate or enterprise overall strategy. Another way of looking at it is the Data Analytics process should enable "Think business backwards, not data forward." That is, do not develop Data Analytics first and start to look for customers…. Other way around: "start with a customer pain-point or opportunity and address it by developing Data Analytics".

Operating Model

Operating Model is the underlying governance in which the insights value chain lives. Core matters to be addressed include deciding where the analytics unit will sit within the organization and how it will function and interact with BUs (e.g., centralized, decentralized, hybrid).

The CRISP-DM Process

The Cross-Industry Standard Process for Data Mining (CRISP-DM) is the model [11] [12] (Figure 7) that is designed as a general model and can be applied to a wide variety of industries and business problems. Figure 7 provides the several techniques in CRISP-DM:[118]

Business Understanding

In my experience, this is one of the most important phases in an ML process. The success of initiative and expectations to stakeholders and project outcomes are derived from clear understanding of the business requirements. This phase also helps to uncover current as-is process and the proposed machine learning objectives.

Some of the steps that will help in this phase include:

- Establish good understanding with a Data & M L champion.
- Understand the business process—especially, accumulate as much tacit knowledge as possible from experts and champions involved.
- Make sure the M L process and outcome are directly connected to strategic drivers and the future roadmap of the organization.
- Assess the ML process is for cost synergies vs. monetization & long-term competitive advantage.
- Finally, understand the objectives are driven from cost vs. revenue business lines.

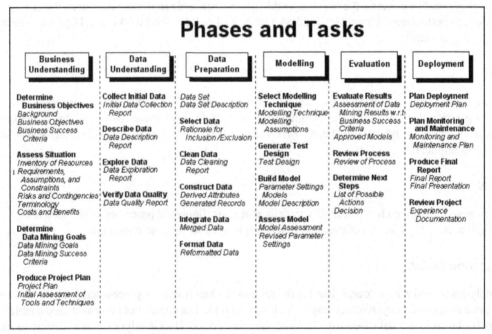

Figure 7: CRISP-DM [11]

---

[118] Introduction to IBM SPSS Modeler and Data Mining, ftp://ftp.software.ibm.com/software/analytics/spss/support/Modeler/Documentation/14/UserManual/CRISP-DM.pdf

Data Understanding

The preamble or pre-requisite for the phase is the business understanding phase. During the Data Understanding phase, involve data scientists and data engineers, as they will be creating data models and ML algorithms. Generally, this phase accounts for 30 to 40 % of the project time.

Lifecycle—Industry Standards[119] (Refer Figure 8).

Task (Proportion of Effort)	Subtasks	Business	Stakeholder Data Scientist	IT/ Operations
1. Problem Understanding (5% to 10%)	a) Determine Objective	X	X	
	b) Define Success Criteria	X	X	
	c) Assess Constraints	X	X	X
2. Data Understanding (10% to 25%)	a) Assess Data Situation	X	X	X
	b) Obtain Data (Access)		X	X
	c) Explore Data	X	X	X
3. Data Preparation (20% to 40%)	a) Filter Data		X	X
	b) Clean Data		X	X
	c) Feature Engineering	X	X	
4. Modeling (20% to 30%)	a) Select Model Approach		X	
	b) Build Models		X	
5. Evaluation of Results (5% to 10%)	a) Select Model		X	
	b) Validate Model		X	
	c) Explain Model	X	X	
6. Deployment (5% to 15%)	a) Deploy Model		X	X
	b) Monitor and Maintain	X	X	X
	c) Terminate	X	X	X

Figure 8: ML Lifecycle Standards [12]

Data Preparation Phase

Data preparation phase involves working with several origination units to collect the required data for developing Machine Learning algorithms. During this phase, we need to work with biasness and product groups in order to develop data processing and transformation rules. These rules will subsequently be deployed at the Edge and at the Cloud level.

Major work areas under this phase include:

- Select Data
  - o Based on business objectives, identify key data attributes—these attributes, later, will be input to ML models.
  - o Not all attributes are intrinsically available within the data—need to develop synthetic variables that explains the models' outcome.
- Clean Data
  - o Data is gold and no data is bad data; all we need to develop is a process to cleanse the data to be utilized under modeling and understanding purposes.
- Construct Data
  - o Based on the business objectives, construct the data to fully describe the ML charter.
- Integrate Data
  - o Integrate data involves working with several business units within the organization or working with 3rd party providers to collect necessary data.

---

[119] Preparing and Architecting for Machine Learning - https://www.gartner.com/binaries/content/assets/events/keywords/catalyst/catus8/preparing_and_architecting_for_machine_learning.pdf

- Format Data
  - o Format data to be pluggable usable for model development.

Data Modelling

This phase is also known as Machine Learning phase. In this phase, the actual mission learning algorithm is built, trained, validated and tested to meet the biasness objectives. During this phase, both data scientists and business Subject Matter Experts (SMEs) are closely involved and worked with Users. The successful development of algorithms is more of a collaborative process than siloed approaches.

This phase comprises the following major tasks:

- Select Modeling Technique
  - o Supervised:
    - Classification
      - Decision Trees, Naïve Bayesians, Support Vector Machines, K-Nearest Neighbor
    - Regression
      - Linear Regression, Non-linear Regression
  - o Unsupervised
    - Clustering
      - Hierarchical (Agglomerative, divisive)

Table 3: Difference between Supervised and Unsupervised Algorithms.

	Supervised Learning	Unsupervised Learning
Input Data	Uses known and labeled data	Uses unknown data
Computational Complexity	Very computational intense	Less computational
Real Time	Use off-line analysis	Use Real-time analysis
Number of Classes	Known	Unknown
Accuracy of results	Accurate and reliable	Reliable

- Generate Test Design
  - o Develop algorithm (on paper or with small datasets)
- Build Model
  - o Model description
  - o Codification of Model
- Assess Model
  - o Train Model
  - o Validate Model and
  - o Test Model

Multi-Label Cyber Classification for Identification of Cyber-related Events in RSS Use Case

RSS[120] stands for "Really Simple Syndication". It is a way to easily distribute a list of headlines, update notices, and sometimes content to a wide number of people. It is used by computer programs that organize those headlines and notices for easy reading. Risk Assessment Companies apply Machine Learning and AI to detect relevant cyber-attack[121] related articles. The use case entails the following key outcomes:

• Extraction of cyber loss magnitude, date of attack, company impacted, information compromised if service down, location of the attack, type of attack (cloud, DDOS, data breach)

• Extraction of dependencies of Subject-Object from one/group of sentences relating to a cyber Event, the event being a cyber-attack.

To derive the above mentioned objectives, generally multi-label classification is applied to RSS feeds. The goal is to identify the company that was the target of cyber event, presence of cyber event and severity of cyber magnitude.

Data:

Please see Table 4: RSS Cyber Feeds

Ingest RSS feeds in real-time to process the cyber data extraction.

Machine Learning:

Application of Natural Language Processing (NLP), Spacy,[122] real-time Classification engine, Cybersecurity risk model entity extraction and adaptive entity definitions as data grow.

 Best to Divide data into Training and Validation:

Assuming you have enough data to do proper held-out test data (rather than cross-validation), the following is an instructive way to get a handle on variances:

• Split your data into training and testing (80/20 is indeed a good starting point)

• Split the training data into training and validation (again, 80/20 is a fair split).

• Subsample random selections of your training data train the classifier with this, and record the performance on the validation set.

• Try a series of runs with different amounts of training data: Randomly sample 20% of it, say, 10 times and observe performance on the validation data, then do the same with 40%, 60%, 80%. You should see both greater performance with more data, but also lower variance across the different random samples.

• To get a handle on variance due to the size of test data, perform the same procedure in reverse. Train on all of your training data, then randomly sample a percentage of your validation data a number of times and observe performance. You should now find that the mean performance on small samples of your validation data is roughly the same as the performance on all the validation data, but the variance is much higher with smaller numbers of test samples.

---

[120] What is RSS? - https://rss.softwaregarden.com/aboutrss.html

[121] Dixons Carphone data breach, 5.9 million payment cards exposed - https://securityaffairs.co/wordpress/73479/data-breach/dixons-carphone-hacked.html

[122] Industrial-Strength Natural Language Processing - https://spacy.io/

Table 4: RSS Cyber Feeds

Label	Title	Link	Pub Date	Category	Content
0	Trump s Big Military Parade Will Reportedly Cost Way, Way More Than Previously Expected	https://taskandpurpose.com/trumps-military-parade-cost-estimate/	Thu Aug 16 22:48:05 IST 2018	[News, Bullet Points, Donald J. Trump, Military parade, military parades, news, parade, parades]	The grand military parade that President Donald Trump wants to hold in the nation capital in November is now expected to cost 92 million 80 million more than the original estimate.
1	Popular Social Media App Timehop Hit With Huge Data Breach	https://securingtomorrow.mcafee.com/consumer/consumer-threat-notices/timehop-data-breach/	Tue Jul 10 22:11:47 IST 2018	[Consumer Threat Notices, Data Breach, social media]	The Fourth of July is characterized by barbeques, fireworks, and patriotism and now cyberattacks!
0	Security expert discovered a bug that affects million Kaspersky VPN users	https://securityaffairs.co/wordpress/75222/hacking/kaspersky-vpn-flaw.html	Fri Aug 10 11:50:41 IST 2018	[Breaking News, Hacking, Kaspersky VPN, Pierluigi Paganini, Security Affairs]	A security issue exists in Kaspersky VPN = v1.4.0.216 which leaks your DNS Address even after you connected to any virtual server.
1	Ticketmaster UK Breached Via Supplier	https://www.infosecurity-magazine.com:443/news/ticketmaster-uk-breached-via/	Thu Jun 28 14:23:00 IST 2018	Null	Firm says less than 5% of global customer base hit
0	Eight men arrested in alleged $15 million hacking scheme	https://www.cyberscoop.com/business-email-compromise-15-million-ghana/	Tue Jun 26 05:36:05 IST 2018	[Government, business email compromise]	Eight men across the U.S. and Ghana have been arrested and charged with a 15 million hacking and fraud scheme.

Data Holdout

If a lot of data are available, simply take two independent samples and use one for training and one for testing. The more training, the better the model. The more test data, the more accurate the error estimate.

Problem: Obtaining data is often expensive and time consuming. Example: Corpus annotated with word senses, Dairy Sensor data, experimental data on sub-cat preferences.

Solution: Obtain a limited data set and use a holdout procedure. Most straightforward: Random split into test and training set. Typically, between 1/3 and 1/10 held out for testing.

Data Stratification

Problem: The split into training and test set might be unrepresentative, e.g., a certain class is not represented in the training set, thus, the model will not learn to classify it.

Solution: Use stratified holdout, i.e., sample in such a way that each class is represented in both sets.

Example: Data set with two classes A and B. Aim: Construct a 10% test set. Take a 10% sample of all instances of class A plus a 10% sample of all instances of class B. However, this procedure doesn't work well on small data sets.

Deployment

The successful ML algorithms are deployed for business and user purposes. The deployment phase consists of working with HPC, data engineers and IT groups.

Building Blocks of AI—Major Components of AI

The most important part of any AI project is to acquire datasets for solving problem or address the use case (please see Figure 9). As indicated in CRISP-DM, one preamble step to the acquisition of data is identification of business objective or classify the problem:

1. Classify the Problem

This step entails understanding the customer needs or identification of customer pain-point or fulfilling a market need or developing an innovative blue-ocean product. The successful outcome of this step is clear and concise business requirements with projected financial metrics in terms of market capture or maintain the

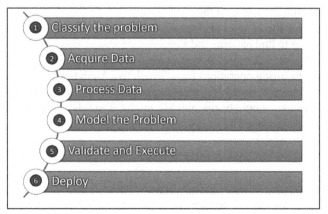

Figure 9: Stages of ML Process

existing market position. From Software Product Life-cycle point of view, the success of this step is measured in terms product requirements that are clear and validated.

Once requirements are businesses validated, the outcome of the step decides the downstream AI & ML processes from effort and manpower number point of view. For instance, if we are developing AI solutions as an incremental step, i.e., fixing issues in the existing product or releasing a point-release, the downstream operations from data point of view would be non-exhaustive as compared to a green field solution.

 Green Field vs. Brown Field Opportunities

In a classical Software Development paradigm, the definition of a Green Field Software project is the one that has no legacy code to worry about or Service Level Agreements (SLA) to maintain. In fact, the development team would be writing the first line of code as part of Green Field Software project.

Brown Field Software development project is the one in which the software development team would create a code branch from existing source code to start the development activity. In other words, the brown field software development projects need to maintain the existing codebases and must perform backward compatibility tests and must validate the existing requirements, one the development is complete, to make sure no code is broken.

On the other hand, the outcome of the step indicated a green field opportunity, the whole data life-cycle (CRISP-DM) and Software Development Life-cycle have to be carved out before finalizing budget for the business or technical executive approvals.

2. Acquire data

Data is the central for AI projects. Lack of having representative data to address the step 1 is a showstopper from Data Science and Data Engineering operations. Data for the ML process can be structured, unstructured and semi-structured. The data types could be audio, image, text, time series, video and structured/semi-structured.

3. Process the data

Based on the Business Goals (Step 1), and sometimes on heuristics, identify how to prepare for the ML execution. The steps include transformation, normalization and cleansing.

4. Model the Problem

The algorithm selection depends upon the type of problem or charter of the project that needs to be delivered. For instance, if we are planning to deliver a forecasting engine for the company's sales operations, the algorithm more likely falls into forecasting models; on the other hand, if the charter of the project is to deliver insights for optimizing work hours for the IT operation teams, the model work would involve time-series analysis with existing HR time systems to find patterns and suggest recommendation based on workforce optimization. Pattern mining is the best case. Likewise, if customers use case calls for explicit predictive analysis of margin prediction for manufacturing operations then "Predictive Analytics" is the choice. Similarly, clinical analytics model for identification of spread of disease in an epidemiological use case needs perform causation analysis—association rule mining would be optimal ML model. The following are list of ML techniques [13] and problem types:[123] (please see Figure 10).

---

[123] Visualizing the uses and potential impact of AI and other analytics
   Open interactive popup - https://www.mckinsey.com/featured-insights/artificial-intelligence/visualizing-the-uses-and-potential-impact-of-ai-and-other-analytics

Techniques	Problem Type
Feed Forward Neural Networks (AI)	Anomaly Detection
Recurrent Neural Network (AI)	Classification
Convolutional Neural Networks (AI)	Clustering
Tree-based ensemble learning	Data generation
Dimensionality Reduction	Ranking
Classifiers	Recommender System

Figure 10: AI Techniques & Problem Type

5. Validate and Execute

The execution process will likely comprise many cycles of running the ML routine and tuning and refining results [14]. One important aspect to be considered as part of validation process is model interpretability (Figure 12). AI algorithms have been notorious[124] for being "black-boxes", providing no way to understand inner workings and making it difficult to explain business users and stakeholders. Here are high-level mode interpretability techniques:[125]

6. Deploy

Finally, the output of the ML process is deployed to provide some form of business value. From tactical implementation point of view, the following are major components [15] AI:[126] (Refer Figure 11)

- Data & Knowledge Representation
- AI Programming Languages & Tools
  - o Popular Frameworks: Pytorch, Tensorflow, Keras, ONNX, Scikit-Learn, MXNet, Chainer, Microsoft CNTK, Spark ML, Python, R, NLTK
  - o IDEs: PyCharm, Jupyter, Visual Studio Code, Visual Studio, Anaconda
- AI Hardware
  - o HaHardware for AI systems vary based on type and place model constructed. Generally, CPU, GPUs, FPGA (embedded systems) are common hardware
- Heuristic Search:
  - o Heuristic search[127] refers to a search strategy that attempts to optimize a problem by iteratively improving the solution based on a given heuristic function [14] or a cost measure.[128]

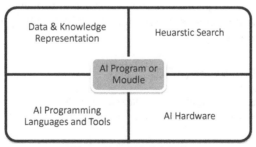

Figure 11: AI Module [14]

---

[124] DataRobot – Model Interpretability - https://www.datarobot.com/wiki/interpretability/

[125] Model interpretability in Azure Machine Learning service - https://docs.microsoft.com/en-us/azure/machine-learning/service/how-to-machine-learning-interpretability

[126] Foundations of Artificial Intelligence and Expert Systems by V S Janakiraman, P Gopalakrishnan and K Sarukesi Laxmi Publications © 2017

[127] Heuristic Search - https://link.springer.com/referenceworkentry/10.1007%2F978-1-4419-9863-7_875

[128] The *Handbook of Artificial Intelligence* Vols. 1 and 2, William Kaufmann, Los Altos,CA 1981.

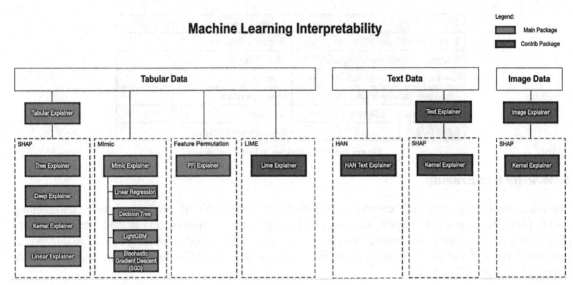

Figure 12: ML Interpretability

AI Reference Architectures

At its core, AI process involves conversion of data into actionable insights. There are several parts involved in the AI process. Understanding business requirements, understanding existing data, building models that extracts information or patterns from the data and building behavioral or predictive models. The AI process outlines the following major steps:[129,130]

- Business Understanding
- Data Acquisition and Understanding
- Modeling
  - o Constrained Modeling for performance against limited resources
- Deployment
  - o Pre-model evaluation for deploying in edge or Turing device (please see Figure 12)
- Customer Acceptance.

From the learning point of view, two processes, generally, ensemble AI: from development work scope, Training and Testing Phase and from production deployment point of view, register model and deploy the model. To understand the several interconnected AI processes, the following major reference architectures are followed (please see Figure 13):

Knowledge Discovery in Databases (KDD)

The Knowledge Discovery (please see Figure 14) in Databases are traditionally applied to unravel information in large databases. KDD involves the following important steps [16] [17]:

- Cleaning and Integration
- Selection and Transformation

---

[129] Team Data Science Process for data scientists - https://docs.microsoft.com/en-us/azure/machine-learning/team-data-science-process/team-data-science-process-for-data-scientists

[130] What is the Team Data Science Process? - https://docs.microsoft.com/en-us/azure/machine-learning/team-data-science-process/overview

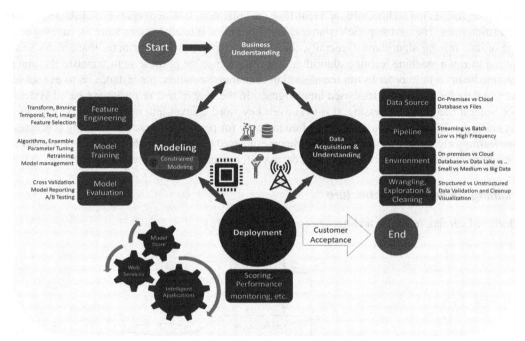

Figure 13: Data Science Process with Edge device

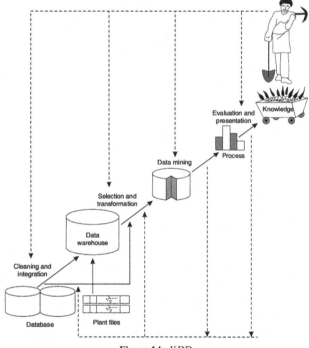

Figure 14: KDD

- Data Mining
- Evaluation and Presentation.

At the base of the KDD [16] (see Figure 14), we have data sources or databases. The data from data sources are cleansed and integrated in order to be stored in data bases or data warehouses. For instance, we are developing foundation architecture for a real-time data platform, this step represents cleansing of data for data validation rules. The next step, Selection and Transformation is to enable data store for running machine learning or data mining algorithms. Generally, the selection represents key attributes that are derived from or required to run a machine learning algorithm. The transformation purpose is to translate the input data into intermediator form in order to run the algorithm. The transformation, for instance, is to extract feature engineering data for audio or video-based input signals. In the case of text or document-based systems, for instance Natural Language Processing, it is to convert key word indexes into matrix form. The transformed input is either persisted on database or loaded into memory for performing machine learning processes. The machine learning algorithms vary based on the use case or based on the type of the data. The processed output of the machine learning process is stored into a database for visualization and reporting purposes.

Data Mining Reference Architecture

The Machine Learning reference architecture (see Figure 15) [16]:

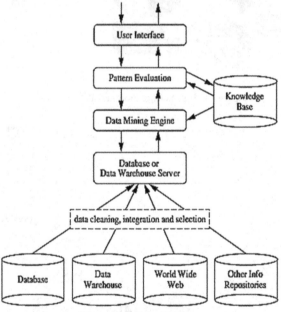

Figure 15: ML Architecture

Streaming Processing Reference Architecture

Applications, sensors, monitoring devices, and gateways broadcast continuous event data known as data streams. Streaming data is high volume, with respect to number of sensor events, and has a lighter payload, generally text or binary (BASE64 converted or Telemetry Transport, than non-streaming systems. In Stream processing architecture (please see Figure 16), any number of streams can enter the system, many or time function bound or event triggered. Each stream can provide elements at its own schedule; they need not have the same data rates or data types, and the time between elements of one stream need not be uniform.[131] The fact that the rate of arrival of stream elements is not under the control of the system distinguishes stream processing from the processing of data that goes on within a database-management system. The latter system controls the rate at which data is read from the disk, enterprise queues, or in-memory, and therefore never has to worry about data getting lost as it attempts to execute queries.

---

[131] Mining of Massive Datasets 2nd Edition, Cambridge University Press; 2 edition (December 29, 2014)

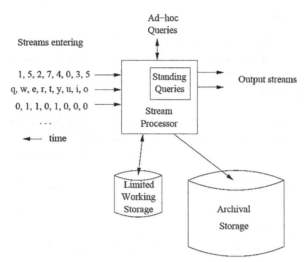

Figure 16: Streaming System Architecture [19]

Streams may be archived in a large archival store (please note: archival strategy depends upon workload architectures: yes for batch processing and may be for real-time), but we assume it is not possible to answer queries from the archival store. It could be examined only under special circumstances using time-consuming retrieval processes. There is also a working store, into which summaries or parts of streams may be placed (for instance, in cloud based architectures, the working store could be high responsive premium Solid State Drives[132] (SSD) or Ultra SSD), and which can be used for answering queries. The working store might be disk, or it might be main memory, depending on how fast we need to process queries. But either way, it is of sufficiently limited capacity that it cannot store all the data from all the streams.

Stream Data Examples

Stream data sources are generally real-time and embedded devices are either non-constrained, constrained, or extremely constrained devices (xCDs). Nevertheless, the data emit from these devices classically fall into stream category with periodicity, data and latency processing needs. Examples of stream data source include Dairy Sensor data,[133] Satellite Image data,[134] Weather data[135] and Traffic data.[136]

Sensor data

Imagine a temperature sensor bobbing about in the ocean, sending back to a base station a reading of the surface temperature each hour. The data produced by this sensor is a stream of real numbers. It is not a very interesting stream, since the data rate is so low. It would not stress modern technology, and the entire stream could be kept in main memory, essentially forever.

Image data

Satellites often send streams down to earth consisting of many terabytes of images per day. Surveillance cameras produce images with lower resolution than satellites, but there can be many of them, each producing a stream of images at intervals like one second. London is said to have six million such cameras, each producing a stream.

---

[132] Pricing Managed Disks - https://azure.microsoft.com/en-us/pricing/details/managed-disks/

[133] Dairy Sensor Data - http://hanuinnotech.com/

[134] A planetary-scale platform for Earth science data & analysis - https://developers.google.com/earth-engine/datasets

[135] Past Weather by Zip Code - Data Table - https://www.climate.gov/maps-data/dataset/past-weather-zip-code-data-table

[136] Data.Gov - https://catalog.data.gov/dataset/traffic-data

Internet and Web Traffic data

Click stream and Web traffic fall into Stream data sources. For instance, traffic pattern recognition or safety and weather data and the effect on safety can be used as analytics from the data source.

Streaming Processing Rules

The stream processing architectures must obey stream processing rules for successful stream analytics. The velocity and data of the streams are so high, without following the rules, that many encounter availability and resilient issues. Additionally, stream processing architectures may encounter stream impurities issue – BLAST model.

Streaming processing architectures follow the eight rules [20] of streaming:[137]

- Rule 1: Keep the Data Moving
- Rule 2: Query using SQL on Streams (StreamSQL)
- Rule 3: Handle Stream Imperfections (Delayed, Missing and Out-of-Order Data)
- Rule 4: Generate Predictable Outcomes
- Rule 5: Integrate Stored and Streaming Data
- Rule 6: Guarantee Data Safety and Availability
- Rule 7: Partition and Scale Applications Automatically
- Rule 8: Process and Respond Instantaneously

BLAST—Stream Model

Stream Engineering systems (i.e., IoT Devices, Edge Architectures) exhibit asymmetric behavior when the underlying Sensor or Custom Purpose-built device experience health or device issues. Generally, the symptoms of asymmetric behavior include:

- **B**urst & Queuing
- Low **L**atency
- **A**vailability
- **S**calability
- Fault **T**olerance

When developing Architectures for the Stream and Edge Architectures, it is prudent to analyze the system architectures from BLAST point of view.

AI Data Pipeline

The term "pipeline" refers to movement of huge Big Data workloads. AI Data pipeline refers to the moving of high velocity and high frequency big data. There are several processes involved in creating a robust data pipeline. For a big data pipeline, the data (raw or structured) is ingested into the Data Store through Data orchestration tools in batches, or streamed near real-time using enterprise bus or distributed stream technologies. This data lands in a data lake for long-term persisted storage. As part of your analytics workflow, use in-memory compute engine to read data from multiple data sources and turn it into breakthrough insights using Machine Learning Analytics Engine (please see Figure 17).

---

[137] The 8 Requirements of Real-Time Stream Processing - http://cs.brown.edu/~ugur/8rulesSigRec.pdf

Figure 17: Reference Pipeline Process

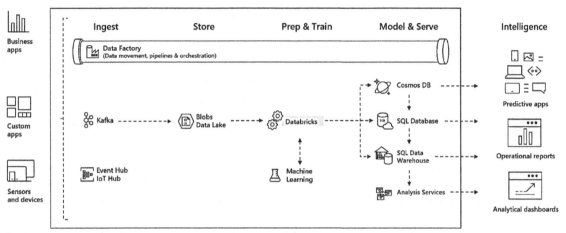

Figure 18: Data Pipeline

For example, consider following Microsoft Azure Data Pipeline processing Databricks, a Spark based stream processing engine. Azure Databricks is a fast, easy, and collaborative Apache Spark-based analytics service. For a big data pipeline (see Figure 18), the data (raw or structured) is ingested into Azure through Azure Data Factory in batches, or streamed near real-time using Kafka, Event Hub, or IoT Hub. This data lands in a data lake for long-term persisted storage in Azure Blob Storage or Azure Data Lake Storage. As part of your analytics workflow, use Azure Databricks to read data from multiple data sources such as Azure Blob Storage, Azure Data Lake Storage, Azure Cosmos DB, or Azure SQL Data Warehouse and turn it into breakthrough insights using Spark.[138]

AI Data Pipeline, generally, refers to Big Data workloads and is performed on High Compute Engines such as Cloud or Stream servers such Apache Spark.

As of writing of this book, there is a huge movement of data processing shifting from Cloud to Edge processing.

Edge Processing

At a high level, an end-to-end model deployment consists of three stages through which the data travels: edge (data ingest), core (training clusters, data lake), and cloud (data archival). This movement of data is very typical in applications such as IoT, where the data spans all three phases of the data pipeline. With the growth of IoT applications and consumer devices, a constant and significant amount of data is being generated and ingested at the edge. As an example, the Edge can consist of an army of sensors that gather raw data, ranging from a few gigabytes (GB) to a few terabytes (TB) a day, depending on the application.

Moving this volume of raw data from thousands of Edge locations over the network at scale is impractical and is prone to performance issues. Shifting analytics processing and inference at the Edge is an ideal way to reduce data movement and to provide faster results [24,25]. In such advanced solution cases, the data that is fed into the data lake is being operated upon from an analytics and an AI/DL perspective, and the trained AI/DL models are pushed back to the edge for real-time inferencing [25].

---

[138] What is Azure Databricks? - https://docs.microsoft.com/en-us/azure/azure-databricks/what-is-azure-databricks

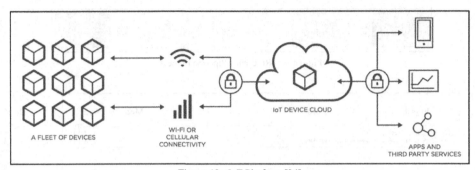

Figure 19: IoT Platform [24]

IoT Platform (see Figure 19) provides an end-to-end connectivity to bring it online and brings intelligence to "Things" [22].

At the source level, you have several types of devices that generate data at several different frequencies and velocities. The connectivity layer is the one that collects data from the sensors and uploads to either edge level or to the Cloud level. On the Cloud level, the IoT data is processed through Analytics frameworks and ML engines to derive useful insights. Finally, the end insights are stored in data base or looped back to the users.

Here is the end-to-end platform view from Intel (see Figure 20): The goal of the System Architecture Specification is very straightforward—to connect any device of any type to the Cloud. The specification is also a guideline for third party developers, Original Equipment Manufacturers (OEM), and other developer communities to develop products for IoT [23].

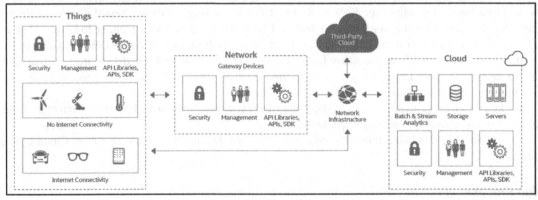

Figure 20: Intel IoT Platform

The tenets for the application developers to deploy IoT solutions are as follows:

- Services to monetize the IoT infrastructure
- Seamless data ingestion
- Intelligence deployment to any-form-factor devices by running analytics on Cloud or on to the device.

In the grand scheme of IoT, the philosophy is very straightforward: *Deployment of distributed data collection, contextual intelligence, centralized data platform, and holistic integration of devices to the Enterprise Data architecture systems.*

End-to-End Platform

End-to-End platform is generally used by businesses or service providers as a completely vertically integrated service offering or product architecture. For instance, Samsung offers IoT platform to integrate all data sources.

End-to-End platform offers all parts of architecture: (a) Hardware or sensor product, (b) Gateway or Edge collector, (c) Mobile app as an Edge device, (d) Middleware component, (e) Message Bus or Message Queue injection, (f) Compute or Analytics platform, and (g) UI or Insights recommendation.

The reference architecture for End-to-End IoT is as follows:

The reference architecture for IoT is very much like that of the architecture of data processing systems except that the IoT architectures consume the data before the platform can massage or compact the data.

On the left handbside (see Figure 21), the data is generated by sensors, human activity, motors or sensors/actuators. The sensors could be of active, passive or a hybrid of the data generators. The data from the generators are collected by the main processing units' data controllers (microcontroller units). The data controllers could be in-memory memories or sensor units. The sensor units comprise silicon circuit board with RTOS (real-time operating system) that collects data either by polling a particular sensor hardware address or triggered by interrupt routine. The data is then processed at the edge level or sensor unit level for immediate action or closed loop behavior. This is achieved by combining events from several data sources. Complex Event Processing. The processed data is stored or relayed to traditional Cloud platforms.

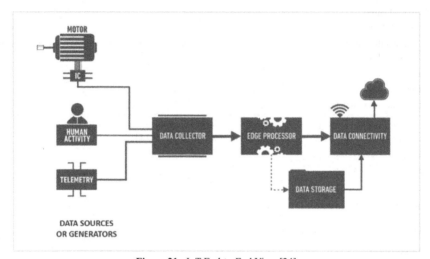

Figure 21: IoT End-to-End View [24]

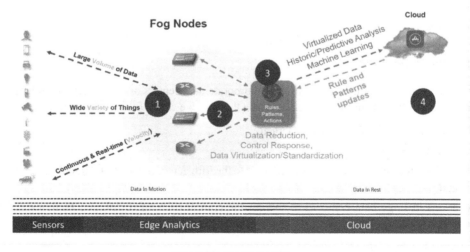

Figure 22: Fog Nodes [24]

The End-to-End architecture contains sensors or data generators that feed the data into Edge data nodes processors. The purpose of the Edge node is to evaluate rules on incoming high velocity multi-sensor data. The data is validated against any anomalies for alerting the user. In addition, the context-based analytics can be run on the in-coming data in order to derive contextual intelligence. Once the data is analyzed, it can be compressed in order to reduce the data footprint and data transmissions costs.

As shown in the above diagram (see Figure 22):

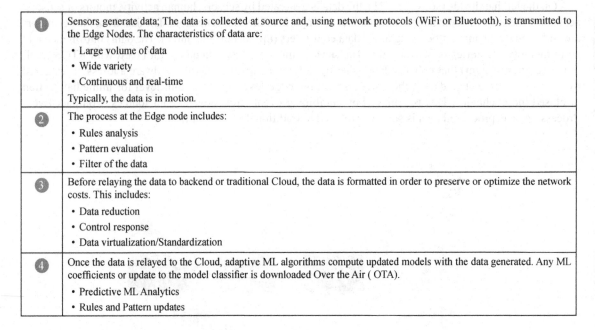

①	Sensors generate data; The data is collected at source and, using network protocols (WiFi or Bluetooth), is transmitted to the Edge Nodes. The characteristics of data are: • Large volume of data • Wide variety • Continuous and real-time Typically, the data is in motion.
②	The process at the Edge node includes: • Rules analysis • Pattern evaluation • Filter of the data
③	Before relaying the data to backend or traditional Cloud, the data is formatted in order to preserve or optimize the network costs. This includes: • Data reduction • Control response • Data virtualization/Standardization
④	Once the data is relayed to the Cloud, adaptive ML algorithms compute updated models with the data generated. Any ML coefficients or update to the model classifier is downloaded Over the Air (OTA). • Predictive ML Analytics • Rules and Pattern updates

The end-to-end platform provides several benefits:

• Over the Air firmware (OTA) upgrade
• Device management
• Cloud connection
• Cellular connectivity
• Analytics
• User interface

Data Shifting to the Edge

Data generation and processing are shifting from corporate data centers or central servers to Edge servers or machines [26]. Four big technology drivers underlie the massive shift of data to the edge:[139,140]

1. AI has become a cost-effective and practical,
2. Billions of IoT devices are being deployed,
3. Wireless operators are upgrading their networks to the fifth generation of cellular mobile communications (5G), and
4. Innovations in Edge data centers are solving the complexities of distributed facilities and unit cost economics.

---

[139] Data at the Edge - https://www.seagate.com/enterprise-storage/what-is-it-4-0/
[140] State of the Edge - https://www.seagate.com/www-content/enterprise-storage/it-4-0/images/Data-At-The-Edge-UP1.pdf

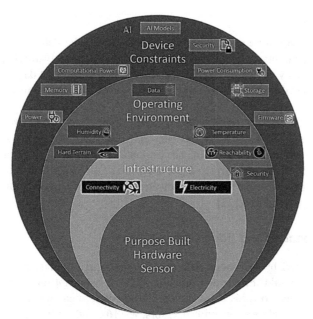

Figure 23: Edge Device Operating Environment

Nevertheless, the processing at the Edge comes with its own processing and constraints. Here is the list of conditions that the Edge devices should endure to provide resilient technologies (please see Figure 23):

Harsh Conditions

Unlike centralized hyperscale data centers which can be carefully and methodically located and have complete resilience/data backups, Edge data centers often need to go near the data, no matter how harsh the location (i.e., the data sensors could be located hundreds of feet down in reservoirs or located at center of ocean collecting data at oil rigs or attached to Cattle Necklace collecting data). Edge architectures, therefore, must consider higher or lower sustained temperatures, varying humidity, terrain dust, stricter shock and vibration requirements, flood and earthquake risks, air pollution, and other environmental factors that aren't usually a concern in centralized locations [25].

Remote Locations

The more the edge expands or wider the edge devices get deployed, the more it will need facilities in distant locations that are hard to service and secure, such as on an oil field, dairy remote fields, or at the base of a cell tower. These types of Edge locations present higher operational costs and greater risks to data [25]. Generally, higher service costs or truck rolls will result in maintaining the device.

Quality of Service (QoS)/Low Latency

Edge devices perform data processing closer to the source. To provide closed loop feedback, the edge processing required to be high resilient and low latency. The data processing responses need to be SLA compliant.

No-Touch/Self-Healing

Remote locations require the optimal combination of resiliency, redundancy, and flexibility that can be delivered via no-touch/self-healing technologies. Ideally, a remote data center can detect problems and heal them without any user intervention and be able to guarantee a certain quality of service. This capability can be brute forced through hardware redundancy and resource overprovisioning, but it can be ideally delivered with softwares that intelligently re-balances resources using orchestration and automation.

Global Data Experience

While Edge nodes allow data to be stored and acted upon locally, the data should be treated as a global entity—its location should not matter to the user. This calls for storage software (file systems) that support a global namespace across many devices and which can modify the location of data as user behavior changes. Granular multitenancy control allows multiple applications to coexist in the same location and service device across a shared, distributed cluster.

Despite harsh conditions, device constraints, operating limitations and environmental factors, moving to Edge is not a choice but a necessity. It's a necessary one, especially for highly secure or mission critical data that needs to be processed in real-time. IDC has estimated that by 2020, 10% of the world's data will be processed on the Edge.[141] I wouldn't be surprised if it proves to be more. As the IoT continues to grow, and the opportunities for using it become even more sophisticated, we can only assume companies won't just move to the Edge—they will be pushed to it [27].

Nomenclature of Embedded and Edge Devices

Embedded devices powered by microcontrollers are becoming popular and every year billons[142] of microcontrollers are getting sold. Embedded processing is evolving not only towards higher capabilities and processing speeds, but also towards allowing a multitude of applications to run autonomously. There is a growing market for small-scale embedded processors; 8-, 16-, and 32-bit microcontrollers with on-chip RAM and flash memory, I/O capabilities, and networking interfaces such as low-rate wireless personal area networks (LR-WPANs—standards IEEE 802.15.4), bluetooth, and Wi Fi, which are increasingly integrated as tiny System-on-a-Chip (SoC) solutions [28]. SoCs enable the design of devices with small physical footprints of a few mm^2 with very low power consumption, e.g., in the milli- to microwatt range, but which are capable of hosting complete communication protocol stacks, including small web servers.

Embedded devices, generally, would have the following characteristics:

- Computational capabilities: 8-bit,16-bit,32-bit working memory and storage
- Power supply: Wired electricity powered, battery, or hybrid
- Sensor/Actuator: Environmental data attributes through sensors and control of devices using actuators
- Communication Interface: Wireless or wired communication connectivity-device to internet connectivity.
- Operating System (OS): Main loop, event-based, real-time or full-featured OS
  - o Main Loop OS constantly consumes power
  - o Using a Real-Time Operating System (RTOS) can allow you to do more work for the same amount of power consumed simply by freeing up the Central Processing Unit (CPU) in time.[143] That is done by ensuring everything is event driven and absolutely no CPU cycles are wasted by polling anything
  - o Full featured OS utilities full power
- User Interface: Buttons, UI and Full Windows based access.

---

[141] Moving To The Edge: Evolving IoT Data Storage In 2018 - https://www.forbes.com/sites/danielnewman/2018/01/23/moving-to-the-edge-evolving-iot-data-storage-in-2018/#6055c30141a5

[142] Why Microcontrollers are important? - https://www.tensorflow.org/lite/microcontrollers

[143] FreeRTOS and power consumption - https://www.freertos.org/FreeRTOS_Support_Forum_Archive/June_2014/freertos_FreeRTOS_and_power_consumption_8849ff7cj.html

	CPU	Memory	Power	Communication	OS, EE
Basic	8-bit PIC 8-bit 8051, 32-bit Cortex-M	Kilobytes	Battery	802.15.4,[144] 802.11, Z-Wave	Main loop, Contiki, RTOS
Advanced	32-bit ARM9, Intel Atom, More powerful Intel and AMD Processors	Megabytes, Gigabytes	Electrical Fixed	802.11, LTE, 3G, 4G, 5G, GPRS, Wired	Linux, Java, Python

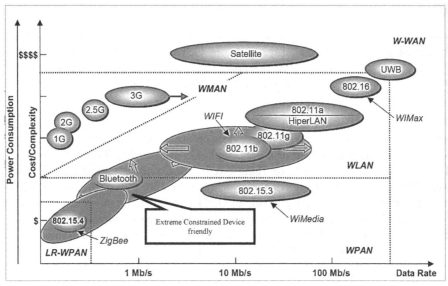

Figure 24: Standard Technology Map [29]

Microcontrollers [28]

Microcontrollers typically provide several ports (physically accessible pins) that allow interaction with sensors and actuators. These include General Purpose I/O (GPIO), support for digital sensors that communicate using standard interfaces such as Serial Peripheral Interface (SPI) and Inter-Integrated Circuit (I²C) and Analog Digital Circuits (ADCs) for supporting analog input. For certain actuators, such as motors, Pulse-Width Modulation (PWM), a method of reducing the average power delivered by an electrical signal, can be easily implemented. As low-power operation is paramount to battery-powered devices, the microcontroller hosts functions that facilitate sleeping and operations in different sleep modes and interrupts that can wake up the device on external and internal events, e.g., when there is activity on a GPIO port or the radio (hardware interrupt), as well as timer-based wake-ups. Some devices even go as far as harvesting energy from their environment, e.g., solar and thermal energy. Where the application allows, specifically in the case of periodic monitoring as opposed to continuous monitoring of phenomena, devices tend to be highly duty-cycled, remaining in ultra Low-Power Mode (LPM) for as long as possible [28].

 IoT Scale

There will be more than 24 billion IoT devices by 2020. That's approximately four devices for every human being on the planet [26]. Data gravity bends the fabric of Cloud, extending it to the edge and forming the Edge Cloud.

---

[144] 802.15.4 - https://www.microchip.com/design-centers/wireless-connectivity/embedded-wireless/802-15-4

To interact with peripherals such as external on-board storage or display, it is common to use serial interfaces such as SPI, I2C, or Universal Asynchronous Receiver Transmitter (UART). These interfaces can also be used to communicate with another microcontroller on the device (e.g., a separate controller managing the energy subsystem, performing power point tracking). This is common when there is a need for offloading certain tasks, or when in some cases the entire application logic is put on a separate host processor. It is not unusual for the microcontroller, Radio Frequency Integrated Circuit (RFIC) or SoC to host hardware security processors, e.g., to accelerate the Advanced Encryption Standard (AES) operations. This is necessary to allow encrypted communication over the radio link without the need for a host processor and is far more energy-efficient than using software implementations of cryptographic ciphers [28].

Connectivity

Created to support low-cost, low-power networks, the IEEE's 802.15.4 standard defines the MAC and PHY layer used by, but not limited to, networking specifications such as Zigbee®, 6LoWPAN, Thread, WiSUN and MiWi™ protocols. As shown in Figure 24, the low power consumption connectivity protocols such as Zigbee and IEEE 802.15.4 for constrained power devices (please see Figure 24).

Operating System

Owing to limited computational resources, basic devices do not use OS in the traditional sense. Something as simple as a single-threaded main-loop or a lightweight OS such as FreeRTOS, Atomthreads, AVIX-RT, ChibiOS/RT, ERIKA Enterprise, TinyOS, or Thingsquare Mist/Contiki are most often implemented [28]. These lightweight OSs offer significant functionality, including memory and concurrency model management, sensor, actuator, and radio drivers, (multi-) threading, TCP/IP, and higher-level protocol stacks. The purpose of the book is to provide more details into ML models in constrained resources, the book only introduces high level architectures of IoT OS. Please refer to classical OS books for more details into IoT OS workings.

Free Real Time Operating Systems (RTOS) features[145] [30]:

- Pre-emptive or cooperative operation
- Very flexible task priority assignment
- Flexible, fast and light weight task notification mechanism
- Queues
- Binary semaphores
- Counting semaphores
- Mutexes
- Recursive mutexes
- Software timers
- Event groups
- Tick hook functions
- Idle hook functions
- Stack overflow checking
- Trace recording
- Task run-time statistics gathering
- Optional commercial licensing and support
- Full interrupt nesting model (for some architectures)
- A tick-less capability for extreme low power applications
- Software managed interrupt stack when appropriate (this can help save RAM).

---

[145] Mastering the FreeRTOS – Real Time Kerney - https://freertos.org/Documentation/161204_Mastering_the_FreeRTOS_Real_Time_Kernel-A_Hands-On_Tutorial_Guide.pdf

The following are some of the commercially available IoT OS [31,32,33,34]:

- Free RTOS
- Contiki
- Android Things
- TI-RIOT
- TinyOS
- RIOT

Free RTOS

FreeRTOS[146] is a real-time operating system kernel for embedded devices designed to be small and simple. It been ported to 35 microcontrollers and it is distributed under the GPL. Enriched software libraries make it easy to connect with small IoT devices. Written in C it provides methods for multiple threads or tasks, mutexes, semaphores and software timers. This IoT Operating System uses the Cloud service of Amazon Web Service called AWS IoT Core to run the IoT applications. The memory footprint is only 6–15 kb[147] which makes it a more adaptable small powered microcontroller [30,31,32,33].

TinyOS

TinyOS[148] is an open source, BSD-licensed operating system designed for low-power wireless devices, such as those used in sensor networks, ubiquitous computing, personal area networks, smart buildings, and smart meters. A worldwide community from academia and industry use, develop and support the operating system as well as its associated tools, averaging 35,000 downloads a year. The core language of TinyOS is nesC which is a dialect of C language. TinyOS is popular among developers for its memory optimization characteristics. A component of TinyOS neutralizes some abstractions of IoT systems, for example, sensing, packet communication, routing, etc. The developer group of this IoT Operating System is TinyOS Alliance [30,31,32,33].

The design principle of TinyOS[149] is key to Low Duty Cycle Operations [34]:

- Sleep—majority of time
- Wakeup—quickly start processing
- Active—minimize work and return to sleep

Power consumption[150] [34]: Please see the following table for TinyOS power consumption. It is not scalar of the unit that contributes overall power consumption but number of wakeups or interrupts make huge difference (please see Table 5).

Table 5: Tiny OS Power Consumption

Operation	Telos	Mica2	MicaZ
Minimum Voltage	1.8V	2.7V	2.7V
Mote Standby (RTC on)	5.1 μA	19.0 μA	27.0 μA
MCU Idle (DCO on)	54.5 μA	3.2 mA	3.2 mA
MCU Active	1.8 mA	8.0 mA	8.0 mA
MCU + Radio RX	21.8 mA	15.1 mA	23.3 mA
MCU + Radio TX (0dBm)	19.5 mA	25.4 mA	21.0 mA
MCU + Flash Read	4.1 mA	9.4 mA	9.4 mA
MCU + Flash Write	15.1 mA	21.6 mA	21.6 mA
MCU Wakeup	6 μs	180 μs	180 μs
Radio Wakeup	580 μs	1800 μs	860 μs

Credits: https://sceweb.uhcl.edu/yang/public/WSN_modules/TinyOS.ppt

---

[146] FreeRTOS - https://www.freertos.org/index.html
[147] Free RTOS - https://www.cs.unc.edu/~anderson/teach/comp790/powerpoints/freertos-proj.pdf
[148] TinyOS - http://www.tinyos.net/
[149] The mote revolution: Low Powr Wireless Sensor Network Devices, Hot Chips 2004
[150] TinyOS - https://sceweb.uhcl.edu/yang/public/WSN_modules/TinyOS.ppt

Contiki

Invented in 2002, Contiki[151] is an open-source IoT operating system particularly popular for low power microcontrollers and other IoT devices to run effectively using internet protocol IPv6, and IPv4. These operating systems support wireless standard CoAP, 6lowpan, RPL. Mostly this IoT OS is suitable for low-powered internet connectivity.

Contiki is designed to run on classes of hardware devices that are severely constrained in terms of memory, power, processing power, and communication bandwidth [31,32]. For example, in terms of memory, despite providing multitasking and a built-in TCP/IP stack, Contiki only needs about 10 kB of RAM and 30 kB of ROM. A typical Contiki system has memory of the order of kilobytes, a power budget of the order of milliwatts, processing speed measured in megahertz, and communication bandwidth of the order of hundreds of kilobits/second. This class of systems includes various types of embedded systems as well as a number of old 8-bit computers [32].

Low Power Operations

Many Contiki systems are severely power-constrained. Battery operated wireless sensors may need to provide years of unattended operation, often with no way to recharge or replace batteries. Contiki provides a set of mechanisms for reducing the power consumption of the system on which it runs. The default mechanism for attaining low-power operation of the radio is called ContikiMAC.

Threads

To run efficiently on memory-constrained systems, the Contiki programming model is based on protothreads. A protothread is a memory-efficient programming abstraction that shares features of both multi-threading and event-driven programming to attain a low memory overhead. The kernel invokes the protothread of a process in response to an internal or external event. Examples of internal events are timers that fire, or messages being posted from other processes. Examples of external events are sensors that trigger, or incoming packets from a radio neighbor.

Protothreads are cooperatively scheduled. This means that a Contiki process must always explicitly yield control back to the kernel at regular intervals. Contiki processes may use a special protothread construct to avoid waiting for events while yielding control to the kernel between each event invocation. Finally, Contiki supports per-process optional preemptive multi-threading, inter-process communication using message-passing events, and an optional GUI subsystem with either direct graphic support for locally connected terminals or networked virtual display with virtual network computing or over Telnet.

☀ Protothreads

Protothreads are an extremely lightweight, stackless threads that provides a blocking context on top of an event-driven system, without the overhead of per-thread stacks. The purpose of protothreads is to implement sequential flow of control without using complex state machines or full multi-threading. Protothreads provides conditional blocking inside a C function.

In memory constrained systems, such as deeply embedded systems, traditional multi-threading may have a too large memory overhead. In traditional multi-threading, each thread requires its own stack, that typically is over-provisioned. The stacks may use large parts of the available memory.

The main advantage of protothreads over ordinary threads is that protothreads are very lightweight: a protothread does not require its own stack. Rather, all protothreads run on the same stack and context switching is done by stack rewinding. This is advantageous in memory constrained systems where a stack for a thread might use a large part of the available memory. A protothread requires only **two bytes of memory per protothread**. Moreover, protothreads are implemented in pure C and do not require any machine-specific assembler code. [34]

---

[151] Contiki - http://contiki-os.org/

Code Size [35]

An operating system for constrained devices must be compact in terms of both code size and RAM usage to leave room for applications running on top of the system [35]. The code size of Contiki is larger than that of TinyOS, but smaller than that of the Mantis system. Contiki's event kernel is significantly larger than that of TinyOS because of the different services provided. While the TinyOS event kernel only provides a FIFO event queue scheduler, the Contiki kernel supports both FIFO events and poll handlers with priorities. Furthermore, the flexibility in Contiki requires more run-time code than for a system like TinyOS, where compile time optimization can be done to a larger extent [35] [36] (Please see Table 6).

Table 6: Code Size

Module	Code Size (AVR)	Code Size (MSP430)	RAM Usage
Kernel	1044	810	+ 4e + 2p
Service layer	128	110	0
Program loader	-	658	8
Multi-threading	678	582	8+s
Timer library	90	60	0
Replicator stub	182	98	4
Replicator	1752	1558	200
Total	3874	3876	

Power Optimization

Power consumption can be decreased significantly by placing the processor into a low power state each time the idle task runs. FreeRTOS also has a special tick-less mode. Using the tick-less mode allows the processor to enter a lower power mode than would otherwise be possible and remain in the low power mode for longer [30]. Another option is, configUSE_TICKLESS_IDLE, which also affects the scheduling algorithm, as its use can result in the tick interrupt being turned off completely for extended periods. The full state machine (please see Figure 25) consists of Not Ready, Ready, Running, Suspended and Blocked Modes. Task in FreeRTOS executes at a time. Tasks have no knowledge of scheduler activity. Task states [30,31,32,33]:

- Running: Actively executing and using the processor
- Ready—Able to execute, but not because a task of equal or higher priority is in the running state.
- Blocked—Waiting for a temporal or external event, e.g., queue and semaphore events, or calling vTaskDelay() to block until delay period has expired. Always have a "timeout" period, after which the task is unblocked.
- Suspended—Only enter via vTaskSuspend(), depart via xTaskResume() API calls.

configUSE_TICKLESS_IDLE is an advanced option provided specifically for use in applications that must minimize their power consumption [30].

In MCU powered applications, the application logic at the device level is usually implemented on top of the OS, calling existing functions provided, as a main application, or within a main loop for simpler cases [27]. A typical task for the application logic might be to read values from sensors and to provide these over the radio interface to a listening gateway or sink node at a predefined frequency. This may or may not be done in a semantically correct manner with the correct units. In the case of raw data (e.g., only numeric values with no units, no location or time of sensor data), the gateway device or a web-hosted application may be needed to perform a transformation on the data such that it becomes semantically correct, appends relevant metadata, etc.

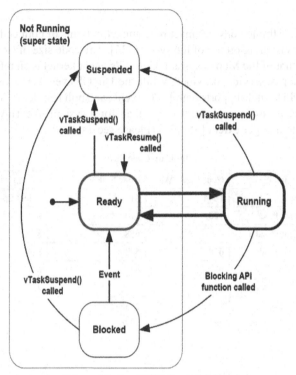

Figure 25: Free RTOS State Machine

📠 **IEEE 802.15.4**

Created to support low-cost, low-power networks, the IEEE's 802.15.4 standard defines the MAC and PHY layer used by, but not limited to, networking specifications such as Zigbee®, 6LoWPAN, Thread, WiSUN and MiWi™ protocols.

IEEE 802.15.4-based products from Microchip are deployed today in a wide range of applications from battery-free, energy-harvesting wireless light switches, to alarm sensors with several years of battery life, to high-performance mesh utility networks supporting smart meters and street lighting. Our complete line of IEEE 802.15.4 transceivers, RF microcontrollers (MCUs) and regulatory-certified modules help you deliver the functionality you need with the low-power performance demanded by your customers.

Chapter Summary:

After reading the chapter, you should be able to fairly answer

• The role of Digital Transformation in new products

• The Digital Feedback Loop

• The purpose of Insights Value Chain

• CRISP-DM Model

• Reference architectures for Data Mining and Stream Processing Architectures

• AI Reference architecture and Edge Processing

References

1. Prabhat Kumar, Artificial Intelligence: Reshaping Life and Business, Chapter 2, BPB Publications, 1 edn (August 17, 2019), ISBN-10: 9388511077.
2. Louis Columbus. 10 Charts That Will Change Your Perspective on Artificial Intelligence's Growth. Jan. 12, 2018, URL: https://www.forbes.com/sites/louiscolumbus/2018/01/12/10-charts-that-will-change-your-perspective-on-artificial-intelligences-growth/#1c94aee04758, Access Date: September 18, 2019.
3. Statista. Revenues from the Artificial Intelligence for Enterprise Applications Market Worldwide, from 2016 to 2025. Published by Statista Research Department, Nov 25, 2019, URL: https://www.statista.com/statistics/607612/worldwide-artificial-intelligence-for-enterprise-applications/Access Date: December 25, 2019.
4. Michael Shirer and Marianne Daquila. Worldwide Spending on Artificial Intelligence Systems Will be Nearly $98 Billion in 2023, According to New IDC Spending Guide. 04 Sept. 2019, URL: https://www.idc.com/getdoc.jsp?containerId=prUS45481219, Access Date: 18 Sep 2019.
5. James Manyika, Michael Chui, Jacques Bughin and et al. Disruptive Technologies: Advances That Will Transform Life, Business, and the Global Economy. May 2013, https://www.mckinsey.com/~/media/McKinsey/Business%20Functions/McKinsey%20Digital/Our%20Insights/Disruptive%20technologies/MGI_Disruptive_technologies_Full_report_May2013.ashx, Access date: 09/18/2018.
6. Yoav Shoham, Raymond Perrault, Erik Brynjolfsson, Jack Clark and Calvin LeGassick. Artificial Intelligence Annual Index – 2017. November 2017, URL: http://aiindex.org/2017-report.pdf, Access Date: Jan 08. 2019.
7. David Rogers. The Digital Transformation Playbook: Rethink Your Business for the Digital Age. March 15, 2016, ISBN-10: 9780231175449, Columbia Business School Publishing © 2016.
8. Steve Guggenheimer. AI and Digital Feedback Loops. July17, 2018, URL: https://docs.microsoft.com/en-us/archive/blogs/stevengu/ai-and-digital-feedback-loops, Access Date: December 20, 2019.
9. Bernie Thibeault. A Complete Guide to Microsoft's Digital Feedback Loop. September 11, 2018, URL: https://www.aerieconsulting.com/blog/microsofts-digital-feedback-loop, Access Date: Sept. 18, 2019.
10. Niko Mohr and Holger Hürtgen. Achieving Business Impact With Data. April 2018, URL: https://www.mckinsey.com/~/media/mckinsey/business%20functions/mckinsey%20analytics/our%20insights/achieving%20business%20impact%20with%20data/achieving-business-impact-with-data_final.ashx, Access Date: Sept. 18, 2019.
11. Pete Chapman, Julian Clinton, Randy Kerber, et al. "CRISP-DM 1.0" March 2000, ftp://ftp.software.ibm.com/software/analytics/spss/support/Modeler/Documentation/14/UserManual/CRISP-DM.pdf , Access Date: Nov. 22, 2018.
12. Carlton E. Sapp. Preparing and Architecting for Machine Learning. 17 January 2017, https://www.gartner.com/binaries/content/assets/events/keywords/catalyst/catus8/preparing_and_architecting_for_machine_learning.pdf,   Access   Date: November 22, 2018.
13. McKinsey Global Institute. Visualizing the Uses and Potential Impact of AI and Other Analytics. April 2018, URL: https://www.mckinsey.com/featured-insights/artificial-intelligence/visualizing-the-uses-and-potential-impact-of-ai-and-other-analytics, Access Date: Sept. 18, 2019.
14. William Kaufmann, The Handbook of Artificial Intelligence Vols. 1 and 2, Los Altos, CA 1981, ISBN-10: 0201168863, Addison-Wesley, Ist Edn (January 1, 1986).
15. V. S. Janakiraman, P. Gopalakrishnan and K. Sarukesi, Foundations of Artificial Intelligence and Expert Systems, Laxmi Publications Pvt Ltd., Ist Edn (March 15, 2019), ASIN: B07PPLMRS2.
16. Jaiwei Han and Macheline Kambler, Data Mining Concepts and Techniques, Morgan Kaufmann; 3rd Edn (June 15, 2011), ISBN-10: 9780123814791.
17. Bernard Marr. What Is Digital Twin Technology - And Why Is It So Important? March 6, 2017, https://www.forbes.com/sites/bernardmarr/2017/03/06/what-is-digital-twin-technology-and-why-is-it-so-important/#2db6e6432e2a.
18. Sam Lucero. Enterprise IoT Leading IoT, Driving Industry Digital Transformation. March 2017, https://e.huawei.com/us/solutions/technical/iot , Access date: December 14, 2018.
19. Jure Leskovec, Anand Rajaraman, and Jeffrey David Ullman, Mining of Massive Datasets, Cambridge University Press 2nd Edn (December 29, 2014), ISBN-10: 9781107077232.
20. Michael Stonebraker, Uğur Çetintemel, and Stan Zdonik. The 8 Requirements of Real-Time Stream Processing", URL: http://cs.brown.edu/~ugur/8rulesSigRec.pdf, Access Date: Nov. 22, 2019.
21. Sundar Ranganathan. Edge to Core to Cloud Architecture for AI. August 2018, URL: https://www.netapp.com/us/media/wp-7271.pdf, Access Date: December 25, 2019.
22. Jeffrey Lee. How to Choose the Right IoT Platform: The Ultimate Checklist. April 25, https://hackernoon.com/how-to-choose-the-right-iot-platform-the-ultimate-checklist-47b5575d4e20 , Access Date: August 6, 2018.
23. David McKinney. Intel Champions Internet of Things and The Intel IoT Reference Architecture. April 23, 2015,https://www.intel.com/content/www/us/en/internet-of-things/white-papers/iot-platform-reference-architecture-paper.html, Access Date: August 6, 2018.

24. Chandrasekar Vuppalapati, Building Enterprise IoT Applications, CRC Press, 1st edition (December 17, 2019), ISBN-10: 0367173859.

25. Kelvin Lui and Jeff Karmiol. AI Infrastructure Reference Architecture - IBM Systems. June 2018, URL: https://www.ibm.com/downloads/cas/W1JQBNJV, Access Date: December 25, 2019.

26. John Morris, Matt Trifiro and Jacob Smith. State of the Edge. URL: https://www.seagate.com/www-content/enterprise-storage/it-4-0/images/Data-At-The-Edge-UP1.pdf, Access Date: December 25, 2019.

27. Daniel Newman. Moving to the Edge: Evolving IoT Data Storage In 2018. Date: Jan. 23, 2018, URL: https://www.forbes.com/sites/danielnewman/2018/01/23/moving-to-the-edge-evolving-iot-data-storage-in-2018/#6055c30141a5, Access Date: December 19, 2019.

28. Vlasios Tsiatsis, Stamatis Karnouskos, Jan Holler, David Boyle and Catherine Mulligan, Internet of Things: Technologies and Applications for a New Age of Intelligence, Second Edition, Academic Press, 2nd Edn (December 14, 2018), ISBN-10: 0128144351.

29. Jose A. Gutierrez. IEEE Std. 802.15.4. 2005. URL: https://people.eecs.berkeley.edu/~prabal/teaching/cs294-11-f05/slides/day21.pdf, Access Date: Jan 08, 2019.

30. Richard Barry. Mastering the FreeRTOS Real Time Kernel. 2016, URL: https://freertos.org/Documentation/161204_Mastering_the_FreeRTOS_Real_Time_Kernel-A_Hands-On_Tutorial_Guide.pdf, Access Date: Sep 18, 2019.

31. Simone Cirani, Gianluigi Ferrari, Marco Picone and Luca Veltri, Internet of Things: Architectures, Protocols and Standards, Wiley, 1st Edn (November 5, 2018), ISBN-10: 1119359678.

32. Rakib Chowdhury. Top 15 Best IoT Operating System For Your IoT Devices in 2020, URL: https://www.ubuntupit.com/best-iot-operating-system-for-your-iot-devices/, Access Date: Jan. 12, 2020.

33. C. Sabri, L. Kriaa and S.L. Azzouz. Comparison of IoT Constrained Devices Operating Systems: A Survey," *2017 IEEE/ACS 14th International Conference on Computer Systems and Applications (AICCSA)*, Hammamet, 2017, pp. 369–375. doi: 10.1109/AICCSA.2017.187, URL: http://ieeexplore.ieee.org/stamp/stamp.jsp?tp=&arnumber=8308310&isnumber=8308243.

34. Adam Dunkels and Oliver Schmidt. About Protothreads. URL: http://dunkels.com/adam/pt/about.html, Access Date: March 21, 2020.

35. Yang. TinyOS Architecture. URL: https://sceweb.uhcl.edu/yang/public/WSN_modules/TinyOS.ppt, Access Date: 01/12/2020.

36. Adam Dunkels, Bjorn¨Gron¨vall, Thiemo Voigt. Contiki—a Lightweight and Flexible Operating System for Tiny Networked Sensors. URL: http://www.dunkels.com/adam/dunkels04contiki.pdf, Access Date: December 25, 2019.

SECTION-II

Data Sources and Engineering Tools

CHAPTER 3

Data—Call for Democratization

"People who are really serious about software should make their own hardware."

Alan Kay

The chapter starts with the perennial issue in democratizing the Artificial Intelligence models—"the lack of data" from the world's most important and largest employer, the agriculture industry and data from developing economies. The need for capturing the data and the development of Computer systems that enables the World Food Organization mandate to create a sustainable food future by 2050. The chapter introduces the Edge device reference architecture with engineering design characteristics of Edge device that includes AI/ML models, Connectivity, and Hardware design characteristics.

Big data surrounds us. Every minute, our smartphones collect billions of data points on our locations, search histories, lifestyle, and habits. The collected data is mined to provide exclusive recommendations that have business value for the companies and marketers. Such collected data has become one of the world's most valuable commodities for companies, a data revolution indeed. The $6-billion-dollar sale of the game Candy Crush[152] was not so pricey because of the app itself, but because of the massive amounts of user data that came with it [1]. Similarly, Monsanto's aggressive move into the Data era was perhaps punctuated in October 2013 with their announced acquisition of the Climate Corporation for $930 million. One may ask why would a firm with its roots in fertilizers and pesticides spend nearly $1 billion on an information technology (IT) company? This aggressive acquisition demonstrates the evolution of the industry [1]. The real valuation is not due to tangible assets but based on data infused insights—"the Climate Corporation is focused on unlocking new value for the farm through data science", commented Hugh Grant, the chairman and chief executive officer for Monsanto [1]. Data has surpassed oil as the world's most valuable resource.[153] Data is giving rise to a new economy[154] and has become a competitive and strategic asset for companies and nations alike.

---

[152] Activision completes $5.9 billion purchase of Candy Crush makers King - https://www.wired.co.uk/article/king-candy-crush-activision-acquisition

[153] The world's most valuable resource is no longer oil, but data - https://www.economist.com/leaders/2017/05/06/the-worlds-most-valuable-resource-is-no-longer-oil-but-data

[154] Data is giving rise to a new economy - https://www.economist.com/briefing/2017/05/06/data-is-giving-rise-to-a-new-economy

> According to Gartner, by 2022, as a result of digital business projects, 75% of enterprise-generated data will be created and processed outside the traditional,[155,156] enterprise data center or Cloud, up from less than 10% in 2018 [2,3,4].

However, the data revolution has not reached every economic sector and every country yet. Agriculture sector in developing countries has been largely passed over. Not only is this a huge missed opportunity for big data companies, it's a significant obstacle in the path towards sustainable food [5]. Richer countries are benefitting more from the new Information Communication Technologies (ICT) possibilities than poorer countries that lack the resources for investment, training and experimentation [5]. According to McKinsey consulting group, African countries spend about 1.1% of their Gross Domestic Product (GDP) on investment in and use of internet services, less than a third of what, on an average, is spent by richer countries—meaning that the gap in the internet availability and use is growing every year, as some regions accelerate ahead [6]. The graph below (Figure 1) shows how advanced economies are ahead of the rest of the world on every indicator of access to, use of, and impact of the use of digital technologies, Information Communication Technologies (ICT) [7].

Huge disparities in ICT spending is a reality between developed economies and poor nations, ironically, the ICT plays a significant role in overcoming societal issues such poverty and education. There is direct correlation between ICT penetration and decrease in global poverty [8], i.e., a clear correlation between falling global absolute poverty and rising ICT penetration (please see Figure 2) [8]. The report finds that technological progress, measured as the share of ICT capital stock has a statistically significant impact on inequality, and the effect of technological change was greater than that of financial globalization [8]. In other words, the improvement of life is more profound for the poor communities in the world by penetration of ICT than disbursement of loans via the International Monetary Fund. The deeper the ICT penetration, the greater is the access to compute resources and, thus, address the perennial issue in democratization of AI—data capture. Having deeper ICT technological penetration is a win-win situation for global communities.

In the same vein, the use of ICT in agriculture sector will improve standard of living for the earth, as the agriculture industry sector (please see Figure 3) is the largest employer in the world, e.g., most of the world's more than 570 million farms are small and family-run [9]. Additionally, agricultural economists and

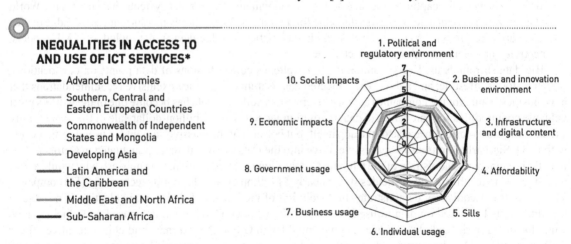

Figure 1: Inequalities in Access To and Use of ICT [7]

---

[155] Hewlett Packard Enterprise Commits $4 Billion to Accelerate the Intelligent Edge - https://www.hpe.com/us/en/newsroom/press-release/2018/06/hewlett-packard-enterprise-commits-4-billion-to-accelerate-the-intelligent-edge.html

[156] What Edge Computing Means for Infrastructure and Operations Leaders - https://www.gartner.com/smarterwithgartner/what-edge-computing-means-for-infrastructure-and-operations-leaders/

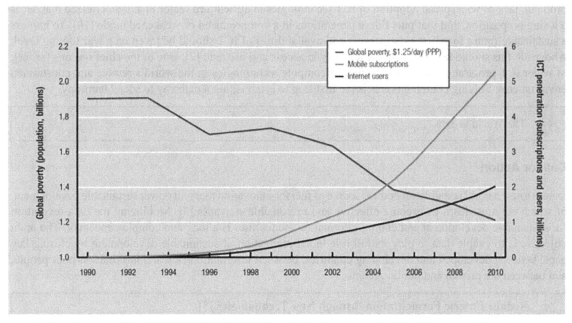

Sources: World Bank *PovCal* database (1990, 1993, 1996, 1999, 2002, 2005, 2008, 2010): authors' calculations and interpolation, ITU *World Telecommunication/ICT Indicators* database June 2013.

Figure 2: Falling Global Absolute Poverty and Rising ICT Penetration [8]

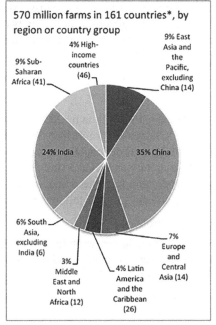

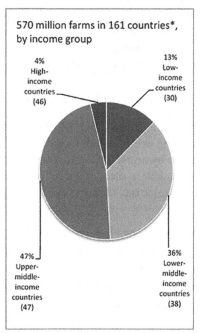

Figure 3: Small Farms [9]

other development specialists generally agree that investing in agriculture is an effective strategy for reducing poverty, inequality and hunger, especially in countries where the agriculture sector employs a large share of the population [9].

Not having data from the most employed sector of the world is a limiting factor for the future generations from the employment and sustainability point of view. Additionally, climate change, importantly, is a data problem, lack of widespread adoption of ICT technologies in agriculture sector in a democratized manner is a losing proposition, and that puts future generations in a compromised or weakened mode [10]. To harness a sustainable future for all, the data access and availabilities of ICT should be taken on a war footing level. Above all, this should be a revolution for equity in access and use data [7]. One of the chief reasons for lack of wider ICT penetration is the availability of compute technologies in the world's remote and constrained environments. Solving compute issues is the first step to establishing a pathway to social harmony.

 The World is data.

Call for Action

Data is a new natural resource, an endless source of fuel for innovation that will power sustainable development, of which we must learn to become effective and responsible stewards [7]. Mobilizing the data revolution for sustainable development and ending information inequalities is a long and complex endeavor. The main objective is to enable data to play its full role in the realization of sustainable development by closing key gaps: between developed and developing countries, between information-rich and information-poor people, and between the private and public sectors.

 Reduce Enteric Fermentation through New Technologies [5]

IoT Sensors play an important role in identifying and providing information to Dairy Farmers on Rumination. Cow necklace sensors collect rumination threshold count that are indicators of cow health.[157] Using threshold counts, we can monitor and control agricultural greenhouse gas emissions. For instance, ruminant livestock were responsible for around half of all agricultural production emissions in 2010. Of these emissions, the largest source is "enteric methane", or cow burps. Increasing productivity of ruminants also reduces methane emissions, mainly because more milk and meat is produced per kilogram of feed. In addition, new technologies can reduce enteric fermentation.

Creating Sustainable Food Future by 2050

As the global population grows from 7 billion in 2010 to a projected 9.8 billion by 2050, and incomes grow across the developing world, overall food demand is on course to increase by more than 50%, and demand for animal-based foods by nearly 70% [11]. Yet today, hundreds of millions of people remain hungry, agriculture already uses almost half of the world's vegetated land, and agriculture and related land-use change generate one-quarter of annual greenhouse gas (GHG) emissions.[158] There is a huge need and scope for agriculture productivity improvement. The menu for a sustainable food future calls for the following [11]:

1. Reduce Growth in Demand for Food and Other Agricultural Products

Globally, food loss and waste results in nearly $1 trillion in economic losses, contribute to food insecurity in some developing countries, squander away agricultural land and water resources, and generate roughly one-quarter of all agricultural Green House Gas (GHG) emissions.[159] Actions [11]:

a. Reduce food loss and wastage

b. Shift to healthier and more sustainable diets

c. Avoid competition from bioenergy for food crops and land

d. Achieve replacement-level fertility rates

---

[157] Rumination - http://hanuinnotech.com/dairyanalytics.html

[158] Can we feed the world without destroying the planet? - https://wrr-food.wri.org/

[159] Course 1 Reduce Growth in Demand for Food and other Agricultural Products (Synthesis) - https://wrr-food.wri.org/course/reduce-growth-demand-food-and-other-agricultural-products-synthesis

2. Increase Food Production without Expanding Agricultural Land[160]

Productivity gain is the name of the game. What I mean by it is without exerting additional pressure on agriculture lands and on livestock, improving productivity efficiencies through ICT and technological democratization will help to feed 10 billion people in 2050 [11]. The single most important need for a sustainable food future is boosting the natural resource efficiency of agriculture that is, producing more food per hectare, per animal, per kilogram of fertilizer, and per liter of water. Such productivity gains reduce both the need for additional land and the emissions from production processes. ICT plan is an important role.

> Optimistically, farmers have so far continued to steadily boost yields by farming smarter in a variety of ways, and new technologies are opening up new potential. Whatever the degree of optimism, the policy implications are the same: Going forward, the world needs to make even greater efforts to boost productivity than in the past to achieve a sustainable food future [11].

3. Protect and Restore Natural Ecosystems and Limit Agricultural Land-Shifting [11]

Focus on the land-management efforts that must complement food demand-reduction efforts and productivity gains to avoid the harms of agricultural land expansion. One guiding principle is the need to make land-use decisions that enhance efficiency for all purposes—not just agriculture but also carbon storage and other ecosystem services. Another principle is the need to explicitly link efforts to boost agricultural yield gains with protection of natural lands.[161]

4. Increase Fish Supply

Fish, including finfish and shellfish, provide only small percentages of total global calories and protein, but they contribute 17% of animal-based protein, and are particularly important for more than 3 billion people in developing countries. We project fish consumption to rise by 58% between 2010 and 2050, but the wild fish catch peaked at 94 million tons in the mid-1990s and has since stagnated or perhaps declined.

5. Reduce Greenhouse Gas Emissions from Agricultural Production

Agricultural production emissions arise from livestock farming, application of nitrogen fertilizers[162] (please see Figure 4), rice cultivation, and energy use [12]. These production processes are traditionally regarded as hard to control. In general, our estimates of mitigation potential in this are more optimistic, partly because many analyses have not fully captured the opportunities for productivity gains and partly because we factor

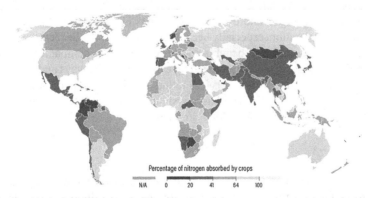

Percentage of nitrogen absorbed by crops

N/A 0 20 41 64 100

Figure 4: The Percentage of Applied Nitrogen That is Absorbed by Crops Varies Widely Across the World [11]

---

[160] Course 2 Increase Food Production without Expanding Agricultural Land - https://wrr-food.wri.org/course/increase-food-production-without-expanding-agricultural-land-synthesis

[161] Course 3 Protect and Restore Natural Ecosystems and Limit Agricultural Land-Shifting (Synthesis) - https://wrr-food.wri.org/course/protect-and-restore-natural-ecosystems-and-limit-agricultural-land-shifting-synthesis

[162] Nitrogen and Phosphorus Pollution Data Access Tool - https://www.epa.gov/nutrient-policy-data/nitrogen-and-phosphorus-pollution-data-access-tool

in promising potential for technological innovations.[163] Sensor technologies and ML can play an important role in addressing this issue [11].

 Nitrogen and Phosphorus Pollution Data Access Tool (NPDAT)

Over the last 50 years, the amount of nitrogen and phosphorus pollution entering our nation's waters has escalated dramatically [12]. For example, 30% of U.S. streams have high levels of nitrogen (N) and phosphorus (P) pollution. Reported drinking water violations for nitrates have doubled in the last eight years. N and P pollution threatens human health, aquatic ecosystems, and economic prosperity. The U.S. Environmental Protection Agency's new tool,[164] NPDAT,[165] is designed to help states develop effective state N/P reduction strategies to combat this serious and growing environmental problem [12].

Data Access Tool - NPDAT:[166]

In order to meet the menu for a sustainable food future, currently, the agriculture data is not available for many small farms across the world. Agriculture data is key for sustainability [13]. Even more, the agriculture data enable small farms to fight and withstand climate change effects and are critical for our future generations as 80% of food produced by the small farms. Mapping and data collection should be a priority for national governments, international agencies, and even private parties [11]. A good example for better mapping is peatland conservation and restoration, especially because peatlands cannot be identified from satellite imagery.

Farm data would provide immense benefits for agricultural-tech companies and farmers. Companies could use data on fertilizer usage, irrigation and seed type to improve the ways and regions where they market agricultural products, and better tailor their services to the individual needs of farmers [13]. Farmers would benefit from the increased efficiency that data-based insights could provide. With more access to data, would provide farmers the ability to improve agriculture production.

Investments

There is already considerable interest in data from the agribusiness industry. Monsanto bought Climate Corp, a company using big data for weather services, for almost $1 billion in 2013, the biggest agriculture-

---

[163] Course 5 Reduce Greenhouse Gas Emissions from Agricultural Production (Synthesis) - https://wrr-food.wri.org/course/reduce-greenhouse-gas-emissions-agricultural-production-synthesis

[164] Nitrogen and Phosphorus Pollution Data Access Tool - https://www.epa.gov/nutrient-policy-data/nitrogen-and-phosphorus-pollution-data-access-tool

[165] Nitrogen and Phosphorus Pollution Data Access Tool (NPDAT) - https://www.epa.gov/sites/production/files/documents/npdatfactsheet.pdf

[166] Data Access Tool - https://gispub2.epa.gov/npdat/

technology deal in history. Ag-tech start-ups received more than $10 billion dollars in investments in 2017. These investments, however, have been concentrated in wealthier countries where the immediate prospect for revenue is higher. According to the United Nation's Food and Agriculture Organization [14], "the world's 500 million smallholder farmers risk being left behind in structural and rural transformations."[167] There are promising new markets for farming technology in the developing world, which hold lots of underproductive agricultural land and the biggest potential for yield increases that could close the world's sustainable food gap [14]. The chasm is development compute technologies that can work efficiently in constrained environments [15]; Current technologies are better suited for resourceful environments with continuous electrical power with availability of connected networks. The same compute systems are not suitable to operate in constrained environments and extremely constrained environments [15][16].

Designing AI that Uses Less Energy

More than ever before, the revolution triggered by the development of digital technologies and their widespread adoption tend to obscure its impact on the environment.[168] Nevertheless, there is an urgent need to take this on board. Two years ago [17], the American Association of Semi-Conductor Manufacturers predicted that by 2040, the global demand for data storage capacity, which grows at the pace of the progress of AI, will exceed the available world production of silicon.[169]

Furthermore, by 2040 the energy required for computation will equally have exceeded world energy production [16]; the progress of the blockchain may also cause our energy requirements to skyrocket. At a time when global warming is a scientific certainty, it is no longer possible to pursue technological and societal developments if those are completely detached from the need to preserve our environment.[170] Designing AI that uses less energy is the way to achieve sustainable computation that balances compute needs with environmental and societal needs [18].

⚡ **The World Energy Production**

The Semiconductor Industry Association (SIA) has predicted that, with current engineering approaches, the required energy use for computation will exceed the estimated world energy supply by the year 2040. By 2040 the energy required for computation will equally have exceeded world energy production [17].

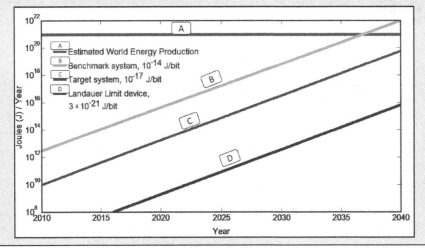

---

[167] FAO – The State of Food and Agriculture - http://www.fao.org/3/a-i7658e.pdf

[168] See the Greenpeace report Clicking Clean: Who is winning the race to build a green internet? - https://secured-static.greenpeace.org/austria/Global/austria/dokumente/Clicking%20Clean%202017.pdf

[169] See the 2015 American Semi-Conductor Industry's report: rebooting the IT revolution, a call to action - https://www.src.org/news-room/rebooting-the-it-revolution.pdf

[170] For a meaningful Artificial Intelligence - https://www.aiforhumanity.fr/pdfs/MissionVillani_Report_ENG-VF.pdf

> Shown here is the annual energy consumption of computation over time. The upward trends are due to SIA predicted increases in global raw-bit transitions (which are strongly correlated with capacity to compute information). It is important to note that system level energy per bit values included are functions not only of information computation/logic circuits, but also other computational device components such as storage and interface circuitry. The "benchmark" curve shows the growing energy demand for the system level energy per bit values of mainstream systems. The target system curve uses the practical lower limit system level energy per bit value set by factors such as materials. The Landauer limit curve uses the minimal device energy per bit value provided by the Landauer's Principle [17].

Our proposed frameworks in this book enables the democratization of compute and AI to the gross roots of the societies [15] bringing balance of energy needs for the future of compute [17], and, thus, bridging the ICT gap. The democratization of AI is no longer a choice but an essential tactics for the survival and betterment of humanity.

The Last Mile—Constrained Compute Devices & "AI Chasm"

The Fourth wave of compute has spurred the development of Edge devices. Edge devices come in various forms and shapes with varying compute capacities.[171] The general classification of Edge devices includes: Class 0, Class 1 and Class 2. Class 0 and Class 1 devices collect vast amount of environmental & geolocation data on a periodic basis [18]. Due to constrained environments, the class 0 devices require external devices such as gateways and cell phones to relay/connect to the internet.

Classes of Constrained Devices

Despite the overwhelming variety of internet-connected devices that can be envisioned, it may be worthwhile to have some succinct terminology[172] for different classes of constrained devices (Table 1) [18].

The constrained devices[173,174] are end nodes with sensors/actuators that can handle a specific application purpose [19,20]. They are usually connected to gateway-like devices, low power lossy network, and in-turn communicates with the IoT cloud platforms. Typically, they communicate via low power wireless protocols like BLE, 802.15.4 (6LoWPAN, Zigbee, Thread, WirelessHART, etc.), LPWAN, etc., and mostly battery powered with low data rate.

Table 1: Classes of Constrained Devices (KiB = 1024 bytes)

Name	Data size (e.g., RAM)	Code Size (i.e., Flash)
Class 0	<< 10 KiB	<< 100 KiB
Class 1, C1	~ 10 KiB	~ 100 KiB
Class 2, C2	~ 50 KiB	~ 250 KiB

---

[171] RFC 7228 - https://tools.ietf.org/html/rfc7228
[172] RFC 7228 https://tools.ietf.org/html/rfc7228
[173] Constrained devices - https://www.cisoplatform.com/profiles/blogs/classification-of-iot-devices
[174] Neural Network / Machine Learning in resource constrained environments - https://medium.com/@raghu.madabushi/neural-network-machine-learning-in-resource-constrained-environments-c934ff1f522

Class 0 devices are *extremely constrained devices (xCDs), i.e.,* sensor-like motes (data size less than 10 KB and code size less than 100 kb). They are so severely constrained in memory and processing capabilities that most likely they will not have the resources required to communicate directly with the internet in a secure manner[18].

Class 0 devices will participate in internet communications with the help of larger devices acting as proxies, gateways, or servers. In democratization of AI, the larger devices play an important role of collecting data from Class 0 devices and relay to central enterprise or cloud data centers. Class 0 devices generally cannot be secured or managed comprehensively in the traditional sense. They will most likely be preconfigured (and will be reconfigured rarely, if at all) with a very small data set. For management purposes, they could answer/keep alive signals and send on/ off or basic health or heartbeat indications.

Class 1 devices are *quite constrained* in code space (less than 10 KB) and processing capabilities (less than 100 kb), such that they cannot easily talk to other internet nodes employing a full protocol stack such as using HTTP, Transport Layer Security (TLS), and related security protocols and XML-based data representations. However, they are capable enough to use a protocol stack specifically designed for constrained nodes (such as the Constrained Application Protocol (CoAP) over UDP [COAP]) and participate in meaningful conversations without the help of a gateway node. Therefore, they can be integrated as fully developed peers into an IP network, but they need to be frugal with state memory, code space, and often power expenditure for protocol and application usage. Deployment of AI on these devices has to follow a stringent and strict performance to feature vs. code space to power requirements.

Class 2 devices are *less constrained* and fundamentally capable of supporting most of the same protocol stacks as used on notebooks or servers. However, even these devices can benefit from lightweight and energy-efficient protocols and from consuming less bandwidth [16].

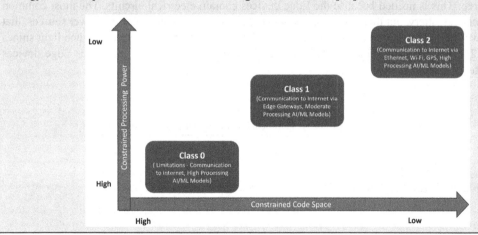

Class 1 and Class 2 devices are less constrained and can perform enough compute to sustain with good connectivity. In the following section, we will expand more on Class 0 devices, which are mostly micro-controller based and have very limited resources. The goal is having an Edge device that can operate in a constrained environment and relay data to the central connected network or Internet alleviate two issues [21]: one, data collection and second, application of AI[175] and we can deploy the AI applications where required the most.

Edge Device Architecture

The following diagram (Figure 5) contains a high level view of Edge devices:

---

[175] Democratizing AI - https://insidebigdata.com/2019/08/06/democratizing-ai/

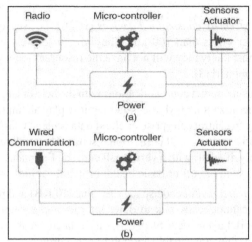

Figure 5: Edge Reference Model

- **Communication module:** This gives the Edge device its communication capabilities. It is typically either a radio transceiver (Bluetooth Low-Energy, CoAP) with an antenna or a wired connection (Wi Fi).
- **Microcontroller:** It is a small microprocessor that runs the software of the Edge device. The brain of the device is its software or firmware that is burned during the device construction. Many of the ML and AI algorithms are part of firmware of the device.
- **Sensors or actuators:** These give the Edge devices a way to sense and interact with the physical world.
- **Power source:** This is needed because the Edge devices contain electrical circuits. The most common power source is a battery, but there are other examples as well, such as piezoelectric power sources, that provide power when physical force is applied, or small solar cells that provide power when light shines on them [22]. The following table (Table 2: Power Source) contains power source for Edge devices [22—Chapter 7]:

Table 2: Power Source

Power source	Typical maximum current (mA)	Typical charge (mAh)
CR2032 button cell	20	200
AA alkaline battery	20	3000
Solar cell	40	Limitless
RF power	25	Limitless

Microcontrollers have two types of memory: Read-only memory (ROM) and random-access memory (RAM). ROM is used to store the program code that encodes the behavior of the device and RAM is used for temporary data the software needs to do its task. For example, temporary data includes storage for program variables and buffer memory for handling radio traffic.

AI Model

Tiny Edge devices have limited hardware capacity that includes small memory, limited storage, small footprint processing power, and finite power source. The choice of AI model is very much limited by the capacity. Nevertheless, there are two ways of deploying ML models on the tiny devices.

- Model Development and Deployment for custom built hardware
- ML models based on microcontroller packaged frameworks—for example, TensorFlow Lite.

> -☀- TensorFlow can be used for both network training and inference, whereas TensorFlow Lite is
> specifically designed for inference on devices with limited compute (phones, tablets and other
> embedded devices). TensorFlow for C is aimed at Deep Learning models for microcontrollers—Tiny &
> Constrained devices.

Custom Built Hardware ML Models

Custom Built Hardware ML models can be supervised, unsupervised or reinforcement models based. The design consideration is the size and model complexity to be validated before deploying the model into hardware. Second, on the model design side, the model could be built on traditional machine learning development environments such as Anaconda, Spyder, Microsoft Visual Studio or Arduino Scratch. The development language is C.

K-Means Clustering

The following code contains a simple K-means cluster.[176] The goal of the console application is to arrange the data points into cluster by calculating the distance based on two pre-defined cluster centroids.

K-Means Cluster header file

```
#include<stdlib.h>
/*
* Simple k-means implementation for arbitrary data structures
*
* Since k-means partitions based on inter-object "distance" the same
* machinery can be used to support any object type that can calculate a
* "distance" between pairs.
*
* To use the k-means infrastructure, just fill out the kmeans_config
* structure and invoke the kmeans() function.
*/
/*
* Threaded calculation is available using pthreads, which practically
* means UNIX platforms only, unless you're building with a posix
* compatible environment.
*
* #define KMEANS_THREADED
*/
#defineKMEANS_NULL_CLUSTER -1
/*
* If the algorithm doesn't converge within this number of iterations,
* it will return with a failure error code.
*/
#defineKMEANS_MAX_ITERATIONS 1000
/*
* The code doesn't try to figure out how many threads to use, so
* best to set this to the number of cores you expect to have
* available. The threshold is the value of k*n at which to
* move to multi-threading.
*/
```

---

[176] KMeans Cluster - https://github.com/pramsey/kmeans

```
#ifdef KMEANS_THREADED
#define KMEANS_THR_MAX 4
#define KMEANS_THR_THRESHOLD 250000
#endif
#definekmeans_malloc(size) malloc(size)
#definekmeans_free(ptr) free(ptr)
typedefvoid* Pointer;
typedefenum {
        KMEANS_OK,
        KMEANS_EXCEEDED_MAX_ITERATIONS,
        KMEANS_ERROR
} kmeans_result;
/*
* Prototype for the distance calculating function
*/
typedefdouble (*kmeans_distance_method) (constPointer a, constPointer b);
typedefvoid (*kmeans_centroid_method) (constPointer* objs, constint* clusters, size_t num_objs, int
cluster, Pointer centroid);
typedefstructkmeans_config
{
        /* Function returns the "distance" between any pair of objects */
        kmeans_distance_method distance_method;
        /* Function returns the "centroid" of a collection of objects */
        kmeans_centroid_method centroid_method;
        /* An array of objects to be analyzed. User allocates this array */
        /* and is responsible for freeing it. */

        Pointer* objs;
        /* Number of objects in the preceding array */
        size_t num_objs;
        /* An array of inital centers for the algorithm */
        /* Can be randomly assigned, or using proportions, */

        Pointer* centers;
        /* Number of means we are calculating, length of preceding array */
        unsignedint k;
        /* Maximum number of times to iterate the algorithm, or 0 for */
        /* library default */
        unsignedint max_iterations;
        /* Iteration counter */
        unsignedint total_iterations;
        /* Array to fill in with cluster numbers. User allocates and frees. */
        int* clusters;
} kmeans_config;
/* This is where the magic happens. */
kmeans_result kmeans(kmeans_config* config);
```

Cluster Calculation

```c
#include<assert.h>
#include<float.h>
#include<math.h>
#include<stdlib.h>
#include<stdio.h>
#include<string.h>
#include"kmeans.h"
#ifdef KMEANS_THREADED
#include<pthread.h>
#endif
staticvoid
update_r(kmeans_config* config)
{
        int i;
        for (i = 0; i <config->num_objs; i++)
        {
                double distance, curr_distance;
                int cluster, curr_cluster;
                Pointer obj;
                assert(config->objs != NULL);
                assert(config->num_objs > 0);
                assert(config->centers);
                assert(config->clusters);
                obj = config->objs[i];
                /*
                * Don't try to cluster NULL objects, just add them
                * to the "unclusterable cluster"
                */
                if (!obj)
                {
                        config->clusters[i] = KMEANS_NULL_CLUSTER;
                        continue;
                }
                /* Initialize with distance to first cluster */
                curr_distance = (config->distance_method)(obj, config->centers[0]);
                curr_cluster = 0;
                /* Check all other cluster centers and find the nearest */
                for (cluster = 1; cluster <config->k; cluster++)
                {
                        distance = (config->distance_method)(obj, config->centers[cluster]);
                        if (distance < curr_distance)
                        {
                                curr_distance = distance;
                                curr_cluster = cluster;
                        }
                }
                /* Store the nearest cluster this object is in */
                config->clusters[i] = curr_cluster;
        }
```

```
}
staticvoid
update_means(kmeans_config* config)
{
        int i;
        for (i = 0; i <config->k; i++)
        {
                /* Update the centroid for this cluster */
                (config->centroid_method)(config->objs, config->clusters, config->num_objs, i, config-
>centers[i]);
        }
}
#ifdef KMEANS_THREADED
#endif/* KMEANS_THREADED */
kmeans_result
kmeans(kmeans_config* config)
{
        int iterations = 0;
        int* clusters_last;
        size_t clusters_sz = sizeof(int) * config->num_objs;
        assert(config);
        assert(config->objs);
        assert(config->num_objs);
        assert(config->distance_method);
        assert(config->centroid_method);
        assert(config->centers);
        assert(config->k);
        assert(config->clusters);
        assert(config->k <= config->num_objs);
        /* Zero out cluster numbers, just in case user forgets */
        memset(config->clusters, 0, clusters_sz);
        /* Set default max iterations if necessary */
        if (!config->max_iterations)
                config->max_iterations = KMEANS_MAX_ITERATIONS;
        /*
        * Previous cluster state array. At this time, r doesn't mean anything
        * but it's ok
        */
        clusters_last = kmeans_malloc(clusters_sz);
        while (1)
        {
                /* Store the previous state of the clustering */
                memcpy(clusters_last, config->clusters, clusters_sz);
#ifdef KMEANS_THREADED
                update_r_threaded(config);
                update_means_threaded(config);
#else
                update_r(config);
                update_means(config);
#endif
```

```
                    /*
                     * if all the cluster numbers are unchanged since last time,
                     * we are at a stable solution, so we can stop here
                     */
                    if (memcmp(clusters_last, config->clusters, clusters_sz) == 0)
                    {
                            kmeans_free(clusters_last);
                            config->total_iterations = iterations;
                            returnKMEANS_OK;
                    }
                    if (iterations++ >config->max_iterations)
                    {
                            kmeans_free(clusters_last);
                            config->total_iterations = iterations;
                            returnKMEANS_EXCEEDED_MAX_ITERATIONS;
                    }
            }
        kmeans_free(clusters_last);
        config->total_iterations = iterations;
        returnKMEANS_ERROR;
}
staticdouble d_distance(constPointera, constPointerb)
{
        double da = *((double*)a);
        double db = *((double*)b);
        return fabs(da - db);
}
staticvoid d_centroid(constPointer* objs, constint* clusters, size_tnum_objs, intcluster, Pointercentroid)
{
        int i;
        int num_cluster = 0;
        double sum = 0;
        double** doubles = (double**)objs;
        double* dcentroid = (double*)centroid;
        if (num_objs<= 0) return;
        for (i = 0; i <num_objs; i++)
        {
                /* Only process objects of interest */
                if (clusters[i] != cluster)
                        continue;
                sum += *(doubles[i]);
                num_cluster++;
        }
        if (num_cluster)
        {
                sum /= num_cluster;
                *dcentroid = sum;
        }
        return;
}
```

```
int main(intnargs, char** args)
{
        // https://github.com/pramsey/kmeans
        double v[10] = { 1.0, 2.0, 3.0, 4.0, 5.0, 6.0, -7.0, 8.0, 9.0, 10.0 };
        double c[2] = { 2.0, 5.0 };
        kmeans_config config;
        kmeans_result result;
        int i;
        config.num_objs = 11;
        config.k = 2;
        config.max_iterations = 100;
        config.distance_method = d_distance;
        config.centroid_method = d_centroid;
        config.objs = calloc(config.num_objs, sizeof(Pointer));
        config.centers = calloc(config.k, sizeof(Pointer));
        config.clusters = calloc(config.num_objs, sizeof(int));
        /* populate objs */
        for (i = 0; i < config.num_objs - 1; i++)
        {
                config.objs[i] = &(v[i]);
        }
        config.objs[10] = NULL;
        // config.objs = objs;
        /* populate centroids */
        for (i = 0; i < config.k; i++)
        {
                config.centers[i] = &(c[i]);
        }
        /* run k-means */
        result = kmeans(&config);
        /* print result */
        for (i = 0; i < config.num_objs; i++)
        {
                if (config.objs[i])
                        printf("%g [%d]\n", *((double*)config.objs[i]), config.clusters[i]);
                else
                        printf("NN [%d]\n", config.clusters[i]);
        }
        free(config.objs);
        free(config.clusters);
        free(config.centers);
        return 0;
}
```

The Output:

Please see Figure 6 for K-means cluster output.

   Finally, the model gets integrated into hardware firmware code and gets deployed as part of the Firmware and flashed into the hardware (please see Figure 7).

Figure 6: K-means Cluster Output

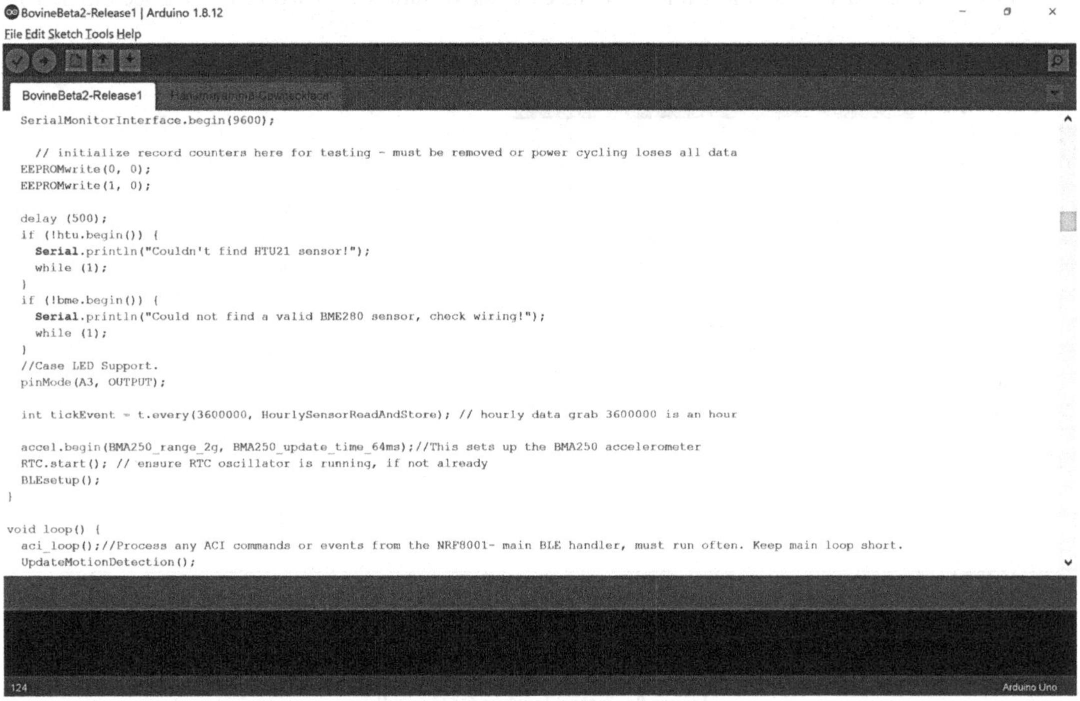

Figure 7: IDE

ML Models based Packaged Frameworks—TensorFlow for C

TensorFlow is a full-fledged deep learning framework. The TensorFlow Lite is for mobile devices. In order to deploy TensorFlow on Tiny and microcontroller devices, TensorFlow for C[177] is the way to go.

TensorFlow for C is supported on the following systems:

- Linux, 64-bit, x86
- macOS X, Version 10.12.6 (Sierra) or higher
- Windows, 64-bit x86

---

[177] TensorFlow for C - https://www.tensorflow.org/install/lang_c

Download TensorFlow for C

The TensorFlow for C can be downloaded from:

TensorFlow C library	URL
Linux	
Linux CPU only	https://storage.googleapis.com/tensorflow/libtensorflow/libtensorflow-cpu-linux-x86_64-1.15.0.tar.gz
Linux GPU support	https://storage.googleapis.com/tensorflow/libtensorflow/libtensorflow-gpu-linux-x86_64-1.15.0.tar.gz
macOS	
macOS CPU only	https://storage.googleapis.com/tensorflow/libtensorflow/libtensorflow-cpu-darwin-x86_64-1.15.0.tar.gz
Windows	
Windows CPU only	https://storage.googleapis.com/tensorflow/libtensorflow/libtensorflow-cpu-windows-x86_64-1.15.0.zip
Windows GPU only	https://storage.googleapis.com/tensorflow/libtensorflow/libtensorflow-gpu-windows-x86_64-1.15.0.zip

Once downloaded, extract TensorFlow for C. In this case, on Windows 10 machine (please see Figure 8):

Figure 8: TensorFlow Lite for C Install

Building First TensorFlow for C Model—Hello World

The following code is a simple Hello World-based console application that uses Tensor Flow for C to deploy a simple C application (please see Figure 9).

```c
#include<stdio.h>
#include<tensorflow/c/c_api.h>
int main() {
    printf("Hello from TensorFlow C library version %s\n", TF_Version());
return 0;
}
```

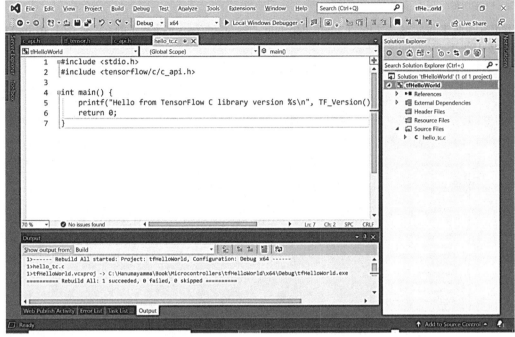

Figure 9: Microsoft Visual Studio - TensorFlow IDE

Compiler Setup

To compile the application, please add "Additional Include Directory" as part of the C/++ General settings and include Library folder (please see Figure 10).

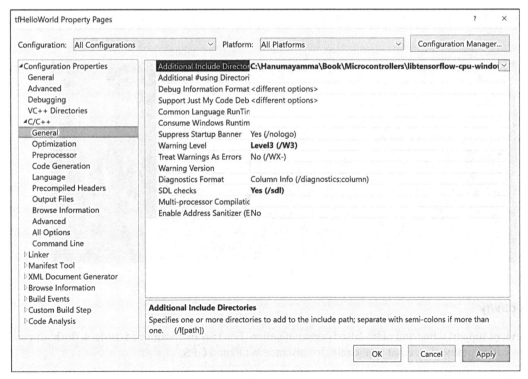

Figure 10: Linker Options

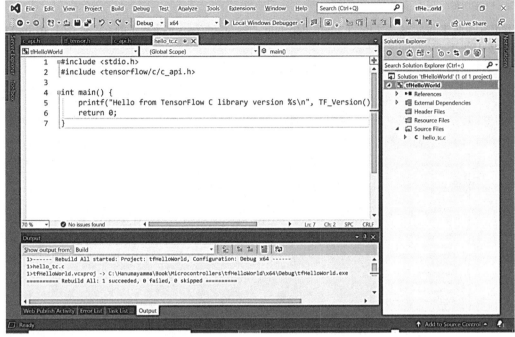

Linker Dependencies

Add TensorFlow Libraries to Linker Input dependencies: Add TensorFlow Library & DLL to build the application (please see Figure 11).

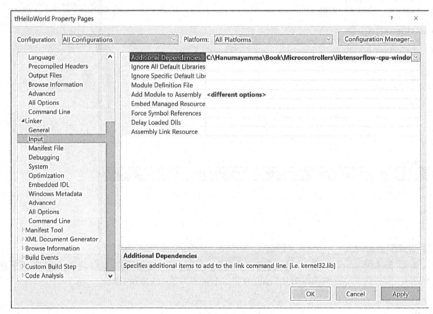

Figure 11: Liner Library

Output

Once successfully compiled, the console application generates output shown in Figure 12).

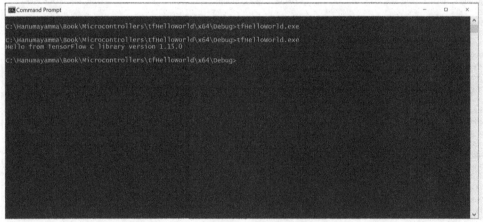

Figure 12: Output

Connectivity

Edge devices support wired and radio-based communication. For Tiny or constrained devices, the high-power communication protocols are not applicable, for instance Wi Fi and GPS.

Bluetooth Low Energy (BLE)

The key features of Bluetooth wireless technology are robustness, low power consumption, and low cost [23]. Many features of the core specification are optional, allowing product differentiation.[178]

There are two forms of bluetooth wireless technology systems: Basic Rate (BR) and Low Energy (LE). Both systems include device discovery, connection establishment and connection mechanisms (See Figure 13) [24]. Bluetooth wireless technology is a short-range communications system intended to replace the cable(s) connecting portable and/or fixed electronic devices

The Basic Rate system, for instance Bluetooth available on Mac systems, includes optional Enhanced Data Rate (EDR), Alternate Media Access Control (MAC) and Physical (PHY) layer extensions. The Basic Rate system offers synchronous and asynchronous connections with data rates of 721.2 kb/s for Basic Rate, 2.1 Mb/s for Enhanced Data Rate and high-speed operation up to 54 Mb/s with the 802.11 AMP [24].

The LE system, in iOS and Android phones, includes features designed to enable products that require lower current consumption, lower complexity and lower cost than BR/EDR. The LE system is also designed for use cases and applications with lower data rates and has lower duty cycles. Depending on the use case or application, one system including any optional parts may be more optimal than the other [24].

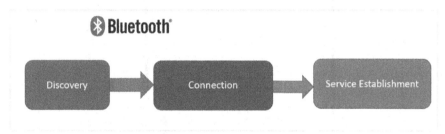

Figure 13: Bluetooth Connection

Bluetooth Core

The Bluetooth Core system consists of a host and one or more controllers. A host is a logical entity defined as all the layers below the non-core profiles and above the Host Controller Interface (HCI) (see Figure 14). A controller is a logical entity defined as all the layers below HCI. An implementation of the host and controller may contain the respective parts of the HCI. Two types of controllers are defined in this version of the Core Specification: Primary Controllers and Secondary Controllers.

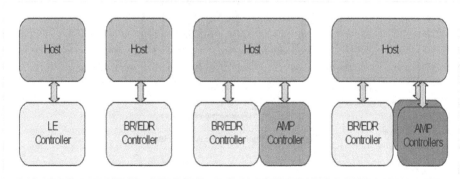

Figure 14: Bluetooth Host and Controller Combination [10]

---

[178] Bluetooth Specification - https://www.bluetooth.com/specifications/protocol-specifications

Bluetooth Core System Architecture

The following figure (see Figure 15) provides the Bluetooth Core architecture:[179] The core blocks, each with its associated communication protocol [24].

- Link Manager, Link Controller and BR/EDR Radio blocks comprise a BR/EDR controller.
- AMP PAL, AMP MAC, and AMP PHY comprise an AMP controller.
- Link Manager, Link Controller and LE Radio blocks comprise an LE Controller. L2CAP, SDP and GAP blocks comprise a BR/EDR host.
- L2CAP, SMP, Attribute protocol, GAP and Generic Attribute Profile (GATT) blocks comprise an LE host.
- A BR/EDR/LE host combines the set of blocks from each respective host.

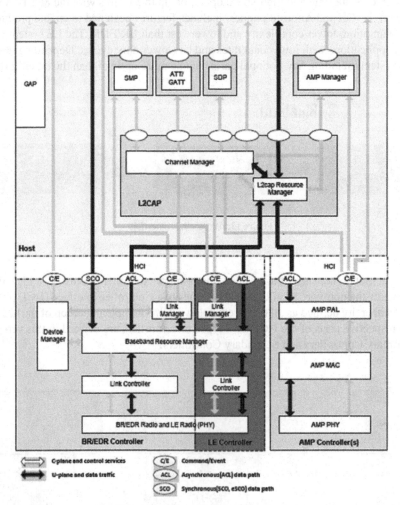

Figure 15: Bluetooth Core Architecture

Bluetooth Low Energy

The LE radio operates in the unlicensed 2.4 GHz industrial, scientific and medical radio (ISM) band (see Figure 16)[180] [25].

---

[179] Bluetooth Specification - https://www.bluetooth.com/specifications/protocol-specifications
[180] RF Basics for Non-RF Engineers - http://www.ti.com/lit/ml/slap127/slap127.pdf

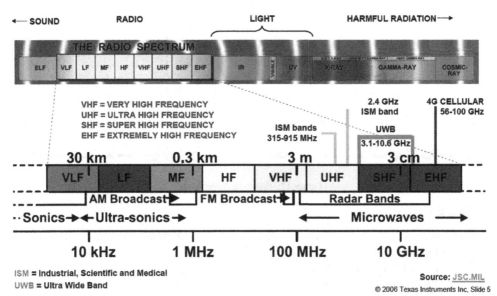

Figure 16: Electromagnetic Spectrum

Unlicensed ISM & Short-Range Device (SRD)

Here are the details of Unlicensed ISM & Short-Range Device (SRD) [11]:

- **USA/Canada:**
  - o 260 – 470 MHz (FCC Part 15.231; 15.205)
  - o 902 – 928 MHz (FCC Part 15.247; 15.249)
  - o 2400 – 2483.5 MHz (FCC Part 15.247; 15.249)

- **Europe:**
  - o 433.050 – 434.790 MHz (ETSI EN 300 220)
  - o 863.0 – 870.0 MHz (ETSI EN 300 220)
  - o 2400 – 2483.5 MHz (ETSI EN 300 440 or ETSI EN 300 328)

- **Japan:**
  - o 315 MHz (Ultra low power applications)
  - o 426 – 430, 449, 469 MHz (ARIB STD-T67)
  - o 2400 – 2483.5 MHz (ARIB STD-T66)
  - o 2471 – 2497 MHz (ARIB RCR STD-33)

LE employs two multiple access schemes: Frequency division multiple access (FDMA) and time division multiple access (TDMA). Forty (40) physical channels, separated by 2 MHz, are used in the FDMA scheme. Three (3) are used as primary advertising channels and 37 are used as secondary advertising channels and as data channels. A TDMA-based polling scheme is used, in which one device transmits a packet at a predetermined time and a corresponding device responds with a packet after a predetermined interval [24].

Devices that transmit advertising packets on the advertising PHY channels are referred to as advertisers. Devices that receive advertising packets (see Figure 17) on the advertising channels without the intention to connect to the advertising device are referred to as scanners. Transmissions on the advertising PHY channels occur in advertising events. At the start of each advertising event, the advertiser sends an advertising packet corresponding to the advertising event type. Depending on the type of advertising packet, the scanner may make a request to the advertiser on the same advertising PHY channel, which may be followed by a response from the advertiser on the same advertising PHY channel. The advertising PHY channel changes on the next advertising packet sent by the advertiser in the same advertising event. The advertiser may end the advertising event at any time during the event. The first advertising PHY channel is used at the start of the next advertising event.

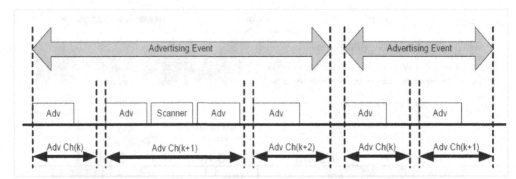

Figure 17: Bluetooth Advertising

Bluetooth Low Energy Protocol

Bluetooth Low Energy (BLE, also marketed as Bluetooth Smart) started as part of the Bluetooth 4.0 Core Specification. It's tempting to present BLE as a smaller, highly optimized version of its bigger brother, classic Bluetooth, however, BLE has an entirely different lineage and design goals.

While Bluetooth Low Energy is a good technology on its own merit, what makes BLE genuinely exciting—and what has pushed its phenomenal adoption rate so far so quickly—is that it's the right technology, with the right compromises, at the right time. For a relatively young standard (it was introduced in 2010), BLE has seen an uncommonly rapid adoption rate, and the number of product designs that already include BLE puts it well ahead of other wireless technologies at the same point of time in their release cycles.

The Bluetooth Protocol stack (see Figure 18) has GATT, ATT, and L2CAP.

Key Terms	Description
Generic Attribute Profile (GATT)	The GATT profile is a general specification for sending and receiving short pieces of data known as "attributes" over a BLE link. All current Low Energy application profiles are based on GATT.
Attribute Protocol (ATT)	GATT is built on top of the Attribute Protocol (ATT). This is also referred to as GATT/ATT. ATT is optimized to run on BLE devices. To this end, it uses as few bytes as possible. Each attribute is uniquely identified by a Universally Unique Identifier (UUID), which is a standardized 128-bit format for a string ID used to uniquely identify information.
L2CAP	L2CAP is used within the Bluetooth protocol stack. It passes packets to either the Host Controller Interface (HCI) or on a hostless system, directly to the Link Manager/ACL link.

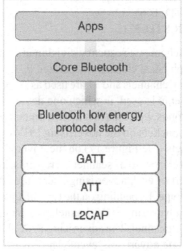

Figure 18: Bluetooth Protocol Stack

Compared to other wireless standards, the rapid growth of BLE is relatively easy to explain: BLE has gone further faster because its fate is so intimately tied to the phenomenal growth in smartphones, tablets, and mobile computing. Early and active adoption of BLE by mobile industry heavyweights like Apple and Samsung broke open the doors for wider implementation of BLE. Apple has put significant effort into producing a reliable BLE stack and publishing design guidelines around BLE. This, in turn, pushed silicon vendors to commit their limited resources to the technology they felt was the most likely to succeed or flourish in the long run, and the Apple stamp of approval is clearly a compelling argument when you need to justify every research and development dollar invested. While the mobile and tablet markets become increasingly mature and costs and margins are decreasing, the need for connectivity with the outside world on these devices has a huge growth potential, and it offers peripheral vendors a unique opportunity to provide innovative solutions to problems people might not even realize that they have today.

There are two major players involved in all bluetooth low energy communication: the central and the peripheral (see Figure 19). Based on a somewhat traditional client-server architecture, a peripheral typically has data that is needed by other devices. A central typically uses the information served up by peripherals to accomplish some task. As Figure 19 shows, for example, a heart rate monitor may have useful information that your Mac or iOS app may need in order to display the user's heart rate in a user-friendly way. The following code set up BLE interface on the hardware:

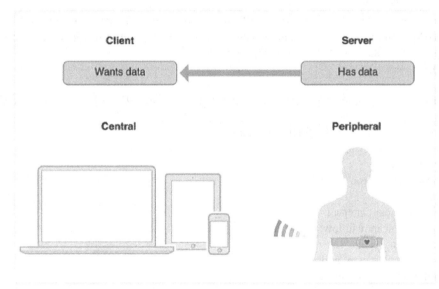

Figure 19: Central and Peripheral

```
void setup(void)
{
Wire.begin();
SerialMonitorInterface.begin(9600);
// initialize record counters here for testing - must be removed or power cycling loses all data
EEPROMwrite(0, 0);
EEPROMwrite(1, 0);
delay (500);
if (!htu.begin()) {
Serial.println("Couldn't find HTU21 sensor!");
while (1);
}
if (!bme.begin()) {
Serial.println("Could not find a valid BME280 sensor, check wiring!");
while (1);
}
//Case LED Support.
pinMode(A3, OUTPUT);
int tickEvent = t.every(3600000, HourlySensorReadAndStore); // hourly data grab 3600000 is an hour
accel.begin(BMA250_range_2g, BMA250_update_time_64ms);//This sets up the BMA250 accelerometer
RTC.start(); // ensure RTC oscillator is running, if not already
BLEsetup();
}
tBleStatus aci_gap_start_observation_procedure(uint16_t scan_interval, uint16_t scan_window, uint8_t
scan_type,
uint8_t own_address_type, uint8_t filter_duplicates)
{
struct hci_request rq;
gap_start_observation_proc_cp cp;
uint8_t status;
cp.scan_interval = scan_interval;
cp.scan_window = scan_window;
cp.scan_type = scan_type;
cp.own_address_type = own_address_type;
cp.filter_duplicates = filter_duplicates;
Osal_MemSet(&rq, 0, sizeof(rq));
rq.ogf = OGF_VENDOR_CMD;
rq.ocf = OCF_GAP_START_OBSERVATION_PROC;
rq.cparam = &cp;
rq.clen = sizeof(cp);
rq.event = EVT_CMD_STATUS;
rq.rparam = &status;
rq.rlen = 1;
if (hci_send_req(&rq, FALSE) < 0)
return BLE_STATUS_TIMEOUT;
return status;
}
```

Hardware—Storage

Constrained devices provide storage like C based file storage. Actual storage is on EPROM.

Files in C

Embedded and SCADA based devices use Input/Output (IO) operations that are driven from C device drivers. In nutshell, the IO operations are directly tied to C based IO.

In this section, I would like to introduce very high-level C IO File operations. In the subsequent sections, the IO that is performed on EPROM, Android and iOS use basics of C style IO functions [3].

 C File operations File IO operations are performed in Embedded Storage software such as File IO and embedded data base SQLite.

File Operations

In C the following file operations are performed:

1. Creating file
2. Opening an existing file
3. Reading and Writing into the file
4. Closing the file

When working with files, you need to declare a pointer of type file. This declaration is needed for communication between the file and program.

FILE *fptr;

The above line defines a pointer to a file.
Opening a file—for creation or edit
Opening a file is performed by calling fopen—defined stdio.h

ptr = fopen("fileopen","mode")

File Operation modes:

File Mode	Meaning of Mode	During Inexistence of file
r	Open for reading.	If the file does not exist, fopen() returns NULL.
rb	Open for reading in binary mode.	If the file does not exist, fopen() returns NULL.
w	Open for writing.	If the file exists, its contents are overwritten. If the file does not exist, it will be created.
wb	Open for writing in binary mode.	If the file exists, its contents are overwritten. If the file does not exist, it will be created.
a	Open for append. i.e, Data is added to end of file.	If the file does not exists, it will be created.
ab	Open for append in binary mode. i.e, Data is added to end of file.	If the file does not exist, it will be created.
r+	Open for both reading and writing.	If the file does not exist, fopen() returns NULL.
rb+	Open for both reading and writing in binary mode.	If the file does not exist, fopen() returns NULL.
w+	Open for both reading and writing.	If the file exists, its contents are overwritten. If the file does not exist, it will be created.
wb+	Open for both reading and writing in binary mode.	If the file exists, its contents are overwritten. If the file does not exist, it will be created.
a+	Open for both reading and appending.	If the file does not exist, it will be created.
ab+	Open for both reading and appending in binary mode.	If the file does not exist, it will be created.

Closing a File

The file should be closed after reading/writing. Closing a file is performed using library function fclose().

fclose(fptr); //fptr is the file pointer associated with file to be closed.

Reading a File

For reading and writing to a text file, we use the functions fprintf() and fscanf(). They are just the file versions of printf() and scanf(). The only difference is that, fprint and fscanf expects a pointer to the structure FILE.

In the code below, the scanf is loading a number of sensors in the variable.

```
printf("Enter number of sensors: ");
scanf("%d", &sensor_nums);
    FILE *fptr;
        fptr = (fopen("C:\\Temp\\sensors.txt", "w"));
        if (fptr == NULL)
        {
                printf("Error!");
                exit(1);
        }
```

In the code above, the fptr points to a file on C:\Temp\Sensors.txt

The code below gets the number of sensors from the User and takes the sensor name and sensor readings and writes it to the file. To suppress security warnings in Microsoft Visual Studio environment, please go to Project Properties and Code Generation and Turn off Security Check (see Figure 20 and Table 3).

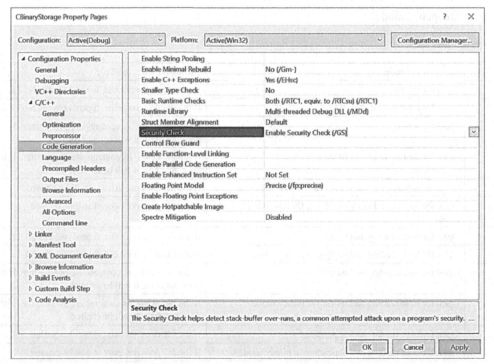

Figure 20: Microsoft Visual Studio - Security Check Feature

Table 3: File Storage

```
#include<stdio.h>
#define_CRT_SECURE_NO_WARNINGS
#include<stdio.h>
int main()
{
        char sensor_name[50];
        int readings, i, sensor_nums;
        printf("Enter number of sensors: ");
        scanf("%d", &sensor_nums);
        FILE *fptr;
        fptr = (fopen("C:\\Temp\\sensors.txt", "w"));
        if (fptr == NULL)
        {
                printf("Error!");
                exit(1);
        }
        for (i = 0; i < sensor_nums; ++i)
        {
                printf("For sensor%d\nEnter name: ", i + 1);
                scanf("%s", sensor_name);
                printf("Enter readings: ");
                scanf("%d", &readings);
                fprintf(fptr, "\nName: %s \nMarks=%d \n", sensor_name, readings);
        }
        fclose(fptr);
        return 0;
}
```

Output

The code above generates the following output (see Figure 21):

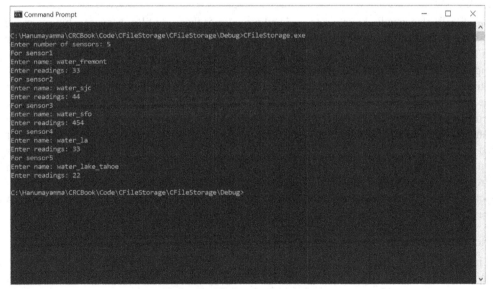

Figure 21: File Storage Console Output

File Output: The file output is written to file (see Figure 22)

Figure 22: File Output

Reading and Writing to a Binary File

Binary files are very similar to arrays of structures, except the structures are in a disk file rather than in an array in memory. Because the structures in a binary file are on disk, you can create very large collections of them (limited only by your available disk space). They are also permanent and always available. The only disadvantage is the slowness that comes from disk access time.

 Sterilization is the process of storing memory objects on to file and reloading them. C File operations make it possible.

Binary files[181] also usually have faster read and write times than text files, because a binary image of the record is stored directly from memory to disk (or vice versa). In a text file, everything has to be converted back and forth to text, and this takes time.

C supports the file-of-structures concept very cleanly. Once you open the file you can read a structure, write a structure, or seek to any structure in the file. This file concept supports the concept of a file pointer. When the file is opened, the pointer points to record 0 (the first record in the file). Any read operation reads the currently pointed-to structure and moves the pointer down one structure. Any write operation writes to the currently pointed-to structure and moves the pointer down one structure. Seek moves the pointer to the requested record.

Keep in mind that C thinks of everything in the disk file as blocks of bytes read from disk into memory or read from memory onto disk. C uses a file pointer, but it can point to any byte location in the file.

Functions fread() and fwrite() are used for reading from and writing to a file on the disk respectively, in case of binary files.

Writing to a Binary File

To write into a binary file, you need to use the function fwrite(). The functions takes four arguments: Address of data to be written in disk, Size of data to be written in disk, number of such type of data and pointer to the file where you want to write.

fwrite(address_data,size_data,numbers_data,pointer_to_file);

---

[181] Binary files - https://computer.howstuffworks.com/c39.htm

Sensor Diag is a simple three integer structure.

```
structsensor_diag
{
        int x, y, z;
};
```

Open file for binary operations:

```
/* create the file of 10 records */
f = fopen("sensor_diag", "w");
```

Write 10 values into the structure and close the file.

```
for (i = 1;i <= 10; i++)
{
        sd_record.x = i;
        fwrite(&sd_record, sizeof(structsensor_diag), 1, f);
}
fclose(f);
```

FSeek to read records

```
f = fopen("sensor_diag", "r");
if (!f)
        return 1;
for (i = 9; i >= 0; i--)
{
        fseek(f, sizeof(structsensor_diag)*i, SEEK_SET);
        fread(&sd_record, sizeof(structsensor_diag), 1, f);
        printf("%d\n", sd_record.x);
}
fclose(f);
printf("\n");
```

Read records in alternate manner (see Figure 23):

```
/* use fseek to read every other record */
f = fopen("sensor_diag", "r");
if (!f)
        return 1;
fseek(f, 0, SEEK_SET);
for (i = 0;i<5; i++)
{
        fread(&sd_record, sizeof(structsensor_diag), 1, f);
        printf("%d\n", sd_record.x);
        fseek(f, sizeof(structsensor_diag), SEEK_CUR);
}
fclose(f);
printf("\n");
```

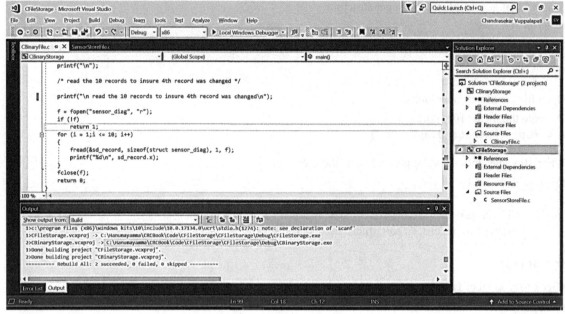

Figure 23: Visual Studio IDE

Code: The full code write to Binary file is listed as part of Table 4.

Table 4: Binary File Write Full Code

```
#include<stdio.h>
#include<stdlib.h>
#include<stdio.h>
/* Sensor records */
structsensor_diag
{
        int x, y, z;
};
/* writes and then reads 10 arbitrary records
from the file "sensor_diag". */
int main()
{
        int i, j;
        FILE *f;
        structsensor_diag sd_record;
        /* create the file of 10 records */
        printf("\n create the file of 10 records \n");
        f = fopen("sensor_diag", "w");
        if (!f)
                return 1;
        for (i = 1;i <= 10; i++)
        {
                sd_record.x = i;
                fwrite(&sd_record, sizeof(structsensor_diag), 1, f);
        }
        fclose(f);
        printf("\n read the 10 records \n");
        /* read the 10 records */
        f = fopen("sensor_diag", "r");
        if (!f)
```

Table 4 contd...

...Table 4 contd.

```
                        return 1;
            for (i = 1;i <= 10; i++)
            {
#include<stdio.h>
#include<stdlib.h>
#include<stdio.h>
/* Sensor records */
structsensor_diag
{
            int x, y, z;
};
/* writes and then reads 10 arbitrary records
from the file "sensor_diag". */
int main()
{
            int i, j;
            FILE *f;
            structsensor_diag sd_record;
            /* create the file of 10 records */
            printf("\n create the file of 10 records \n");
            f = fopen("sensor_diag", "w");
            if (!f)
                        return 1;
            for (i = 1;i <= 10; i++)
            {
                        sd_record.x = i;
                        fwrite(&sd_record, sizeof(structsensor_diag), 1, f);
            }
            fclose(f);
            printf("\n read the 10 records \n");
            /* read the 10 records */
            f = fopen("sensor_diag", "r");
            if (!f)
                        return 1;
            for (i = 1;i <= 10; i++)
            {
                        fread(&sd_record, sizeof(structsensor_diag), 1, f);
                        printf("%d\n", sd_record.x);
            }
            fclose(f);
            printf("\n");
            /* use fseek to read the 10 records in reverse order */
            printf("\n use fseek to read the 10 records in reverse order \n");
            f = fopen("sensor_diag", "r");
            if (!f)
                        return 1;
            for (i = 9; i >= 0; i--)
            {
                        fseek(f, sizeof(structsensor_diag)*i, SEEK_SET);
                        fread(&sd_record, sizeof(structsensor_diag), 1, f);
                        printf("%d\n", sd_record.x);
            }
            fclose(f);
            printf("\n");
            /* use fseek to read every other record */
            printf("\n use fseek to read every other record \n");
            f = fopen("sensor_diag", "r");
            if (!f)
                        return 1;
            fseek(f, 0, SEEK_SET);
            for (i = 0;i<5; i++)
            {
```

Table 4 contd...

...Table 4 contd.

```
                    fread(&sd_record, sizeof(structsensor_diag), 1, f);
                    printf("%d\n", sd_record.x);
                    fseek(f, sizeof(structsensor_diag), SEEK_CUR);
            }
            fclose(f);
            printf("\n");
            /* use fseek to read 4th record, change it, and write it back */
            printf("\n use fseek to read 4th record, change it, and write it back\n");
            f = fopen("sensor_diag", "r+");
            if (!f)
                    return 1;
            fseek(f, sizeof(structsensor_diag) * 3, SEEK_SET);
            fread(&sd_record, sizeof(structsensor_diag), 1, f);
            sd_record.x = 100;
            fseek(f, sizeof(structsensor_diag) * 3, SEEK_SET);
            fwrite(&sd_record, sizeof(structsensor_diag), 1, f);
            fclose(f);
            printf("\n");
            /* read the 10 records to insure 4th record was changed */
            printf("\n read the 10 records to insure 4th record was changed\n");
            f = fopen("sensor_diag", "r");
            if (!f)
                    return 1;
            for (i = 1;i <= 10; i++)
            {
                    fread(&sd_record, sizeof(structsensor_diag), 1, f);
                    printf("%d\n", sd_record.x);
            }
            fclose(f);
            return 0;
}
```

The program above generates the following output (see Figure 24)

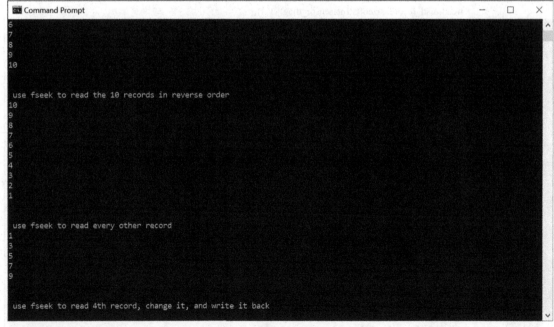

Figure 24: File Output

EPROM Data Storage

The embedded system On-chip ISP (In-System Programing) Flash allows the program memory to be reprogrammed In-System through an SPI (Serial Peripheral Interface) by a conventional nonvolatile memory programmer, or by an On-chip Boot program running on the AVR core. The Boot program can use any interface to download the application program in the Application Flash memory. The application with Kalman filter is developed using C Compilers.

The bootloader[182] is the little program that runs when you turn the embedded system on or press the reset button. Its main function is to load the application into the FLASH. The bootloader is what enables you to program the Arduino using just the USB cable.

Entering the Boot Loader takes place by a jump or call from the application program. This may be initiated by a trigger such as a command received via USART, or SPI interface.

The embedded system Program Flash memory space is divided in two sections, the Boot Program section and the Application Program section. The application Program section contains instructions for starting the connectivity (BLE), EPROM access and other services [4].

EEPROM provides the nonvolatile data storage support. 0-255 record counters will be stored in EEPROM location zero (see Figure 25) [26]. 256–336 record counter with maximum of 2 weeks at 24 records/day will be stored in EEPROM location one. Records will start at EEPROM location two. Additionally, the data arrays store on EEPROM and retrieve command over Bluetooth Low Energy (BLE).

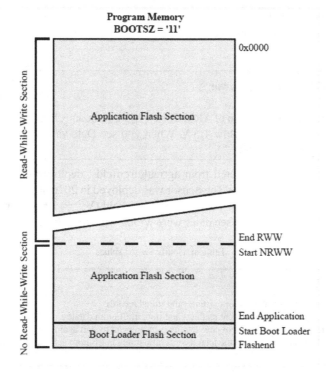

Figure 25: EPROM Program Memory

The following table has the memory initialization for bootloader (please see Table 5):

---

[182] Bootloader - http://arduinoinfo.mywikis.net/w/images/3/35/Atmel-42735-8-bit-AVR-Microcontroller-ATmega328-328P_Datasheet.pdf

Table 5: Memory Loader File

```
MEMORY
{
/* Flash memory region for the program, offset by the bootloader section */
FLASH (rx) : ORIGIN = 64k, LENGTH = 448k
/* 32k (Heap starts here) */
SRAM (rwx) : ORIGIN = 0x10000000, LENGTH = 32k
/* 32k (Globals at bottom, stack on top), heap continues from here */
SRAM_AHB (rwx) : ORIGIN = 0x2007c000, LENGTH = 32k
/* Heap starts at SRAM because if heap starts from higher memory, then
* malloc tracking routines get confused when sbrk() system function
* returns higher heap memory first, and then lower heap memory.
* So we start heap from SRAM, and then move it up to SRAM_AHB if SRAM
* portion runs out completely.
*/
}
```

Setup real-time clock event handler that collects data on an hourly basis:

```
int tickEvent = t.every(3600000, HourlySensorReadAndStore); // hourly data grab 3600000 is an hour
```

The data is collected in the following order:

- Get Current Memory Pointer
- Read Sensor Values from the Store (see Table 6)
- Clear variables.

 Real Time Clock & Battery Power

As per the technical design specification of ARM Cortex-M3 microcontrollers, the Real Time Clock (RTC) typically fails[183] when $V_{i(BAT)}$ drops below 1.6 V. When you see Date values corruptions it could highly possible that the battery power is low.

In the below figure, the data collected from agriculture fields, deployed extremely constrained Dairy Diagnostic device for Animal Husbandry. The Sensor was deployed in 2016 and during the tail end of battery life (March 2017), the battery was below the operating threshold ($V_{i(bat)} < 1.6$ V). The voltage has generated Real Time Clock (RTC) error—you can see dates were wrong with June 30, 2065 (please refer Figure 26).

Table 6: Hourly Sensor Values

```
void HourlySensorReadAndStore ()
{
GetCurrentRecordCount (); // get global count of existing valid stored records
ReadSensorsAndStore(); // get all 5 sensor current readings and store in EEprom datalog
ClearCowActivity (); // clear the running count of cow activity measurement to start again
IncrementCurrentRecordCount (); // increment the total records now stored in EEprom datalog (on EEprom)
}
```

---

183 NXP Semiconductors N.V. 2015 - https://www.nxp.com/docs/en/data-sheet/LPC1759_58_56_54_52_51.pdf

SensorTagID	Date	Time	Humidity
62CB2E1B-E964-4BCE-98BD-107D3D9F039E	12/29/2016	0:09	47.33
62CB2E1B-E964-4BCE-98BD-107D3D9F039E	12/29/2016	1:09	47.44
62CB2E1B-E964-4BCE-98BD-107D3D9F039E	12/29/2016	2:09	47.51
62CB2E1B-E964-4BCE-98BD-107D3D9F039E	12/29/2016	3:10	47.53
62CB2E1B-E964-4BCE-98BD-107D3D9F039E	12/29/2016	4:10	47.66
62CB2E1B-E964-4BCE-98BD-107D3D9F039E	12/29/2016	5:10	47.69
62CB2E1B-E964-4BCE-98BD-107D3D9F039E	12/29/2016	6:10	47.8
62CB2E1B-E964-4BCE-98BD-107D3D9F039E	12/29/2016	7:10	48.03
62CB2E1B-E964-4BCE-98BD-107D3D9F039E	12/29/2016	8:10	48.13
62CB2E1B-E964-4BCE-98BD-107D3D9F039E	1/20/2017	13:46	65.41
62CB2E1B-E964-4BCE-98BD-107D3D9F039E	1/20/2017	14:46	64.97
62CB2E1B-E964-4BCE-98BD-107D3D9F039E	2/6/2017	22:31	61.08
62CB2E1B-E964-4BCE-98BD-107D3D9F039E	6/30/2065	6:00	148
62CB2E1B-E964-4BCE-98BD-107D3D9F039E	6/30/2065	6:15	148
62CB2E1B-E964-4BCE-98BD-107D3D9F039E	6/30/2065	6:30	148
62CB2E1B-E964-...F039E	6/30/2065	6:45	148
62CB2E1B-E964-4BCE-98BD-107D3D9F039E	6/30/2065	9:00	148
62CB2E1B-E964-4BCE-98BD-107D3D9F039E	6/30/2065	9:15	148
62CB2E1B-E964-4BCE-98BD-107D3D9F039E	6/30/2065	9:30	148
62CB2E1B-E964-4BCE-98BD-107D3D9F039E	6/30/2065	9:45	148

Figure 26: RTC Clock error for low battery voltage

Get Current Record Count

The pointer to the record is stored in two register variables: EEPROMread(0) and EEPROMread(1)—see. Table 7.

Table 7: Get Current Record Count

```
// Returns record counter into int from two stored bytes.
void GetCurrentRecordCount () {
totalRecordCount0 = EEPROMread(0); //get lower byte of current stored record count.
totalRecordCount1 = EEPROMread(1); //get upper byte of current stored record count.
if ((totalRecordCount0) < 253) //check if low order byte counter not full yet
{
currentCountBothBytes = (totalRecordCount0); //low order byte not full yet
}
else if ((totalRecordCount1) < 82)
{
currentCountBothBytes = (totalRecordCount0 + totalRecordCount1);
}
}
}
```

EPROM Read:

The following (see Table 8) code provides access to EPROM Read functionality:

Table 8: EPROM Read

```
//Reads out of EEprom and send Part One of the records over ble
//Part One is date time, amb temp, amb humidity, cow temp, cow humidity.
//This is the maximum data transfer possible with this ble arrangement apparently.
//Note that baseRecordByteAddrCount is set to zero on power up, first sw load, cmd "R", and "0" cmd.
void SendRecordPartOne() {
 GetCurrentRecordCount (); //0 to 335 integer count of records in eeprom. At 336 it restarts at 0.
 if ((currentCountBothBytes) > 0)// if no records yet, just exit. Otherwise process just one set.
 {
 //read section and send section
 EEPROMread((baseRecordByteAddrCount),(uint8_t*)timeDateData,int(sizeof(timeDateData)));
 // Send it over bluetooth.
 while (!lib_aci_send_data(PIPE_UART_OVER_BTLE_UART_TX_TX, (uint8_t*)timeDateData, sizeof (timeDateData)))
 {
```

Table 8 contd. ...

...Table 8 contd.

```
for(int i=0; i<50; i++)
aci_loop();
delay (1);
//SerialMonitorInterface.println(F("TX dropped!"));
}
nextRecordByteAddrCount = baseRecordByteAddrCount + (sizeof (timeDateData));
EEPROMread((nextRecordByteAddrCount),(uint8_t*)ambientTemp, int(sizeof (ambientTemp)));
while (!lib_aci_send_data(PIPE_UART_OVER_BTLE_UART_TX_TX, (uint8_t*)ambientTemp, sizeof (ambientTemp)))
{
for(int i=0; i<50; i++)
aci_loop();
delay (1);
// SerialMonitorInterface.println(F("TX dropped!"));
}
while (!lib_aci_send_data(PIPE_UART_OVER_BTLE_UART_TX_TX, (uint8_t*)cowTemp, sizeof(cowTemp)))
{
for(int i=0; i<50; i++)
aci_loop();
delay (1);
// SerialMonitorInterface.println(F("TX dropped!"));
}
nextRecordByteAddrCount5 = nextRecordByteAddrCount4 + (sizeof (cowHumd));
}
}
```

Read Sensor and Store into EPROM

The following code (see Table 9) provides read & write EPROM Sensor data:

Table 9: Read Sensor and Store

```
void ReadSensorsAndStore () {
String timeDateString;
RTC.readTime(); // update RTC library's buffers from chip
timeDateString += (int(RTC.getMonths()));
timeDateString += "/";
timeDateString += (int(RTC.getDays()));
timeDateString += "/";
timeDateString += (RTC.getYears());
timeDateString += " ";
timeDateString += (int(RTC.getHours()));
timeDateString += ":";
timeDateString += (int(RTC.getMinutes()));
timeDateString.toCharArray(timeDateData, sizeof (timeDateData)); // copy into timeDateData buffer
// Get ambient temp into array
floatVal= (htu.readTemperature());
dtostrf(floatVal, 3, 2, ambientTemp); //4 is mininum width, 4 is precision; float value is copied onto buf ambientTemp
// Get cow activity into array, first convert integer count to string
String str = String(activityCount);
// Length (5 digits with one extra character for the null terminator), cowActivity is 6 bytes.
str.toCharArray(cowActivity, sizeof (cowActivity)); // copy into cowActivity buffer
+ (sizeof (cowActivity));
}
```

EPROM Read and Writes

```
byte EEPROMread(unsigned long addr){
uint8_t val=255;
uint8_t I2Caddr=EEPROM_ADDR;
if(addr>0x0000FFFF){
I2Caddr|=0x04;
}
Wire.beginTransmission(I2Caddr);
Wire.write(addr>>8);
Wire.write(addr);
Wire.endTransmission();
Wire.requestFrom(I2Caddr,(uint8_t)1);
while(Wire.available()){
val=Wire.read();
}
return val;
}
byte EEPROMwrite(unsigned long addr, byte val){
uint8_t I2Caddr=EEPROM_ADDR;
if(addr>0x0000FFFF){
I2Caddr|=0x04;
}
Wire.beginTransmission(I2Caddr);
Wire.write(addr>>8);
Wire.write(addr);
Wire.write(val);
Wire.endTransmission();
Wire.beginTransmission(I2Caddr);
unsigned long timeout=millis();
while(Wire.endTransmission() && millis()-timeout<10){
Wire.beginTransmission(I2Caddr);
}
}
```

Current Edge device architectures have operation limitations[184,185] in constrained environments and have resulted in limiting factor[186] with respect to dissemination of AI [18,8000013, 16]. There are multiple facets to the constraints, which often play in combination:

- The maximum code complexity (ROM/Flash),
- The size of state and buffers (RAM),
- The amount of computation feasible in a period ("processing power"),
- The available power,
- The connectivity and highly asymmetric link characteristics
- The deployment accessibility (ability to set security keys, update software, deploy firmware etc.), and
- The cost of the device

---

[184] Classification of IoT Devices - https://www.cisoplatform.com/profiles/blogs/classification-of-iot-devices
[185] IETF 7228 - https://tools.ietf.org/html/rfc7228
[186] Dynamic Computation Offloading for Mobile-Edge Computing with Energy Harvesting Devices - https://arxiv.org/pdf/1605.05488.pdf

- AI/ML Model deployment
- Model interpretability at the inference level

To understand and design AI models that can operate in constrained environments, we have developed Constrained Modeling—Model Accuracy—Connectivity—Hardware Economy Framework. By addressing devices constraints and by designing AI to work on these small footprint devices in a sustainable manner would enable millions of farmers on to the digital revolutions. The next chapter covers Constrained Modeling in detailed aspects.

Chapter Summary:

After reading the chapter, you should fairly answer

• The need for creating Small compute devices to solve data issue.

• The design view of building hardware, looking from three major hardware design characteristics: AI/ML Model Accuracy, Connectivity and Hardware.

References

1. Rob Thomas and Patrick McSharry, Big Data Revolution: What Farmers, Doctors And Insurance Agents Teach Us About Discovering Big Data Patterns, 1st Edition, Wiley (March 2, 2015), ISBN-10: 9781118943717.
2. David Cearley, Brian Burke, Mike Walker, Bob Gill and Thomas Bittman. Top 10 Strategic Technology Trends for 2018: Cloud to the Edge. 8 March 2018, URL: https://www.gartner.com/en/documents/3865403/top-10-strategic-technology-trends-for-2018-cloud-to-the0, Access Date: September 18, 2019.
3. IEEE Innovation. Why Does Edge Computing Matter? August 26, 2019URL: https://innovationatwork.ieee.org/why-does-edge-computing-matter/, Access Date: September 18, 2019.
4. Janet Ranganathan, Richard Waite, Tim Searchinger and Craig Hanson. How to Sustainably Feed 10 Billion People by 2050, in 21 Charts. December 5, 2018, URL: https://www.wri.org/blog/2018/12/how-sustainably-feed-10-billion-people-2050-21-charts, Access Date: September 18, 2019.
5. Thomas Hardjono, David L. Shrier and Alex Pentland. Trust::Data: A New Framework for Identity and Data sharing by Thomas Hardjono. CreateSpace Independent Publishing Platform (October 21, 2016), ISBN-10: 9781539114215.
6. Claire Melamed 2020. A World That Counts," 11 November 2018 URL: https://www.undatarevolution.org/draft-report, Access Date: 22 November 2019.
7. Soumitra Dutta, Thierry Geiger and Bruno Lanvin. The Global Information Technology Report 2015. 2015 URL: http://www3.weforum.org/docs/WEF_Global_IT_Report_2015.pdf, Access Date: 22 November 2019.
8. Sarah K. Lowder, Jakob Skoet and Terri Raney. The Number, Size, and Distribution of Farms, Smallholder Farms, and Family Farms Worldwide, 9 February 2016, URL:https://www.sciencedirect.com/science/article/pii/S0305750X15002703, Access Date: 22 November 2019.
9. John L. Schnase, Tsengdar J. Lee, Chris A. Mattmann, Christopher S. Lynnes, Luca Cinquini, Paul M.Ramirez, Andre F. Hart, Dean N. Williams, Duane Waliser, Pamela Rinsland, W. Philip Webster, Daniel Q. Duffy, Mark A. McInerney, Glenn S. Tamkin, Gerald L. Potter and Laura Carrier. Big Data Challenges in Climate Science. 2015, URL: https://ntrs.nasa.gov/archive/nasa/casi.ntrs.nasa.gov/20160002946.pdf , Access Date: 22 November 2019.
10. Richard Waite, Craig Hanson and Janet Ranganathan. Creating A Sustainable Food Future 2018, URL: https://wrr-food.wri.org, Access Date: Nov. 22 2019.
11. Toby Stover, Izabela Wojtenko and Katherine Bentley. Nitrogen and Phosphorus Pollution Data Access Tool, October 30, 2019. URL: https://www.epa.gov/nutrient-policy-data/nitrogen-and-phosphorus-pollution-data-access-tool#main-content, Access Date: 22 November 2019.
12. Luiz Amaral. The Data Revolution Hasn't Yet Hit Agriculture, March 5, 2019, URL:https://www.wri.org/blog/2019/03/data-revolution-hasnt-yet-hit-agriculture, Access Date: 22 November 2019.
13. Raffaele Bertini, Vito Cistulli, Andre Croppenstedt, Eva Gálvez Nogales, Theresa McMenomy, Ahmad Sadiddin, Jakob Skøt and Thomas, Graeme "The State of Food and Agriculture Leveraging Food Systems for Inclusive Rural Transformation," 2017, URL:http://www.fao.org/3/a-i7658e.pdf, Access Date: 22 November 2019.
14. J. S. Vuppalapati, S. Kedari, A. Ilapakurthy, A. Ilapakurti and C. Vuppalapati. Smart Dairies—Enablement of Smart City at Gross Root Level. Published:2017 IEEE Third International Conference on Big Data Computing Service and Applications (BigDataService), San Francisco, CA, 2017, pp. 118-123. URL:http://ieeexplore.ieee.org/stamp/stamp.jsp?tp=&arnumber=7944928&isnumber=7944900 , Access Date: 22 November 2019.

15. Gabriel Vega. Computation, Energy-Efficiency, and Landauer's Principle. December 5, 2016, URL: http://large.stanford.edu/courses/2016/ph240/vega1/, Access Date: March 16, 2020.
16. Cédric Villani. For a Meaningful Artificial Intelligence. March 8, 2018, URL: https://www.aiforhumanity.fr/pdfs/MissionVillani_Report_ENG-VF.pdf, Access Date: March 16, 2020.
17. C. Bormann, M. Ersue and A. Keranen. Terminology for Constrained-Node NetworksInternet Engineering Task Force (IETF). 2014, URL: https://tools.ietf.org/html/rfc7228, Access Date: November 22, 2019.
18. Simone Cirani, Gianluigi Ferrari, Marco Picone and Luca Veltri. Internet of Things: Architectures, Protocols and Standards. November 5, 2018, URL : https://www.amazon.com/Internet-Things-Architectures-Protocols-Standards/dp/1119359678 , Access Date : November 22, 2019.
19. Nagasai. Classification of IoT Devices, February 18, 2017 URL:https://www.cisoplatform.com/profiles/blogs/classification-of-iot-devices, Access Date : November 22, 2019.
20. Saranyan Vigraham. Democratizing AI. August 6, 2019, URL:https://insidebigdata.com/2019/08/06/democratizing-ai/, Access Date: November 22, 2019.
21. Simone Cirani, Gianluigi Ferrari, Marco Picone and Luca Veltri. Internet of Things: Architectures, Protocols and Standards. John Wiley & Sons Ltd., 1st edition (November 5, 2018), ISBN-10: 1119359678.
22. Robert Davidson, Akiba, Carles Cufí and Kevin Townsend. Getting Started with Bluetooth Low Energy. May 2014, ISBN: 9781491900550, https://www.oreilly.com/library/view/getting-started-with/9781491900550/.
23. Bluetooth, Protocol Specification, https://www.bluetooth.com/specifications/protocol-specifications, 2018.
24. Dag Grini, RF Basics, RF for Non-RF Engineers. 2006. http://www.ti.com/lit/ml/slap127/slap127.pdf, Access Date: 12/02/2018.

## CHAPTER 4

Machine Learning Frameworks and Device Engineering

"Every company is a software company. You have to start thinking and operating like a digital business."
Satya Nadella

The chapter starts with the perennial issue—democratizing the artificial intelligence models—"the lack of data" from the world's most important and largest employer, the agriculture industry and data from developing economies. The need for capturing the data and the development of Compute systems that could fulfill the World Food Organization (WFO) mandate for creating a sustainable food future by 2050. The chapter introduces device constrained modeling framework, the purpose of which is to enable engineering designs to holistically look three major engineering characteristics so as to deploy Machine Learning models in extreme constrained environments. The design characteristics include: AI/ML Model Accuracy, Connectivity and Hardware. Next, the chapter introduces tradeoff modeling between three major hardware design characteristics to fine tune the necessary adjustments in design characteristic to compensate any tradeoff in function or hardware features. And finally, the chapter concludes by introducing factors such as environment & perturbation of data in crossing the Chasm to democratize the AI.

Machine Learning Device Deployments

Artificial Intelligence and Machine Learning (ML) provide several algorithms to enable learning with and without supervision. The learning could be performed on powerful desktops, Edge devices, Cloud instances and hybrid compute cloud instances. In aggregate level, the ML Libraries for compute instances can be divided into dedicated applications for individual computing nodes (for example Weka, scikit-learn, Jupyter Notebook, Anaconda) and for high performance computers (cluster/cloud computing, e.g., Spark, TensorFlow, AutoML, Cloud based Jupyter Notebook). Many large companies offer services that rely on machine learning in public cloud infrastructures. The most popular services of this type are Amazon Machine Learning, Google Prediction, IBM Watson and Microsoft Azure Machine Learning and the dedicated for IoT such as ThingWorx. These solutions analyze data mostly in the cloud and role of IoT devices comes down to software agents providing data for analysis.[187]

Extremely Resource-Constrained (xRC) Systems

In the domain of resource-constrained systems we can find many implementations of ML algorithms on mobile and embedded devices that cooperate with the cloud computing. The architecture of such systems, generally, the learning and inference process are split into two layers: local edge layer constructed with mobile devices and remote server (cloud) layer. For example, in the following Apple Core ML diagram,[188] the Core ML Model

---

[187] Enabling machine learning on resource constrained devices by source code generation of the learned models - https://www.iccs-meeting.org/archive/iccs2018/papers/108610666.pdf

[188] Core ML - https://developer.apple.com/documentation/coreml

Figure 1: Core ML

is constructed from different machine learning models and the learning can be performed on a standalone platform or on a Cloud-based system. The model is bundled in target mobile app (i.e., in the iOS) and can be invoked as part of the mobile application (please see Figure 1).

Deep Learning Device Deployment (please see Figure 2)

Similarly, deep learning models such as TensorFlow, is constructed on Cloud or high-powered Compute systems. The developed model is converted to TensorFlow Lite Buffer (C Language structure) to be deployed to a client device. The TensorFlow Lite FlatBuffer[189] file is then deployed to a client device (e.g., mobile, embedded) and run locally using the TensorFlow Lite interpreter. This conversion process is shown in the diagram below:

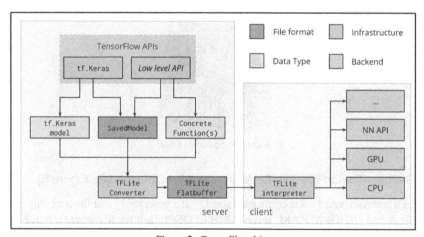

Figure 2: TensorFlow Lite

TensorFlow Lite for Microcontrollers is written in C++ 11 and requires a 32-bit platform. It has been tested extensively with many processors based on the Arm Cortex-M Series architecture and has been ported to other architectures including ESP32. There are example applications available for the following development boards:[190]

Arduino Nano 33 BLE Sense (please see Figure 3)

The Nano 33 BLE Sense[191] (with headers) is Arduino's 3.3V AI enabled board in the smallest available form factor: 45 × 18 mm! The Arduino Nano 33 BLE Sense is a completely new board on a well-known form factor. The Board contains series embedded sensors that include: 9 axis inertial sensor, humidity, temperature, barometric, microphone and gesture, light and proximity.

---

[189] TensorFlow Lite Converter - https://www.tensorflow.org/lite/convert
[190] TensorFlow Lite for Microcontrollers - https://www.tensorflow.org/lite/microcontrollers
[191] Arduino Nano 33 BLE Sense - https://store.arduino.cc/usa/nano-33-ble-sense-with-headers

Figure 3: Nano BLE

SparkFun Edge (please see Figure 4)

SparkFun Edge[192] design utilizes Ambiq Micro's latest Apollo3 Blue microcontroller, whose ultra-efficient ARM Cortex-M4F 48 MHz (with 96 MHz burst mode) processor, is spec'd to run TensorFlow Lite using only 6 uA/MHz. The SparkFun Edge board currently measures ~ 1.6 mA@3V and 48 MHz and can run solely on a CR2032 coin cell battery for up to 10 days. Apollo3 Blue sports all the cutting edge features expected of modern microcontrollers including six configurable I2C/SPI masters, two UARTs, one I2C/SPI slave, a 15-channel 14-bit ADC, and a dedicated Bluetooth processor that supports BLE5. On top of all that the Apollo3 Blue has 1MB of flash and 384KB of SRAM memory—plenty for the vast majority of applications.

Figure 4: SparkFun Edge

Adafruit EdgeBadge—TensorFlow Lite for Microcontrollers (please see Figure 5)

The EdgeBadge is a compact board – it is credit card sized.[193] It's powered by our favorite chip, the ATSAMD51, with 512KB of flash and 192KB of RAM. It has 2 MB of QSPI flash for file storage, handy for TensorFlow Lite files, images, fonts, sounds, or other assets.

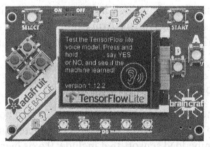

Figure 5: Edge Badge

---

192 SparkFun Edge - https://www.sparkfun.com/products/15170
193 AdaFruit EdgeBadge - https://www.adafruit.com/product/4400

ESP32-DevKitC (please see Figure 6)

ESP32-DevKitC[194] is an AWS-qualified development board. In addition to Espressif's own ESP-IDF SDK, you can use FreeRTOS on ESP32-DevKitC. FreeRTOS provides out-of-the-box connectivity with AWS IoT, AWS Greengrass and other AWS services. ESP32-DevKitC contains the entire basic-support circuitry for ESP32-WROOM-32D, ESP32-WROOM-32U, ESP32-WROVER-B and ESP32-SOLO-1, including a USB-UART bridge, reset- and boot-mode buttons, an LDO regulator and a micro-USB connector. Every important GPIO is available to the developer.

Figure 6: ESP32

For other Edge Machine Learning processors and development boards, please see Appendix A.

 Quantized Models

Quantized classification models[195] offer the smallest model size and fastest performance, at the expense of accuracy.

Extremely Resource-Constrained (xRC) Systems

Extremely resource-constrained (xRC) systems are embedded devices that have small form factor and powered by finite power sources. These devices are microcontroller based with limited processing power, limited memory, intermittent connectivity to the internet or low bandwidth or sporadic connectivity and extremely low storage.[196] The extremely constrained devices[197,198] are end nodes with sensors/actuators that can handle a specific application purpose [4,5]. They are usually connected to gateway-like devices, low power lossy network, and in-turn communicates with the IoT cloud platforms. Typically, they communicate via low power wireless protocols like BLE, 802.15.4 (6LoWPAN, Zigbee, Thread, WirelessHART, etc.), LPWAN, etc., and mostly battery powered with low data rate.

Generally, the goal of machine learning on the extremely constrained devices is not related to theperforming learning process on these devices butis related to the usage of the models learned elsewhere and used on

---

194 ESP32-DevKitC - https://www.espressif.com/en/products/hardware/esp32-devkitc/overview
195 Quantized Classification Models - https://www.tensorflow.org/lite/guide/hosted_models
196 Enabling machine learning on resource constrained devices by source code generation of the learned models - https://www.iccs-meeting.org/archive/iccs2018/papers/108610666.pdf
197 Constrained devices - https://www.cisoplatform.com/profiles/blogs/classification-of-iot-devices
198 Neural Network / Machine Learning in resource constrained environments - https://medium.com/@raghu.madabushi/neural-net-work-machine-learning-in-resource-constrained-environments-c934ff1f522

the devices. One of the limitations of extremely constrained devices is porting the high-level and general-purpose machine learning libraries is not possible. In this situation, the implementation of general-purpose machine learning may consume significant device resources. Thus, the approach in which source code of the model/inference/estimator that expresses the learned model is generated and then compiled into *the device firmware*. The presented concept of the machine learning model source code generation requires three steps to be performed (please see Figure 7):[199]

1. analysis of the machine-learning algorithm and the way how it can be expressed in the source code,

2. analysis of how to get details of machine-learning model from the ones generated by the particular software or library,

3. analysis on how the final code can be optimized for the target embedded architecture regarding its resource constraints.

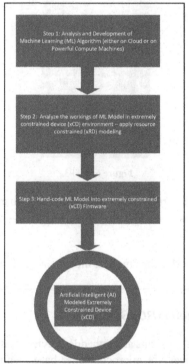

Figure 7: xRC AI-ML Process

To understand and design AI models that can operate in extremely constrained environments, we have developed Constrained Modeling—Model Accuracy—Connectivity—Hardware Economy Framework. The framework analyzes machine learning model, analyzes the device environment constraints, and fits the models into the device firmware. By addressing devices constraints and by designing AI to work on these small footprint devices in a sustainable manner would enable millions of farmers on to the digital revolutions.

xRC Modeling: Model Accuracy-Connectivity-Hardware (MCH) Framework

As we have seen in the above sections, technology dissemination through wider ICT deployment play a pivotal role in creating sustainable planet for the future. Machine Learning and AI are the drivers of sustainable revolution with data infused insights that could potentially reduce the use of pesticides in agriculture to

---

[199] Enabling machine learning on resource constrained devices by source code generation of the learned models - https://www.iccs-meeting.org/archive/iccs2018/papers/108610666.pdf

improve the farm productivity. To realize such future, we need data. The data from developing economies in agriculture hold key for the entire planet for two simple reasons:

1. Small farm agriculture is largest employer of the World in turns number of people livelihoods and

2. Improving productivity of small farms not only brings many people out of poverty but also help to reduce Global Greenhouse Gas (GHG) emissions that help us to fight climate change.

In order to get the data from geographically disparate and distant places with constraints such as connectivity, device operating limiting characteristics, ML model performance attributes, environmental performance issues and operating characteristics, we need to design a system that can function well in such constraint environments. There are three design characteristics (nodes) of the system to balance or tradeoff to make the constrained device operating successfully, with Service Level Agreement (SLA) in mind, in the constrained environment.

 In a constrained environment, a small edge or small custom-built device with higher model accuracy and connectivity will sacrifice hardware performance or hardware economy (i.e., Lower battery life).

The Trade-off Modeling (please see Figure 8)

The Trade-off model is used in finance and manufacturing. The conventional Trade-off model[200] states that unless there is some slack in the system, improving any one of the four basic manufacturing capabilities—Quality, Dependability, Speed and Cost—must necessarily be at the expense of one or more of the other three.

In finance, especially company capital structures, the trade-off theory of capital structure is the idea that the benefit of debt finance and costs it brings.

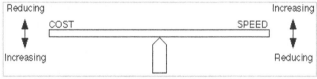

Figure 8: Trade-off Modeling

Hardware Economy—Model Accuracy—Connectivity Trade-off

The constrained model consists of three important nodes: Hardware Economy, ML Model Accuracy and Connectivity. These three design characteristics define the working of custom-built hardware or hardware embedded device.

Hardware Economy

Hardware Economy in a classical term refers to cost of building the hardware given constituents of the device. I would be doing a thorough disservice if I scope hardware economy on Bill of Material (BOM) of the device, i.e., the raw ingredients in making a custom device. BOM is one aspect of it. Nonetheless, for evaluating performance of the device in a constrained environment, computational key attributes are essential. Computational capabilities, of course, are based on design specification of the hardware line item in BOM.

The following Hardware Economy parameters are required to be validated for building devices that operate in constrained environments (please see Figure 9):

• Device Casing
• Processing Power
  o Computational Power

---

[200] Trade-off Models - https://www.ifm.eng.cam.ac.uk/research/dstools/trade-off-models/

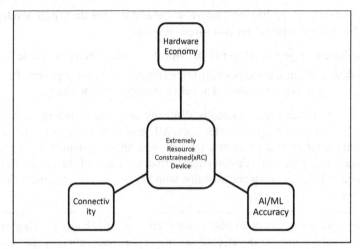

Figure 9: Model-Connectivity-Hardware

- o Offload Computational
- Environmental Perturbations
  - o Extreme Humidity
  - o Extreme Temperature Swings

- Power
  - o Electrical
  - o Battery

- Storage
  - o Offload
  - o Limited
    - Circular
    - Flat-File

- Memory
  - o In Memory Cache

Model (AI/ML) Accuracy

The Model AI/ML Accuracy can be influenced by various factors of hardware and available connectivity options. The factors that influence model accuracy include:

- Hardware Refresh—this parameter predicates the rate of refresh of the AI/ML model on the Constrained Device.
- Over the Air (OTA) Firmware Update: OTA is a direct upgrade of a Firmware on to a target device.
- Eager Learners vs. Lazy Learners ML Model Option: The Machine Learning Learner that performs model inference based on limited set of data vs. fully backed model. Examples of Eager Learner—Decision Tree, Naïve Bayesian and Artificial Neural Networks (ANN) and examples of lazy learner include K-nearest Neighbor. Due to the model construction, eager learners take a long time to train and *less time to predict*. Compared to eager learners, lazy learners have less training time but **more time in predicting**.
- Data Movement needs of the Model: Model performs that needs huge memory transfer between CPU to Memory.
- Model Invocation
  - o In-line AI Model
  - o Stack or Heap Load Models

Connectivity

Connectivity options available to the constrained devices. The options include:

- No Connectivity
- Low Powered Connectivity
  - o CoAP (Constrained Application Protocol)
  - o Bluetooth Low Energy (BLE)
- High Powered Connectivity
  - o Wi-Fi
  - o GPS

The Trade-off model provides a seesaw view of a system by dissecting the influence of one of the key lever modeling/variations on the entire workings/penalties/compromises of the system. Generally, the variation of one key lever with respect to other important design parameters are necessary to develop a resilient, robust and fault-tolerant system. For example, in the above diagram, Trade-off model depicts the performance implications of an ML model, namely: in-memory large dataset plus computational complexity (O (n log (n)) on the battery longevity of the system. Of course, in the above system only two scalar components of a system are depicted (please see Figure 10).

In a real-world production system, multiple interplay of different key design levers are compared and performance studied for deploying ML model. Put it plainly, a cross influence matrix of system performance needs to be considered.

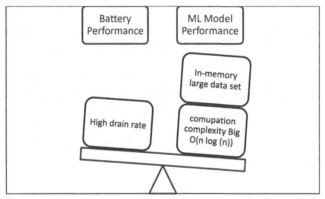

Figure 10: Trade-off Modeling

Application of trade-off modeling on nodes of MCH model provides the following trade-off linkages Please see Figure 11):

- Hardware Economy to Model Accuracy Trade-off
- Model Accuracy to Connectivity Trade-off
- Connectivity to Hardware Economy Trade-off

Hardware Economy—Model Accuracy Trade-off

Hardware Economy and Model Accuracy trade-off modeling analyzes the impacts of connectivity options on the choices of Hardware and Model Accuracy. In this tradeoff modeling, the various connectivity impacts are enumerated for desired operating Hardware and ML model accuracy. In other words, for a given connectivity option, the necessary tradeoff associated with Hardware economy and ML model accuracy can be analyzed. This modeling is essential for IoT devices deployment that are geographically distant and has reachability issue.

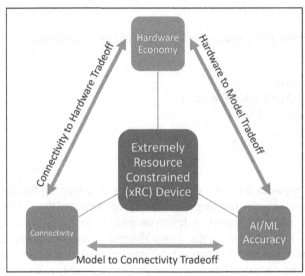

Figure 11: MCH

Modeling no connectivity

AI/ML Model Accuracy: Based on the above connectivity conditions, the trade-off on Hardware economy and Model accuracy needs to be analyzed and designed. For instance, IoT devices that are operated in no connectivity environments have two choices when it comes to ML models update. One, during the device refresh as part of service renewal or new sale. That is, when a new device is deployed, the model contains all the learnings as part of the device. The limitation of this approach, of course due to no connectivity, is the model has staleness with respect to time and learnings.

Similarly, lack of connectivity imposes limitations on the hardware economy. For instance, due to non-connectivity, the ML model operations as well as hardware & in-memory data storage and computations need to be analyzed at finer grade to make sure the desired operating process is achieved. Due to non-connectivity, the device cannot offload the computational processing to devices such as cell phones or Gateway devices. This poses either restrictions in terms of ML models (Active learners vs. Lazy learners) or tax hardware to be self-sufficient (higher storage or memory capacities) and thus increase device costs.

Hardware Economy: The tradeoff attributes Memory, Computational Processing and Storage must compensate for the lack of connectivity to the Internet.

Memory: Memory footprint for the devices that operate under no connectivity could be higher to compensate lack of offloading data to the other devices such as Edge or Gateways. Additionally, based on AI/ML models that are memory intense such as active learner require higher memory.

Storage: Based on device operational SLAs, the storage must be designed with sufficiently large capacity to prevent any data-loss or serviceability issues. Data loss could occur in scenarios if the data generated by hardware due to environmental issues, device health issues, device calibration issues or due to BLAST of sensor streams. The serviceability issues could emerge due to in-time replacement of device upon SLA completion or due to terrain or reachability issues. The most important factor in above two conditions is prevention of data loss due to non-availability of hardware storage capacity. To ensure the success of data capture, it is highly recommended to increase capacity of storage at least by a factor of fifty to seventy percentage.

The storage calculation:

- Let f be the frequency of data collection of the hardware—for instance, a custom IoT device collects temperature of an environment every 30 minutes

Connectivity / No Connectivity		Hardware 🔒 Constraint 🔓 No Constraint			
		Memory	Power Source	Storage	Computational Power
ML Model Accuracy	Self-Contained & Updated only on Hardware Refresh	N/A	N/A	N/A	N/A
	OTA Firmware	N/A	N/A	N/A	N/A
	Model Type: Eager Learner	🔓 In-memory needs are lower compared to Lazy learner as model is inline and/or atomic call.	🔒 Power source could be higher to maintain in-memory model state.	🔒 Due to no connectivity, the device needs to store the data locally instead of offloading to edge device	🔒 The higher capacities of hardware forces higher computational power to maintain the hardware.
	Lazy Learner	🔒 In-memory needs are higher compared to Eager learner as model takes more time to compute the references.	🔒 Power source could be higher to maintain in-memory model state.	🔒 Due to no connectivity, the device needs to store the data locally instead of offloading to edge device	🔒 The higher capacities of hardware forces higher computational power to maintain the hardware.
	Model Invocation	🔓 Inline model invocation exerts lower memory needs.	🔓	N/A	🔒 Inline model invocation requires less computational power compared to stack or heap based.

- Let *s* be the size of the data to be stored on the device—for instance, this could be in bytes or kilobytes based on the data upload
- Let *t* be the time in which the device will offload to an Edge device (like Smartphones) or to the Central Internet
- Let ϵ be the error in connecting device to external Edge devices to offload the data (for instance, the extra number of data capture to be stored before overwriting the data). This data varies from customer to customer and is a variable derived from historical data.

$$Base\ Storage\ size = \sum_{i=1}^{f} s * t$$

Extra storage or cushion required to factor in ϵ would be

$$Storage\ cushion = \sum_{i=1}^{\epsilon} s * t$$

Total storage required: base storage size plus storage cushion

Power: For devices that are operating under constrained environments with no network connectivity, the tax imposed by the connectivity results into higher power requirements. That is, the increased storage and higher memory will result in more power consumption needs.

Modeling Low Power Bluetooth Connectivity

Bluetooth Low Energy (BLE) is a commonly used Wireless Personal Area Network (WPAN) technology for IoT setups. Different researchers have offered a BLE implementation and evaluated its performance as comparable to that of ZigBee/802.15.4. Energy is consumed during a master-slave discovery process, as the master and slave devices are not always in a connected mode [6,7]. A master device searches for available slaves for connection simultaneously along the slave devices,[201] which advertise their availability to the master. Energy used after establishing a connection is also considered, and parameters related to energy, such as transmission and reception, along with the interframe spaces, are analyzed. A neighbor discovery mechanism multicast a high number of messages for IPv6 over BLE, consuming higher energy in BLE-based IoT devices [6,7]. When doing connectable or scan-able advertising, the advertising device consumes significant power.[202] The power per bit can be further improved by increasing the payload to 31 bytes per packet, and by configuring it for broadcast only. We may also take advantage of chips with integrated DC/DC regulators that enable peak current reduction of up to 20% for 3V cells.

Irrespective Frequent BLE advertising with higher computational cycles or in-memory operations to yield higher ML model accuracy is expensive to battery for constrained devices.

AI/ML Model Accuracy: Based on the BLE connectivity conditions, the trade-off on Hardware economy and Model accuracy is as follows:

Connectivity Bluetooth Low Energy (BLE)	Hardware 🔒 Constraint / 🔓 No Constraint			
	Memory	Power Source	Storage	Computational Power
Self-Contained & Updated only on Hardware Refresh	🔓 The data size could be reduced by offloading data to external devices.	🔒 Continuous Bluetooth advertisements could increase power needs.	🔓 Data could be offloaded to external devices.	🔓 The model data and storage are offloaded to Edge devices (Smartphone) and thus could reduce hardware footprint.
OTA Firmware	N/A	N/A	N/A	N/A
Model Type: Eager Learner	🔓 In-memory needs are lower compared to Lazy learner as model is inline and/or atomic call.	🔒 The model data and storage are offloaded to Edge devices (Smartphone) and thus could reduce hardware footprint. Nonetheless, Bluetooth advertisements need higher power.	🔓 Periodic offloading of Sensor data to Edge devices lowers the storage needs.	🔓 Lower Data Size and Code size reduce the needs for more computational power.
Lazy Learner	🔒 Though Data could be offloaded to external devices (Edge), the intrinsic needs of algorithms (i.e., K-means cluster) exerts the need for higher memory.	🔒 Higher hardware footprint (i.e., higher memory footprint) leads to greater power source needs. Additionally, BLE advertisements increase power needs.	🔒 The intrinsic needs of algorithms (i.e., K-means cluster) exerts the need for higher storage.	🔒 The higher capacities of hardware forces higher computational power to maintain the hardware.
Model Invocation	🔓 Inline model invocation exerts lower memory needs.	🔒 Continuous Bluetooth advertisements could increase power needs	N/A	🔒 Inline model invocation requires less computational power compared to stack or heap based.

The leftmost vertical label for the lower rows reads: **ML Model Accuracy**

---

[201] Internet of Things: Challenges, Advances, and Applications (Chapman & Hall/CRC Computer and Information Science Series) 1st Edition

[202] The Importance of Average Power Consumption to Battery Life - https://blog.nordicsemi.com/getconnected/the-importance-of-average-power-consumption-to-battery-life

Hardware Economy: Two factors influence hardware footprint: BLE advertisements and Model accuracy.

Memory: Periodic connectivity to Edge devices and offloading the data could reduce memory and storage needs. Thus, the overall power consumption could be lowered. However, Lazy learner-based AI/ML algorithms require more memory and storage, combining this with BLE advertisements, the memory & power requirements for the hardware increase.

Power Source: Given BLE advertisements, the power source capacities are higher compared to no connectivity devices.

Storage: The storage footprint is lower compared to no-connectivity option as the data from the device is periodically offloaded to Edge device.

Computational Power: Algorithm type dictates the computational power. For instance, eager learners need less computational power compared to lazy learners.

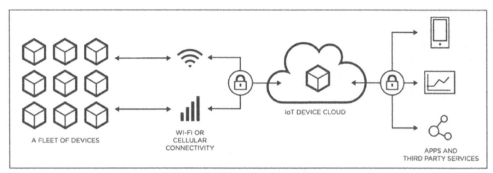

Figure 12: Edge offload Architecture

Edge Offload Architectures (please see Figure 12)

The term offload architecture is referred to architecture processes that reduce the load on the principal architecture by introducing time, location or actor-based intervention to transfer data or control. For instance, the storage space requirements for non-connected storage is much higher than the connected architecture using Wi-Fi or BLE. The advantage of connectivity is the actor can connect to network and transfers application data from Tiny device to Mobile Edge device. The advantage is that the data is refreshed from custom hardware device and the fear of overwrite could be eliminated by the Offload architecture (please see Figure 13).

In the above figure, the left side represents Tiny or constrained devices and the offload architecture connects to the central Edge by Wi-Fi or Bluetooth.

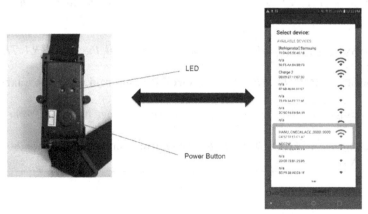

Figure 13: Mobile design with manual BLE Button

Bluetooth Data Service Connection

The Universal Asynchronous Receiver/Transmitter (UART) interface collects data from the Sensor[203] (left in Figure 13). The UART receives the data change event from the device and sends it to the Edge device.

The Bluetooth Low Energy device advertises signals. Mobile device display Bluetooth device and enable User connects to the device.

The BLE Device Connect

```java
package com.example.dairy2;

import android.app.Notification;
import android.app.NotificationManager;
import android.app.PendingIntent;
import android.app.Service;
import android.bluetooth.BluetoothDevice;
import android.content.BroadcastReceiver;
import android.content.Context;
import android.content.Intent;
import android.content.IntentFilter;
import android.os.IBinder;
import android.text.TextUtils;
import android.util.Log;

import androidx.annotation.NonNull;
import androidx.core.app.NotificationCompat;
import androidx.localbroadcastmanager.content.LocalBroadcastManager;

import no.nordicsemi.android.log.Logger;

public class UARTService extends BleProfileService implements UARTManagerCallbacks {
private static final String TAG = "UARTService";

public static final String BROADCAST_UART_TX = "no.nordicsemi.android.nrftoolbox.uart.
BROADCAST_UART_TX";
public static final String BROADCAST_UART_RX = "no.nordicsemi.android.nrftoolbox.uart.
BROADCAST_UART_RX";
public static final String EXTRA_DATA = "no.nordicsemi.android.nrftoolbox.uart.EXTRA_
DATA";

/** A broadcast message with this action and the message in {@link Intent#EXTRA_TEXT} will be sent
t the UART device. */
public final static String ACTION_SEND = "no.nordicsemi.android.nrftoolbox.uart.ACTION_
SEND";
/** A broadcast message with this action is triggered when a message is received from the UART
device. */
private final static String ACTION_RECEIVE = "no.nordicsemi.android.nrftoolbox.uart.
ACTION_RECEIVE";
```

---

[203] Sensor from Hanumayamma Innovations and Technologies, Inc - http://www.hanuinnotech.com

```java
/** Action send when user press the DISCONNECT button on the notification. */
public final static String ACTION_DISCONNECT = "no.nordicsemi.android.nrftoolbox.uart.
ACTION_DISCONNECT";
/** A source of an action. */
public final static String EXTRA_SOURCE = "no.nordicsemi.android.nrftoolbox.uart.EXTRA_
SOURCE";
public final static int SOURCE_NOTIFICATION = 0;
public final static int SOURCE_WEARABLE = 1;
public final static int SOURCE_3RD_PARTY = 2;

private final static int NOTIFICATION_ID = 349; // random
private final static int OPEN_ACTIVITY_REQ = 67; // random
private final static int DISCONNECT_REQ = 97; // random

private UARTManager mManager;

private final LocalBinder mBinder = new UARTBinder();

public class UARTBinder extends LocalBinder implements UARTInterface {
@Override
public void send(final String text) {
mManager.send(text);
}
}

@Override
protected LocalBinder getBinder() {
return mBinder;
}

@Override
protected LoggableBleManager<UARTManagerCallbacks> initializeManager() {
return mManager = new UARTManager(this);
}

@Override
protected boolean shouldAutoConnect() {
return true;
}

@Override
public void onCreate() {
super.onCreate();

registerReceiver(mDisconnectActionBroadcastReceiver, new IntentFilter(ACTION_
DISCONNECT));
registerReceiver(mIntentBroadcastReceiver, new IntentFilter(ACTION_SEND));
}
```

```java
@Override
public void onDestroy() {
// when user has disconnected from the sensor, we have to cancel the notification that we've created
some milliseconds before using unbindService
cancelNotification();
 unregisterReceiver(mDisconnectActionBroadcastReceiver);
 unregisterReceiver(mIntentBroadcastReceiver);

super.onDestroy();
 }

@Override
protected void onRebind() {
// when the activity rebinds to the service, remove the notification
cancelNotification();
 }

@Override
protected void onUnbind() {
// when the activity closes we need to show the notification that user is connected to the sensor
createNotification(R.string.uart_notification_connected_message, 0);
 }

@Override
public void onDeviceConnected(@NonNull final BluetoothDevice device) {
super.onDeviceConnected(device);
//sendMessageToWearables(Constants.UART.DEVICE_CONNECTED, notNull(getDeviceName()));
}

@Override
protected boolean stopWhenDisconnected() {
return false;
 }

@Override
public void onDeviceDisconnected(@NonNull final BluetoothDevice device) {
super.onDeviceDisconnected(device);
//sendMessageToWearables(Constants.UART.DEVICE_DISCONNECTED, notNull(getDeviceName()));
}

@Override
public void onLinkLossOccurred(@NonNull final BluetoothDevice device) {
super.onLinkLossOccurred(device);
//sendMessageToWearables(Constants.UART.DEVICE_LINKLOSS, notNull(getDeviceName()));
}

private String notNull(final String name) {
if (!TextUtils.isEmpty(name))
```

```
    return name;
    return getString(R.string.not_available);
    }

    @Override
    public void onDataReceived(final BluetoothDevice device, final String data) {
    final Intent broadcast = new Intent(BROADCAST_UART_RX);
     broadcast.putExtra(EXTRA_DEVICE, getBluetoothDevice());
     broadcast.putExtra(EXTRA_DATA, data);
     LocalBroadcastManager.getInstance(this).sendBroadcast(broadcast);

    // send the data received to other apps, e.g. the Tasker
    final Intent globalBroadcast = new Intent(ACTION_RECEIVE);
     globalBroadcast.putExtra(BluetoothDevice.EXTRA_DEVICE, getBluetoothDevice());
     globalBroadcast.putExtra(Intent.EXTRA_TEXT, data);
     sendBroadcast(globalBroadcast);
    }

    @Override
    public void onDataSent(final BluetoothDevice device, final String data) {
    final Intent broadcast = new Intent(BROADCAST_UART_TX);
     broadcast.putExtra(EXTRA_DEVICE, getBluetoothDevice());
     broadcast.putExtra(EXTRA_DATA, data);
     LocalBroadcastManager.getInstance(this).sendBroadcast(broadcast);
    }


    /**
    * Cancels the existing notification. If there is no active notification this method does nothing
    */
    private void cancelNotification() {
    final NotificationManager nm = (NotificationManager) getSystemService(NOTIFICATION_
    SERVICE);
     nm.cancel(NOTIFICATION_ID);
    }

    /**
    * This broadcast receiver listens for {@link #ACTION_DISCONNECT} that may be fired by pressing
    Disconnect action button on the notification.
    */
    private final BroadcastReceiver mDisconnectActionBroadcastReceiver = new BroadcastReceiver() {
    @Override
    public void onReceive(final Context context, final Intent intent) {
    final int source = intent.getIntExtra(EXTRA_SOURCE, SOURCE_NOTIFICATION);
    switch (source) {
    case SOURCE_NOTIFICATION:
     Logger.i(getLogSession(), "[Notification] Disconnect action pressed");
    break;
```

```
case SOURCE_WEARABLE:
 Logger.i(getLogSession(), "[WEAR] '" + Constants.ACTION_DISCONNECT + "' message
received");
break;
 }
if (isConnected())
 getBinder().disconnect();
else
stopSelf();
 }
};
```

```
/**
 * Broadcast receiver that listens for {@link #ACTION_SEND} from other apps. Sends the String or int
content of the {@link Intent#EXTRA_TEXT} extra to the remote device.
 * The integer content will be sent as String (65 -> "65", not 65 -> "A").
 */
private BroadcastReceiver mIntentBroadcastReceiver = new BroadcastReceiver() {
@Override
public void onReceive(final Context context, final Intent intent) {
final boolean hasMessage = intent.hasExtra(Intent.EXTRA_TEXT);
if (hasMessage) {
 String message = intent.getStringExtra(Intent.EXTRA_TEXT);
if (message == null) {
final int intValue = intent.getIntExtra(Intent.EXTRA_TEXT, Integer.MIN_VALUE); // how big is the
chance of such data?
if (intValue != Integer.MIN_VALUE)
 message = String.valueOf(intValue);
 }

if (message != null) {
final int source = intent.getIntExtra(EXTRA_SOURCE, SOURCE_3RD_PARTY);
switch (source) {
case SOURCE_WEARABLE:
 Logger.i(getLogSession(), "[WEAR] '" + Constants.UART.COMMAND + "' message received with
data: \"" + message + "\"");
break;
case SOURCE_3RD_PARTY:
default:
 Logger.i(getLogSession(), "[Broadcast] " + ACTION_SEND + " broadcast received with data: \""
+ message + "\"");
break;
 }
mManager.send(message);
return;
 }
 }
// No data od incompatible type of EXTRA_TEXT
if (!hasMessage)
```

```
 Logger.i(getLogSession(), "[Broadcast] " + ACTION_SEND + " broadcast received no data.");
else
Logger.i(getLogSession(), "[Broadcast] " + ACTION_SEND + " broadcast received incompatible
data type. Only String and int are supported.");
}
};
}
```

Offload data

The Mobile User connects the Sensor and collects the data and the collected data relay to the Cloud – an offload process.

The Collected data uploaded to the Cloud (please see Figure 14 and Figure 15):

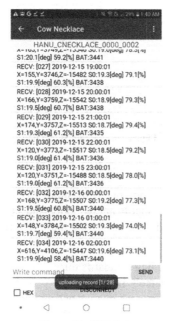

Figure 14: BLE Data

0	[000]	2000-01-10	05:00:01	X=264,Y=-209,Z=-16358	S0:23.3[deg]	46.5[%]	S1:22.9[deg]	45.5[%]	BAT:3220♦
1	[001]	2000-01-10	06:00:01	X=290,Y=-243,Z=-16316	S0:23.7[deg]	45.9[%]	S1:23.7[deg]	44.6[%]	BAT:3233♦
2	[000]	2000-01-10	05:00:01	X=264,Y=-209,Z=-16358	S0:23.3[deg]	46.5[%]	S1:22.9[deg]	45.5[%]	BAT:3220♦
3	[001]	2000-01-10	06:00:01	X=290,Y=-243,Z=-16316	S0:23.7[deg]	45.9[%]	S1:23.7[deg]	44.6[%]	BAT:3233♦
4	[002]	2000-01-10	07:00:01	X=267,Y=-249,Z=-16363	S0:23.7[deg]	46.1[%]	S1:23.7[deg]	44.5[%]	BAT:3236♦
5	[003]	2000-01-10	08:00:01	X=233,Y=-213,Z=-16342	S0:24.0[deg]	46.1[%]	S1:23.8[deg]	44.7[%]	BAT:3236♦
6	[004]	2000-01-10	09:00:01	X=267,Y=-220,Z=-16333	S0:24.3[deg]	45.6[%]	S1:24.0[deg]	45.2[%]	BAT:3231♦
7	[005]	2000-01-10	10:00:01	X=275,Y=-222,Z=-16330	S0:24.6[deg]	45.3[%]	S1:24.4[deg]	43.9[%]	BAT:3244♦
8	[006]	2000-01-10	11:00:01	X=264,Y=-224,Z=-16364	S0:24.9[deg]	50.9[%]	S1:24.6[deg]	49.1[%]	BAT:3238♦
9	[007]	2000-01-10	12:00:01	X=260,Y=-196,Z=-16322	S0:24.9[deg]	47.8[%]	S1:24.8[deg]	46.1[%]	BAT:3229♦
10	[008]	2000-01-10	13:00:01	X=236,Y=-205,Z=-16367	S0:24.8[deg]	46.6[%]	S1:24.7[deg]	45.0[%]	BAT:3230♦
11	[009]	2000-01-10	14:00:01	X=255,Y=-235,Z=-16340	S0:24.5[deg]	46.5[%]	S1:24.4[deg]	44.8[%]	BAT:3229♦
12	[010]	2000-01-10	15:00:01	X=265,Y=-244,Z=-16380	S0:24.3[deg]	45.9[%]	S1:24.2[deg]	44.2[%]	BAT:3229♦
13	[011]	2000-01-10	16:00:01	X=216,Y=-209,Z=-16350	S0:24.3[deg]	46.5[%]	S1:24.2[deg]	44.9[%]	BAT:3228♦
14	[012]	2000-01-10	17:00:01	X=283,Y=-253,Z=-16339	S0:24.3[deg]	46.9[%]	S1:24.2[deg]	45.1[%]	BAT:3224♦
15	[013]	2000-01-10	18:00:01	X=252,Y=-234,Z=-16372	S0:24.0[deg]	45.3[%]	S1:24.0[deg]	43.7[%]	BAT:3223♦
16	[014]	2000-01-10	19:00:01	X=271,Y=-209,Z=-16339	S0:22.9[deg]	45.3[%]	S1:22.3[deg]	44.0[%]	BAT:3178♦

Figure 15: Cow Necklace Data

The Development Environment

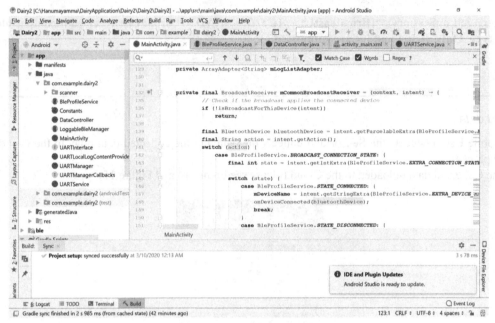

Figure 16: Android Studio IDE

Modeling Wi-Fi Connectivity

Communication technologies are a key component of smart IoT applications. Wireless communication technologies are attractive, because of the significant reduction and simplification in wiring involved. Various wireless standards have been established. One can group these into two main categories, depending on the transmission range [7]:

- **Short-range communication**: including standards for:
  o Wireless LAN, used for Wi-Fi, namely IEEE 802.11
  o Wireless PAN used more widely for measurement and automation applications, such as IEEE 802.15.1 (Bluetooth) (IEEE, 2002) and IEEE 802.15.4 (ZigBee/6LoWPAN) (IEEE, 2003).

All these standards use the instrumentation, scientific and medical (ISM) radio bands, typically operating in the 2.400–2.4835 GHz band.

- **Long-range communication**: including the increasingly important sub-gigahertz IoT communication technologies, such as LoRA, in the 868–870 MHz band. These trade data transmission rates (on the order of hundreds of kbit/s) for longer transmission ranges.

Although Wi-Fi has become the standard for audio and home automation devices that require high throughput rates,[204] power consumption is high, and the devices are usually plugged into power outlets [7]. Most wearables, for example, *should use other standards with a better power efficiency*. For many sub-GHz applications, range is unquestionably the most important design constraint, making this a straightforward determination. Increasing the transmit output power extends the range and coverage but consumes more battery power [8]. Greater range can result in lower system cost since the longer a node's range, the fewer nodes there need to be in the system to provide complete coverage of an area. Wi-Fi on-board device consumes more battery power to maintain high network connectivity (please see Figure 17).

---

[204] The Importance of Average Power Consumption to Battery Life - https://blog.nordicsemi.com/getconnected/the-importance-of-average-power-consumption-to-battery-life

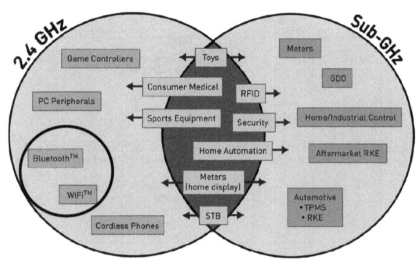

Figure 17: Sub-GHz versus 2.4 GHz Frequency Trends by Application Type [8]

Additionally, Operating life depends upon range, radio sensitivity, data rate and the number of nodes in the network.[205]

Range is determined by the sensitivity of the transceiver and its output power, commonly referred to as link budget. A primary factor affecting radio sensitivity is the data rate. The lower the data rate, the narrower the receive bandwidth is and the greater the sensitivity of the radio. Figure 18 shows how sensitivity is higher at lower data rates.

As seen in the below table (Figure 19), Wi-Fi is better suited with electrical power let constrained devices [7].

For a constrained device with Wi-Fi, on-board chip is expensive. Additionally, if the ML model accuracy needs are higher, i.e., ML model performs higher data movements between Microcontroller and SRAM memory space[206] (Memory-wall & the "von-Neumann bottleneck") the consumption of battery will be higher as computational power goes up [9]. Net-net, it results in a higher battery drain (please see Figure 20).

☀️ Constrained Device on Wi-Fi Connectivity with High Data Movement between Memory and CPU Controllers result in High Load and Drain on Battery Power.

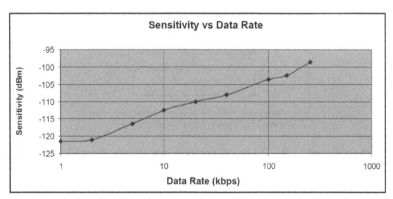

Figure 18: Sensitivity Increases with Lower Data Rates [8]

---

[205] Maximizing Range and Battery Life in Low-Cost Wireless Network - https://www.silabs.com/documents/public/white-papers/Maximize-Wireless-Network-Range-and-Battery-Life.pdf

[206] Energy-Efficient Smart Embedded Memory Design for IoT and AI - https://pdfs.semanticscholar.org/9edf/918044222ae684b7e3d7de434ddf7cbbec3f.pdf

Technology	Latency	Throughput (Kbps)	Peak current Consumption	Comment	Peak power consumption	Power per bit (efficiency)
ANT	5 ms	60	~ 10 mA			
ZigBee	~20 ms	128	~ 30 mA	Consumes 0.035706 W when transferring 24 bytes of data.		0.035706/192 = 185.9 µW/bit
Bluetooth Smart (BLE)	7.5 ms	128 (4.1) 305 (4.2)	~ 10 mA	Packets are broadcasted every 500 ms. Each contain 20 bytes of useful payload	49 µA x 3 V = 0.147 mW	0.147 mW/960 = 0.153 µW/bit
NFC	~polled typically every second, this is manufacturer specific		~ 50 mA			
Wi-Fi	~1.5 ms	6000	~ 116 mA (@1.8 V)	Transmitting a 40 Mbps UDP payload consumes about 116 mA at 1.8 V.	116 mA x 1.8 V = 0.210 W	0.00525 µW/bit

Figure 19: Connectivity-Power Consumption Table

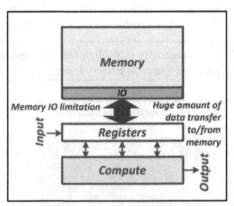

Figure 20: Memory Data Transfer (DX) [9]

AI/ML Model Accuracy: Based on the Wi-Fi connectivity conditions, the trade-off on Hardware economy and Model accuracy is as follows:

Connectivity Wi-Fi		Hardware 🔒 Constraint 🔓 No Constraint			
		Memory	Power Source	Storage	Computational Power
ML Model Accuracy	Self-Contained & Updated only on Hardware Refresh	🔓 The data size could be reduced by offloading data to external devices or central cloud.	🔒 On board Wi-Fi demands more power. For battery powered devices, Wi-Fi option is less favorable.	🔓 Data could be offloaded to the external devices or to the central Cloud.	🔓 The model data and storage are offloaded to the Edge devices (Smartphone) and thus could reduce the hardware footprint.
	Over the Air (OTA) Firmware Update	🔓 OTA allows to deploy the latest and greatest optimized model.	🔒 The Onboard Wi-Fi demands more power , although OTA allows to deploy the latest and greatest optimized model.	🔓 Data could be offloaded to external devices or to the central Cloud.	🔓 OTA allows to deploy the latest and the greatest optimized model.
	Model Type: Eager Learner	🔓 In-memory needs are lower compared to Lazy learner as model is inline and/or require lesser call stack invocations.	🔒 The model data and storage are offloaded to the Edge devices (i.e., Smartphone) and thus could reduce hardware footprint. Nonetheless, Bluetooth advertisements need higher power.	🔓 Periodic offloading of the Sensor data to the Edge devices lowers the on-board storage needs.	🔓 Lower Data Size and Code size reduce the needs for computational power.
	Lazy Learner	🔒 Though the Data could be offloaded to the external devices (Edge Devices), the intrinsic needs of the lazy learning algorithms (i.e., K-means cluster) exerts the need for higher memory.	🔒 Higher hardware footprint (i.e., higher memory footprint) leads to the greater power source needs. Additionally, the Bluetooth advertisements increase the overall power needs.	🔒 The intrinsic needs of lazy algorithms (i.e., K-means cluster) exerts the need for higher storage.	🔒 The higher capacities of hardware forces higher computational power to maintain the power needs of lazy learner.
	Model Invocation	🔓 Inline model invocationsneed lower memory.	🔒 Continuous Bluetooth advertisements increase power needs	N/A	🔒 Inline model invocation requires less computational power compared to lazy learner invocation.

Hardware Economy: Two factors influence Hardware footprint: Wi-Fi power needs and Model accuracy.

Memory: Periodic connectivity to Edge devices and offloading the data could reduce memory and storage needs. Thus, the overall power consumption could be lowered. However, Lazy learner-based AI/ML algorithms require more memory and storage, combining with Wi-Fi, the memory & power requirements for the hardware increase.

Power Source: Given Wi-Fi power needs, the power source capacities are higher compared to no connectivity devices.

Storage: The storage footprint is lower compared to no-connectivity option as the data from the device periodically offloaded to Edge device.

Computational Power: Algorithm type dictates the computational power. For instance, eager learners need less computational power compared to lazy learners.

 In an extremely resource constrained (xRC) environment, a small edge or custom-built device with higher Model Accuracy and Higher Hardware Performance (longer battery life) will sacrifice connectivity frequencies. The Frequency of connectivity must be low powered to achieve desired and balanced hardware to model accuracy (please see Figure 21).

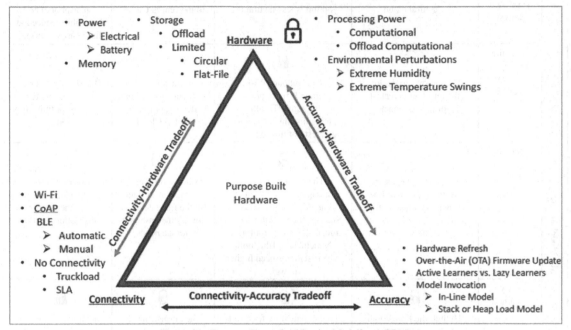

Figure 21: Extreme Constrained Device Modeling (xCDM)

Connectivity—Model Accuracy Trade-off

Connectivity to Model Accuracy trade-off modeling analyzes the impacts of hardware options on the choices of Model Accuracy and Connectivity. In this trade-off modeling, the various hardware impacts are enumerated for desired operating ML Model Accuracy and Connectivity options. In other words, for a given hardware option, the necessary trade-off associated with Connectivity and ML Model Accuracy can be analyzed. This modeling is essential for all IoT devices' deployment that are geographically distant and has reachability issue.

Modeling Memory

Memories are a critical component of any computing system, allowing to store data that need to be accessed for processing. Figure 22 shows the typical memory hierarchy, with increasing speed and decreased energy of access as we move up the pyramid.[207] On the other hand, manufacturing cost decreases, along with increasing density of stored bits, as we move down the hierarchy [9]. Hence, frequently accessed data are stored in faster and more energy efficient embedded memories (caches), situated near the processing engines. Whereas, data that is not accessed very frequently, is stored further away in larger but slower memories, e.g., DRAM, Flash, etc., with larger energy needs [9]. In the following table, the loader file setup the memory space with 64KB and SRAM is set from the offset as described in the following table.

---

[207] Energy-Efficient Smart Embedded Memory Design for IoT and AI - https://pdfs.semanticscholar.org/9edf/918044222ae684b7e3d7d e434ddf7cbbec3f.pdf

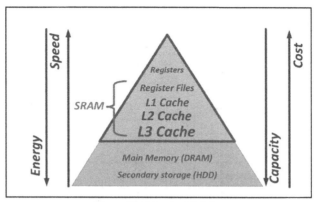

Figure 22: Memory Modeling

```
MEMORY
{
/* Flash memory region for the program, offset by the bootloader section */
FLASH (rx) : ORIGIN = 64k, LENGTH = 448k

/* 32k (Heap starts here) */
SRAM (rwx) : ORIGIN = 0x10000000, LENGTH = 32k

/* 32k (Globals at bottom, stack on top), heap continues from here */
SRAM_AHB (rwx) : ORIGIN = 0x2007c000, LENGTH = 32k

/* Heap starts at SRAM because if heap starts from higher memory, then
 * malloc tracking routines get confused when sbrk() system function
 * returns higher heap memory first, and then lower heap memory.
 * So we start heap from SRAM, and then move it up to SRAM_AHB if SRAM
 * portion runs out completely.
 */
}
```

The Heap pointer location for ARM Cortex 3 starts at 0 × 10000000.

```
/* Provide the start of the initial stack pointer
        * Debugger and ISP may use 32-bytes of space?
        */
       PROVIDE(_vStackTop = ORIGIN(SRAM_AHB) + LENGTH(SRAM_AHB) - 32);
```

Chapter 6 provides details of memory and stack pointer data screen captures.

Static Random-Access Memories (SRAM) are widely used for on-chip caches in modern micro-processors. SRAMs occupy a significant portion of chip area in modern Systems-on-a-Chip (Soc). In the following block diagram[208] of ATmega48A/PA/88A/PA/168A/PA/328/P, a low power CMOS 8-bit microcontrollers, the SRAM is located as internal cache AVR CPU [10] (please see Figure 23 and Figure 24).

Program Flash memory space is divided in two sections, the Boot Program section and the Application Program section. During interrupts and subroutine calls, the return address Program Counter (PC) is stored on the Stack. The Stack is effectively allocated in the general data SRAM, and consequently the Stack size is only limited by the total SRAM size and the usage of the SRAM.[209]

In emerging applications, such as accelerators for machine learning (ML) algorithms like neural networks, embedded memory is even more crucial for improving performance. Google's Tensor Processing Unit or TPU chip is one such example, in which 37% of the chip area is occupied by memory (~ 28MB), to buffer different types of data [9].

---

[208] Microchip ATmega48A/PA/88A/PA/168A/PA/328/P - http://ww1.microchip.com/downloads/en/DeviceDoc/ATmega48A-PA-88A-PA-168A-PA-328-P-DS-DS40002061A.pdf

[209] AVR - http://ww1.microchip.com/downloads/en/DeviceDoc/ATmega48A-PA-88A-PA-168A-PA-328-P-DS-DS40002061A.pdf

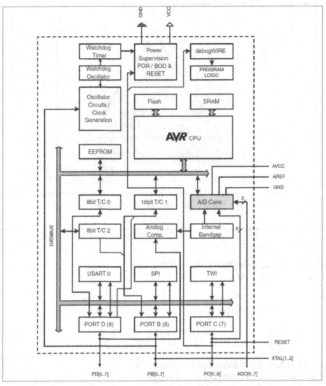

Figure 23: ATmega 48 Block Diagram

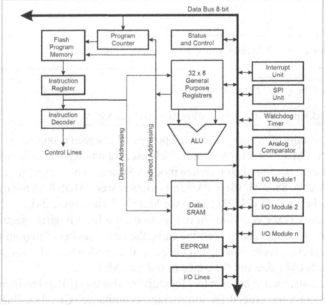

Figure 24: Block Diagram of AVR Architecture

Von Neumann Bottleneck (please see Figure 25)

Early computer systems timings for accessing main memory and for computation were reasonably balanced.

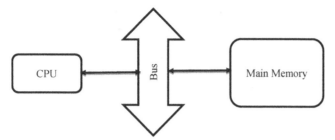

Figure 25: CPU and Memory Bus Architecture

As computational speeds grow at a much faster rate (high powered CPUs) than the main memory access speed, it resuls in a significant performance gap. Von Neumann bottleneck is the discrepancy between CPU compute speed and main memory (DRAM) speed. Consider a following compute example to find maximum capacity of the systems (please see Figure 26):

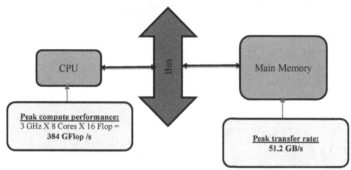

Figure 26: CPU-Memory Computations

   What is neuromorphic technology?

This technology draws its inspiration from the brain's internal organization and is capable of impressive cognitive tasks with less consumption than a light bulb. We speak of "neuromorphic chips". Neuromorphic[210] systems are extremely energy efficient in comparison with processors and graphic cards, due to their exploitation of two strategies. First of all, ***they bring computing and memory as close together as possible, so limiting data exchanges, which are currently the main source of energy consumption in processors***. Second, they carry out computing ***less accurately*** than processors but in a much more energy efficient way, either by using low-precision digital circuits (with small numbers of bits) or using the intrinsic nonlinearities in electronic components, which are an essential part of modern approaches such as neural networks. It should be borne in mind, however, that neuromorphic technologies do not necessarily solve all learning problematics.

Several articles have tried to quantify energy gains obtained via such technologies: IBM's TrueNorth neuromorphic chip, for example, consumes 20 mW/cm2 compared with 100 W/cm2 for conventional computers, for implementation of neural networks. Learning is carried out offline, however. Online learning via these technologies remains a challenge our researchers have yet to resolve, and could therefore be the subject of an innovation challenge.

Source: The Centre for Nanoscience and Technology's contribution to the mission.

A simple DOT product for example: $n = 2^{30}$ (please see output Figure 27)

```
/* Double dotp = 0.0;
 For(int i = 0; i < n; i++) {
 dotp += u[i] * v[i];
 } // assuming u[i] & v[i] is initialized array of integers
 */
double value1, value2;
double n;
/* Assign the values we will use for the pow calculation */
 value1 = 2;
 value2 = 20;
/* Calculate the result of value1 raised to the power of value2 */
 n = pow(value1, value2);
/* Display the result of the calculation */
 printf("%f raised to the power of %f is %f\n", value1, value2, n);
double* u, * v;
 u = malloc(n * sizeof(double));
 v = malloc(n * sizeof(double));
double dotp = 0.0;
for (int i = 0; i < n; i++) {
 u[i] = i * 2;
 v[i] = i * 5;
 }
for(int i = 0; i < n; i++) {
 dotp += u[i] * v[i];
 printf("Value of dot product at index %d is %f\n ", i, dotp);
 }
```

- Computation time: $t_{comp} = \dfrac{2\ GFlop}{384\ GFLOP\ per\ s} = 5.2$ ms

  o Total Operations: $2.n = 2^{31}$ Flops = 2 GFlops

- Data Transfer time: $t_{mem} = \dfrac{16\ GB}{51.2\ GB/s} = 312.5$ ms

  o Amount of data transferred: $2 \times 2^{30} \times 8$ B = 16 GB

- Execution time: $t_{exec} \geq$ **max** (5.2 ms, 312.5 ms) = 312.5 ms

  o Achievable performance: $\dfrac{2\ GFlops}{312.5\ ms} = 6.4$ GFlop/s (< 2% of peak)

  o DOT Product is **memory bound**

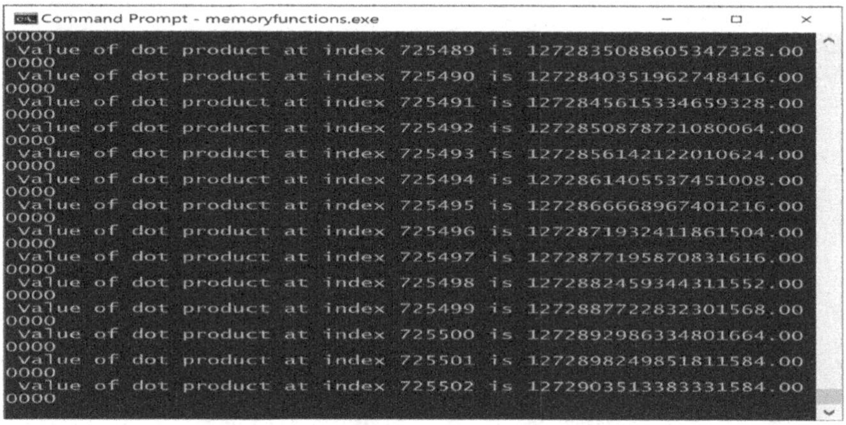

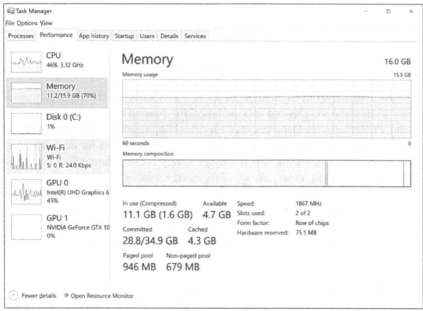

Figure 27: Output

Please find the memory details (Windows Task Manager): please see Figure 28.

Figure 28: Memory Snapshot

Please see CPU snapshot (Windows Task Manager)—see Figure 29.

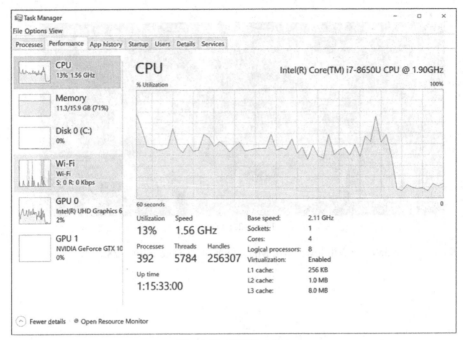

Figure 29: CPU Snapshot

Most machine learning and artificial intelligence applications today use the Von Neumann architecture.[211] "This has a memory to store the weights and data, and the CPU does all of the computation," explains Meng-Fan (Marvin) Chang, Professor in the Department of Electrical Engineering at National Tsing Hua University. "A lot of the data movement is through the bus. Today, they also use GPUs for deep learning that includes convolutions. One of the major problems is that they generally create intermediate data to implement the inference [11]. The data movement, *especially off-chip, causes a lot of penalty in energy and latency*. That is a bottleneck."

Both DRAM (Dynamic Random-Access Memory) and SRAM (Static Random-AccessMemory) are types of Random-Access Memory (RAM).[212] RAM is a semiconductor device internal to the integrated chip that stores the processor that a microcontroller or other processor will use constantly to store variables used in operations while performing calculations. RAM refers to the hardware that provides the memory locations referred to in software as registers. You can think of RAM as working memory where variables are stored while the CPU performs calculations. RAM is much faster to access than external memory and is a critical component to the speed of the processor chip. The architectural difference between the two is that DRAM uses transistors and capacitors in an array of repeating circuits (where each circuit is one bit), whereas SRAM uses several transistors in a circuit to form one bit [12].

Unlike DRAM that uses capacitors to maintain state, SRAM uses several transistors in a cross-coupled flip-flop configuration and does not have leakage issue and does not need to be refreshed. But SRAM still needs constant power to maintain the state of charge and thus is volatile like DRAM [12].

The power consumption of SRAM varies widely depending on how frequently it is accessed; in some instances, it can use as much power as dynamic RAM, when used at high frequencies, and some ICs can consume many watts at full bandwidth. On the other hand, static RAM used at a somewhat slower pace, such as in applications with moderately clocked microprocessors, draws very little power and can have a nearly negligible power consumption when sitting idle—in the region of a few micro-watts[213] (please see Figure 30).

---

[211] AI Architectures Must Change - https://semiengineering.com/ai-architectures-must-change/
[212] What is DRAM (Dynamic Random-Access Memory) vs SRAM? - https://www.microcontrollertips.com/dram-vs-sram/
[213] SRAM – Wikipedia - https://en.wikipedia.org/wiki/Static_random-access_memory

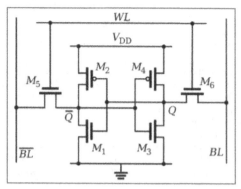

Figure 30: Memory Electronic Circuits [12]

Two aspects are important for SRAM design:[214] dynamic energy consumption and delay of read and write operation. Dynamic energy consumption consumes energy of power supply during read and write operation and determines time battery operation in mobile application.

 In a constrained environment, a small edge or custom-built device with SRAM provides higher speed and efficient energy consumption.

Radically new processor–memory architectures must be developed to address the current mismatch between processor speed and memory access time.[215] The significant energy costs in time and memory to move data will be one of the most difficult barriers for future HPC. As a goal, novel memory management technologies could enhance programmer productivity by using programming methods promoting data locality, by auto-tuning to eliminate the need for low-level optimization, and by managing software layers to ensure reliability.

Connectivity: Connectivity mode influences memory constraint. For no connectivity environments, it is imperative to have higher memory footprint. More importantly, have both SRAM and DRAM to sustain higher footprint and faster access speeds to compensate energy consumption demands. For Wi-Fi and BLE, the type of memory, particularly SRAM, would yield better design.

AI/ML Model Accuracy: Based on the memory constraint, in terms of size & type, the tradeoff on Model accuracy and Connectivity need to be analyzed and designed. For instance, IoT devices that are designed with DRAM based memory must factored in higher power consumption whereas devices designed with SRAM memory may not taxed with higher energy consumption and yields high access speeds. The choice of memory and size play an even important role for no connectivity environments. For Wi-Fi and BLE connectivity modes, the power consumption can be reduced with SRAM or Cache based memories to overcome extra power needs due to Wi-Fi and/or BLE advertisements.

More importantly, memory design plays an important role for the type of model deployed on the constraint device. For Eager learner-based ML designs, i.e., Decision Tree, Naive Bayes, Artificial Neural Networks, the memory needs are lower as eager learners take a long time for train and less time to predict. On the other hand, lazy learners, example: k-nearest neighbor, need higher memory needs to store the model new data points. Having SRAM compensate increased hardware footprint.

---

[214] Delay and Energy Consumption Analysis of Conventional SRAM - http://citeseerx.ist.psu.edu/viewdoc/download?doi=10.1.1.123.6 254&rep=rep1&type=pdf

[215] Rebooting the IT Revolution: A Call to Action - https://www.semiconductors.org/resources/rebooting-the-it-revolution-a-call-to-action-2/

Hardware Memory		Connectivity 🔒 Constraint 🔓 No Constraint		
		No Connectivity	Bluetooth Low Energy (BLE)/ LoRa	Wi-Fi
Model Accuracy	Self-Contained & Updated only on Hardware Refresh	Require higher memory—having both SRAM & DRAM would help the design. SRAM Memory yields more energy efficient design as memory demands are higher given no connectivity. Generally, the memory stores Application Data in a fixed Circular Buffer. RTOS Heap_3 implementation is generally optional.	For energy efficient operations, having SRAM would help. Given Connectivity, the memory data could be offloaded to an external storage or to the cloud storage on a periodic basis. The memory stores Application Data in a fixed Circular Buffer. RTOS Heap_3 implementation is optional.	For energy efficient operations, having SRAM would help. Given Connectivity, the memory data could be offloaded to an external storage or to the cloud on a periodic basis.
	OTA Firmware	Require higher memory—having both SRAM & DRAM would help the design. SRAM Memory yields more energy efficient design as memory demands are higher given no connectivity.	For energy efficient operations, having SRAM would help. Given Connectivity, the memory data could be offloaded to an external storage or to the cloud on a periodic basis.	For energy efficient operations, having SRAM would help. Given OTA, the device is deployed with the most efficient models.
	Model Type: Eager Learner	Having either SRAM or DRAM doesn't make difference as the model call is inline function call.	Having either SRAM or DRAM doesn't make difference as the model call is inline function call.	Having either SRAM or DRAM doesn't make difference as the model call is inline function call.
	Lazy Learner	Needs frequent memory operations and stack calls. Though the Data could be offloaded to external devices (Edge), the intrinsic needs of algorithms (i.e., K-means cluster) exerts the need for higher memory.	Needs higher memory operations and benefited by having higher SRAM.	Needs higher memory operations and benefited by having higher SRAM.
	Model Invocation	Since inline model invocation, exerts lower memory needs.	Since inline model invocation, exerts lower memory needs.	Since inline model invocation, exerts lower memory needs.

Memory Management

The Real Time Operating System (RTOS) kernel needs RAM[216] each time a task, queue, mutex, software timer, semaphore or event group is created. The RAM can be automatically dynamically allocated from the RTOS heap within the RTOS API object creation functions. If RTOS objects are created dynamically then the standard C library malloc() and free() functions can sometimes be used for the purpose, but

1. they are not always available on embedded systems
2. they take up valuable code space
3. they are not thread safe, and
4. they are not deterministic (the amount of time taken to execute the function will differ from call to call)

... so often an alternative memory allocation implementation is required.

One embedded/real time system can have very different RAM and timing requirements to another—so a single RAM allocation algorithm will only ever be appropriate for a subset of applications.

To get around this problem, FreeRTOS keeps the memory allocation API in its portable layer. The portable layer is outside of the source files that implement the core RTOS functionality, allowing an application specific implementation appropriate for the real time system being developed to be provided. When the RTOS kernel requires RAM, instead of calling malloc(), it instead calls pvPortMalloc(). When RAM is being freed, instead of calling free(), the RTOS kernel calls vPortFree().

---

[216] RTOS Memory Management - https://www.freertos.org/a00111.html

FreeRTOS offers several heap managements schemes that range in complexity and features. It is also possible to provide your own heap implementation, and even to use two heap implementations simultaneously. Using two heap implementations simultaneously permits task stacks and other RTOS objects to be placed in fast internal RAM (SRAM), and application data to be placed in slower external RAM-DRAM.

The following heap implementations are possible:

- heap_1 – the very simplest, does not permit memory to be freed
- heap_2 – permits memory to be freed, but does not coalescence adjacent free blocks.
- heap_3 – simply wraps the standard malloc() and free() for thread safety
- heap_4 – coalescences adjacent free blocks to avoid fragmentation. Includes absolute address placement option
- heap_5 – as per heap_4, with the ability to span the heap across multiple non-adjacent memory areas
  Most common, heap_3.

---

 heap_3.c

This implements a simple wrapper for the standard C library malloc() and free() functions that will, in most cases, be supplied with your chosen compiler. The wrapper simply makes the malloc() and free() functions thread safe.
This implementation:

• Requires the linker to setup a heap, and the compiler library to provide malloc() and free() implementations.
• Is not deterministic.
• Will probably considerably increase the RTOS kernel code size.

Note that the configTOTAL_HEAP_SIZE setting in FreeRTOSConfig.h has no effect when heap_3 is used.

Simple Memory Application

The following code exercises a simple C based memory allocation and deallocation exercise.

```
char name[100];
char* description;
 strcpy_s(name,44, "Democratization of Artificial Intelligence");
/* allocate memory dynamically */
 description = malloc(200 * sizeof(char));
 if (description == NULL) {
 fprintf(stderr, "Error - unable to allocate required memory\n");
 }
else {
 strcpy_s(description,65, "Technologies and Innovations for helping humanity and beyond...");
 }
 printf("Book Name = %s\n", name);
 printf("Description: %s\n", description);
int i, n;
int* a;
printf("Number of elements to be entered:");
 scanf_s("%d", &n);
 a = (int*)calloc(n, sizeof(int));
 printf("Enter %d numbers:\n", n);
for (i = 0; i < n; i++) {
 scanf_s("%d", &a[i]);
 }
```

```
printf("The numbers entered are: ");
for (i = 0; i < n; i++) {
printf("%d ", a[i]);
}
free(a);
```

Char name array allocated static 100 char bytes whereas description variable dynamically allocated with 200 bytes. String Copy copies allocated strings into Character buffer.

Next, application scan for number of array elements to be entered. Once entered, the console displays the array. Both malloc & calloc C Runtime functions are used (please see Figure 31 and Figure 32).

```
MALLOC C Runtime
_Check_return__Ret_maybenull__Post_writable_byte_size_(_Size)
_ACRTIMP_CRTALLOCATOR_CRT_JIT_INTRINSIC_CRTRESTRICT_CRT_HYBRIDPATCHABLE
void* __cdecl malloc(
_In__CRT_GUARDOVERFLOWsize_t_Size
);
CALLOC C Runtime
_Check_return__Ret_maybenull__Post_writable_byte_size_(_Count * _Size)
_ACRTIMP_CRT_JIT_INTRINSIC_CRTALLOCATOR_CRTRESTRICT_CRT_HYBRIDPATCHABLE
void* __cdecl calloc(
_In__CRT_GUARDOVERFLOWsize_t_Count,
_In__CRT_GUARDOVERFLOWsize_t_Size
);
FREE C Runtime
_ACRTIMP_CRT_HYBRIDPATCHABLE
void __cdecl free(
_Pre_maybenull__Post_invalid_void* _Block
);
```

IDE

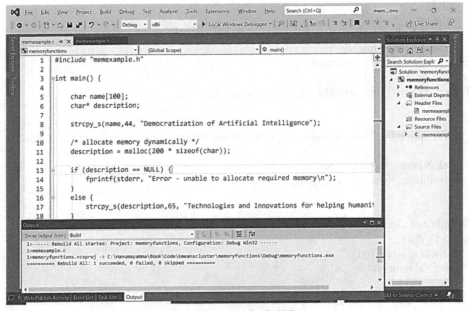

Figure 31: Visual Studio IDE

Output

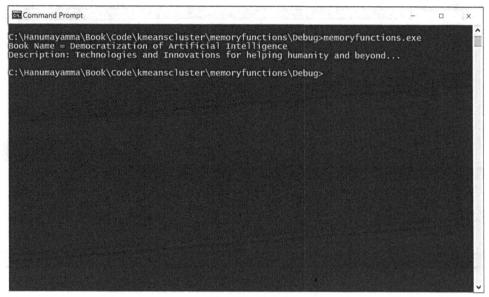

Figure 32: Output

Circular Buffer Design

The data at the edge level is very fast paced. To process data at the edge layer, the computation needs to be quick. Holding events longer due to process cost will pile the events at the source and may result in un-wanted system issues or crashes. The following list provides the most frequently used to process the events at the edge:

- Circular Buffers
- Double Buffering

Circular Buffers

Circular buffers are very useful data structures that are used memory constrained embedded device applications. As the name implies, Circular buffers use predefined fixed size buffers that wrap to the beginning as a close tube (please see Figure 33).

Technically, Circular buffers provide message queue semantics. That is, the data inserted Last in First Out (LIFO) mode.

Circular Buffer C Code

The generic code is used as a library so that User does not worry about code implementation and modification to the library code.

In our library header, we will forward declare the structure:

```
/// Opaque circular buffer structure
typedefstructcircular_buf_tcircular_buf_t;
/// Handle type, the way users interact with the API
typedefcircular_buf_t* cbuf_handle_t;
```

We have defined Circular buffer and defined access data member cbuf_handle_t;

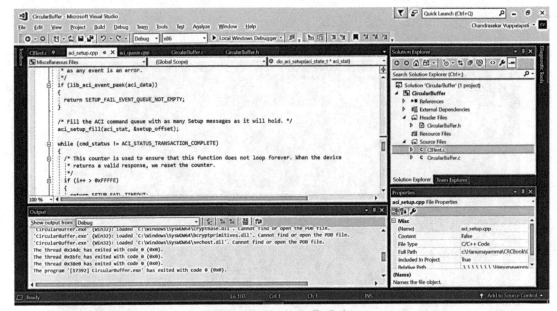

Figure 33: Circular Buffer Code

Circular Buffer Structure:

```
// The definition of our circular buffer structure is hidden from the user
structcircular_buf_t {
        uint8_t * buffer;
        size_t head;
        size_t tail;
        size_t max; //of the buffer
        bool full;
};
```

The Circular buffer has the following data members: Buffer pointer, head, tail, max and Boolean full flag.

The buffer points to unsigned integer of size 8 bytes. Head and Tail are pointers with size_t data length. In essence, both head and tail points to a circular buffer pointer.

Full flag is Boolean with 0 is empty and 1 with full flag.

Construct and Initialize the Buffer

The buffer is constructed with two parameters: Buffer and Size. Please note buffer is to be allocated based on the size and number of entries

```
#defineEXAMPLE_BUFFER_SIZE 10
uint8_t * buffer = malloc(EXAMPLE_BUFFER_SIZE * sizeof(uint8_t));
printf("\n=== C Circular Buffer Check ===\n");
cbuf_handle_t cbuf = circular_buf_init(buffer, EXAMPLE_BUFFER_SIZE);
```

First order of business, first we need to allocate buffer. The Malloc with number of entries in buffer and size of each entry are needed to allocate buffer.

The following code implements allocation of buffer. First things first, in order to initialize buffer, we need to create circular buffer and assign to cbuf variable. Next, assign each data member of cbuf with the newly allocated buffer.

```
cbuf_handle_t circular_buf_init(uint8_t* buffer, size_tsize)
{
        assert(buffer&&size);
        cbuf_handle_t cbuf = malloc(sizeof(circular_buf_t));
        assert(cbuf);
        cbuf->buffer = buffer;
        cbuf->max = size;
        circular_buf_reset(cbuf);
        assert(circular_buf_empty(cbuf));
        return cbuf;
}
```

In our case, buffer and max data members. The following calls initialize cbuf data members.

```
cbuf->buffer = buffer;
cbuf->max = size;
```

Finally, we need to point head, tail and full members (see Figure 34). Since the buffer is not pointing to any active element, we set other parameter by calling circular buffer reset function.

```
void circular_buf_reset(cbuf_handle_tcbuf)
{
        assert(cbuf);
        cbuf->head = 0;
        cbuf->tail = 0;
        cbuf->full = false;
}
```

As shown in Figure 34, both "head" and "tail" points to the same memory address location.

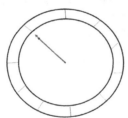

Figure 34: Circular Buffer

Next, once allocated buffer, calling circular buffer initialize method would allocate the buffer to our structure. After this, we can insert entries into the buffer.

```
printf("Buffer initialized. Full: %d, empty: %d, size: %zu\n",
                circular_buf_full(cbuf),
                circular_buf_empty(cbuf),
                circular_buf_size(cbuf));
```

```
=== C Circular Buffer Check ===
Buffer initialized. Full: 0, empty: 1, size: 0
```

Adding elements to buffer

Adding elements to buffer is performed through put operation.

```
printf("\n******\nAdding %d values\n", EXAMPLE_BUFFER_SIZE);
for (uint8_t i = 0; i <EXAMPLE_BUFFER_SIZE; i++)
{

        circular_buf_put(cbuf, i);
        printf("Added %u, Size now: %zu\n", i, circular_buf_size(cbuf));

}
```

After for loop, we have filled all the entries of circular buffer.

```
******
Adding 9 values
Added 0, Size now: 1
Added 1, Size now: 2
Added 2, Size now: 3
Added 3, Size now: 4
Added 4, Size now: 5
Added 5, Size now: 6
Added 6, Size now: 7
Added 7, Size now: 8
Added 8, Size now: 9
Full: 0, empty: 0, size: 9
```

The put operation performs the following: buffer needs to be valid and head points to data.

```
/// Put version 1 continues to add data if the buffer is full
/// Old data is overwritten
/// Requires: cbuf is valid and created by circular_buf_init
void circular_buf_put(cbuf_handle_tcbuf, uint8_tdata)
{
        assert(cbuf&&cbuf->buffer);
        cbuf->buffer[cbuf->head] = data;
        advance_pointer(cbuf);
}
```

Next, we advance the pointer to point to the next buffer position (see Figure 35). The next buffer position is derived by moving the pointer the size of structure. Please note: In circular buffer, all the entries are allocated in a contiguous manner and by moving the pointer to the size of the preceding structure ensures that we are pointing to a contiguous location.

```
staticvoid advance_pointer(cbuf_handle_tcbuf)
{
        assert(cbuf);
        if (cbuf->full)
        {
                cbuf->tail = (cbuf->tail + 1) % cbuf->max;
        }
        cbuf->head = (cbuf->head + 1) % cbuf->max;
        // We mark full because we will advance tail on the next time around
        cbuf->full = (cbuf->head == cbuf->tail);
}
```

Figure 35: Addition of first item

As data are getting populated, both tail and head move one block at a time (see Figure 36).

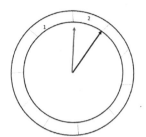

Figure 36: Addition of second item

Please note: head points to new block and tail points to first block.

```
staticvoid advance_pointer(cbuf_handle_tcbuf)
{
        assert(cbuf);
        if (cbuf->full)
        {
                cbuf->tail = (cbuf->tail + 1) % cbuf->max;
        }
        cbuf->head = (cbuf->head + 1) % cbuf->max;
        // We mark full because we will advance tail on the next time around
        cbuf->full = (cbuf->head == cbuf->tail);
}
```

Reading Values

Reading Values from Circular buffer is performed by calling Get operation on Circular Buffer.

```
int circular_buf_get(cbuf_handle_tcbuf, uint8_t * data)
{
        assert(cbuf&&data&&cbuf->buffer);
        int r = -1;
        if (!circular_buf_empty(cbuf))
        {
                *data = cbuf->buffer[cbuf->tail];
                retreat_pointer(cbuf);
                r = 0;
        }
        return r;
}
```

Check Circular buffer is empty first, if not, read the buffer.

```
bool circular_buf_empty(cbuf_handle_tcbuf)
{
        assert(cbuf);
        return (!cbuf->full && (cbuf->head == cbuf->tail));
}
```

Circular buffer empty can be evaluated by checking whether the buffer is full and checking whether head and tail pointers are pointing to the same location.

Driver

```
#include<stdio.h>
#include<stdlib.h>
#include<stdint.h>
#include<stddef.h>
#include<stdbool.h>
#include<assert.h>
#include"CircularBuffer.h"
#defineEXAMPLE_BUFFER_SIZE 10
int main(void)
{
        uint8_t * buffer = malloc(EXAMPLE_BUFFER_SIZE * sizeof(uint8_t));
        printf("\n=== C Circular Buffer Check ===\n");
        cbuf_handle_t cbuf = circular_buf_init(buffer, EXAMPLE_BUFFER_SIZE);
        printf("Buffer initialized. Full: %d, empty: %d, size: %zu\n",
                circular_buf_full(cbuf),
                circular_buf_empty(cbuf),
                circular_buf_size(cbuf));
        printf("\n******\nAdding %d values\n", EXAMPLE_BUFFER_SIZE - 1);
        for (uint8_t i = 0; i < (EXAMPLE_BUFFER_SIZE - 1); i++)
        {
                circular_buf_put(cbuf, i);
                printf("Added %u, Size now: %zu\n", i, circular_buf_size(cbuf));
        }
        printf("\n******\nAdding %d values\n", EXAMPLE_BUFFER_SIZE);
        for (uint8_t i = 0; i <EXAMPLE_BUFFER_SIZE; i++)
        {
                circular_buf_put(cbuf, i);
                printf("Added %u, Size now: %zu\n", i, circular_buf_size(cbuf));
        }
        printf("Full: %d, empty: %d, size: %zu\n",
                circular_buf_full(cbuf),
                circular_buf_empty(cbuf),
                circular_buf_size(cbuf));
        printf("\n******\nReading back values: ");
        while (!circular_buf_empty(cbuf))
        {
                uint8_t data;
                circular_buf_get(cbuf, &data);
                printf("%u ", data);
        }
        printf("\n");
        printf("Full: %d, empty: %d, size: %zu\n",
                circular_buf_full(cbuf),
                circular_buf_empty(cbuf),
                circular_buf_size(cbuf));
        printf("\n******\nAdding %d values\n", EXAMPLE_BUFFER_SIZE + 5);

        free(buffer);
        circular_buf_free(cbuf);
        return 0;
}
```

Modeling Power

Power of constrained device is generally battery powered. There are two basic battery types:

- Single use and
- Rechargeable batteries.

   Single-use batteries have a finite life and need to be replaced. These include alkaline batteries and lithium batteries. Other single-use batteries include silver oxide and miniature lithium specialty batteries and zinc air hearing aid batteries. Rechargeable batteries, of course, can be recharged again and again—some of them up to 1000 times!

   In most of the purpose-built hardware construction for wider global deployment, generally, rechargeable batteries are not the best choice as availability of these batteries are limited and they are prohibitively expensive. Another important factor to consider is the *operating temperature ranges of the battery type.*

 ML Model computational and accuracy are sensitive to temperature variations and needs to be fine-tuned to Geolocations.

Lithium Battery: Please note the scope of the book is not on detailed battery technologies but to cover the impact of battery performance on the AI/ML model inference accuracies. Figure 37 [13] contains cross section view of a Lithium battery.[217]

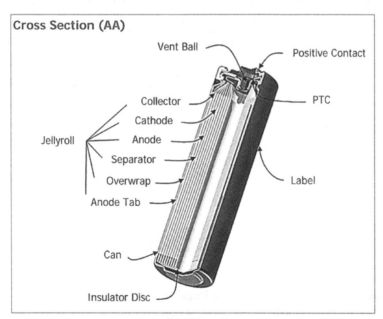

Figure 37: Lithium Battery

- Anode – lithium metal

- Cathode – iron disulfide on an aluminum foil substrate

- Separator – polyolefin

- Electrolyte – lithium salt/organic solvent

- Jellyroll construction – spiral wound multiple layers of

---

[217] Battery Cross Section Diagram - https://data.energizer.com/PDFs/LithiumL91l92_XSection.pdf

- anode/separator/cathode material to produce a high
- surface area for high power cell design
- vent ball – safety mechanism that provides internal
- pressure release
- positive and negative contact surfaces – nickel plated steel
- non-conductive plastic film label – electrically
  - insulates the battery
- Positive Temperature Coefficient (PTC) – over
  - current safety device

Temperature Effect on Performance

Lithium iron disulfide (LiFeS2) batteries have a much *lower sensitivity to temperature compared to other chemical systems.* The recommended operating temperature range is –40°C to +60°C (–40°F to +140°F) – mostly enough for all most all commercial and civil use cases. As with all battery systems, service life is reduced as the discharge temperature is lowered below room temperature (Temperature Effect on Capacity,[218] Figure 38).

Batteries generate power through chemical reactions, and these typically run much more slowly at lower temperatures. However, even at –40°C, the LiFeS2 batteries perform well at the rating drain 200 mA. LiFeS2 batteries can deliver approximately full rated capacity at –40°C if they are discharged at 25 mA. Thus, at these rates, the batteries give comparable performance over the entire 100°C operating range.

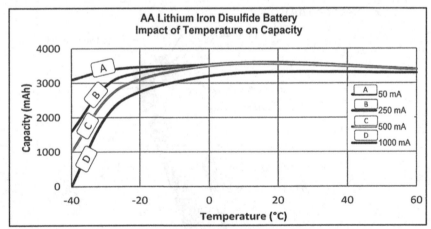

Figure 38: Temperature Performance

Cold Temperatures

Cold temperatures also lower the operating voltage thereby reducing the energy output.[219] Battery capacity is not lost due to cold temperature use, rather it is more difficult to access the battery's full potential due to the slowing of the electrochemical reactions, reducing capacity at high drain rates (please figure 39). The temperature has influence of Volt-Ampere characteristic [14] of semi-conductors and devices.[220]

$$I = Io \left(e \left(\frac{V}{nVt}\right) - 1\right)$$

The Volt equivalent of temperature (V_T) = $T/11{,}600$

[218] Cylindrical Primary Lithium Handbook and A pplication Manual - https://data.energizer.com/pdfs/lithium191192_appman.pdf
[219] Battery Cold Temperature Performance - https://data.energizer.com/pdfs/lithium191192_appman.pdf
[220] Integrated Electronics Analog and Digital - Mcgraw Hill Intl Eds; International Ed edition (June 1975)

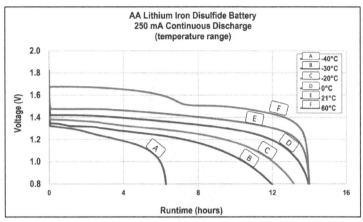

Figure 39: Temperature Performance

At room temperature, $V_T = 0.026$ V or 2.6 mV.

$$\mathbf{I_o}(T) = I_{01} \text{X } 2^{(T-T1)/10}$$

If the temperature is increased at a fixed voltage, the current increases. However, if we reduce V, then I may be bought back to its previous value [14].

Alkaline Battery

Since its commercial introduction in 1959, the Alkaline[221]-Manganese Dioxide battery has advanced to a dominant position in the portable battery market. This came about because the alkaline system is recognized to have several advantages over carbon zinc type batteries. Some of these advantages of alkaline chemistry over the basic carbon zinc chemistry are [15] (please see Figure 40):

- Higher energy density
- Superior service performance at all drain rates
- Superior cold temperature performance
- Lower internal resistance
- Longer shelf life
- Greater resistance to leakage

  Alkaline batteries are designed for long-lasting performance in the broadest range of device applications: A typical cylindrical alkaline battery has the following structure:[222]

- Cathode is a mixture of high purity electrolytic manganese dioxide and carbon conductor.
- Anode is a gelled mixture of zinc powder and electrolyte.
- Separators of specially selected materials prevent migration of any solid particles in the battery.
- Steel can confine active materials and serves as the cathode collector.
- Brass collector serves as the anode collector.
- Positive and negative covers provide contact surfaces of nickel-plated steel.
- Non-conductive plastic film label electrically insulates the battery.
- Nylon seal provides a safety venting mechanism.

---

[221] Alkaline Manganese Dioxide Handbook and Application Manual - https://data.energizer.com/pdfs/alkaline_appman.pdf
[222] Alkaline Battery - https://data.energizer.com/pdfs/alkaline_appman.pdf

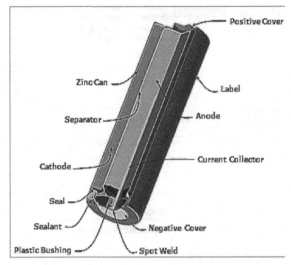

Figure 40: Battery Structure

Temperature Effects on Performance (please see Figure 41 and Figure 42)

The recommended operating temperature range for alkaline batteries is –18°C to 55°C. However, it is important to keep in mind that battery performance is still *impacted by temperature within the recommended range*. Performance of the battery is primarily dependent on how fast critical fuels, water and hydroxyl ions, can move and react in the battery. The mobility of ions is known as diffusion. For example: *maximum battery performance will not be achieved at cold temperatures*. As the temperature decreases, the diffusion of the fuels will decrease resulting in lower performance. Batteries will discharge more efficiently as the operating temperature is increased due to increase diffusion of the fuels. As the temperature is decreased, performance decreases accordingly. The lower temperature limit is determined in part by the temperature at which the electrolyte freezes. The Alkaline-Manganese Dioxide cell can operate at temperatures as low as –20°C , however this performance will be significantly lower in cold temperatures and their subsequent slowing of the chemical kinetics reactions impact the internal resistance (R_i) of the cell. *High drainage rates in cold environments will cause a large voltage drop due to the higher battery R_i.*

Applications using high drain rates (i.e., digital cameras) will be impacted more by cold temperatures than will light drain devices (i.e., MP3 players). No capacity is actually lost due to cold temperatures; rather it is more difficult to access the full potential of the battery due to the slowing of the electrochemical reactions.

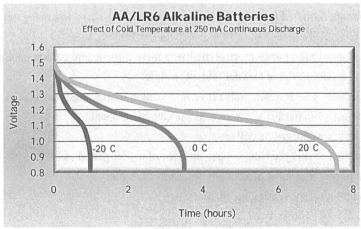

Figure 41: Battery Performance

As an example: if a flashlight is stored in a car on a very cold night, the flashlight would not be as bright or last as long when used. However, if this flashlight is taken indoors and allowed to warm up for several hours, its performance would return to normal. Warm temperatures can increase battery performance since ion mobility and the reaction rates are increased. A boost in battery performance can be observed in very high drain continuous applications that increase the battery temperature.

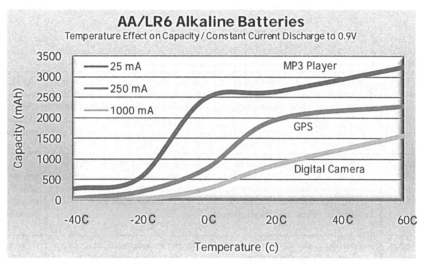

Figure 42: Temperature Effect on Capacity

Here are the characteristics of batteries:

	Lithium	Alkaline
Sizes Available	AA, AAA*, 9V	AA, AAA*, C, D, 9V
Composition	Batteries differ from alkaline batteries in chemistry and construction.	Alkaline batteries are available in a wide variety of sizes to fit most applications.
Shelf Life/ Usable Life	AA, AAA up to 20 years; 9V up to 10 years	AA, AAA, C, D up to 10 years; AAAA, N, 9V, 6V up to 5 years
Leakage Protection	Yes	Yes
Key Features	World's longest lasting AA and AAA batteries in high-tech devices	Protects your devices from leakage of fully used batteries up to 2 years (AA/AAA)
Type of Devices Best Used In	For every day and high-tech devices such as: Home Automation, Handheld Games, Digital Audio, Digital Cameras, and Wireless Game Controllers.	For everyday devices such as: Remote Control, Wireless Mouse, Flashlight, Clock, and Toys.

Connectivity: Connectivity mode influences battery power constraint. For no connectivity environments, it is imperative to have lower battery power consumption rates as compared to Wi-Fi & BLE powered connectivity environments, of course Wi-Fi connectivity enabled devices better served by electrical power. Other factors that influence workings of battery power are temperature sensitivity of a region or location.

Hardware Battery Power		Connectivity		
		No Connectivity	Bluetooth Low Energy (BLE)/LoRa	Wi-Fi
Model Accuracy	Self-Contained & Updated only on Hardware Refresh	Devices that are deployed in temperature sensitive regions , for instance extreme cold locations, the ML model should account for data outliers or perturbations of data or noise in the data due to error in readings that caused by temperature sensitivity.	Battery Power consumption rate is higher due to BLE advertisements.	Battery Power consumption rate is extremely high for Wi-Fi. It's recommended to use grid based electrical power for Wi-Fi.
	OTA Firmware	N/A	N/A	N/A
	Model Type: Eager Learner	AI/ML Model should account for temperature effects on performance of battery and factor in perturbations of sensor readings.	AI/ML Model should account for temperature effects on performance of battery and factor in perturbations of sensor readings.	AI/ML Model should account for temperature effects on performance of battery and factor in perturbations of sensor readings.
	Lazy Learner	The number of data points and the iteration counts will impact battery performance. Additionally, AI/ML Model should account for temperature effects on performance.	AI/ML Model should account for temperature effects on performance of battery and factor in perturbations of sensor readings.	AI/ML Model should account for temperature effects on performance of battery and factor in perturbations of sensor readings.
	Model Invocation	Since inline model invocation, exerts lower battery consumption.	Since inline model invocation, exerts lower battery consumption.	Since inline model invocation, exerts lower battery consumption.

⚡ **Real Time Clock & Battery Power**

As per the technical design specification of ARM Cortex-M3 microcontrollers, the Real Time Clock (RTC) typically fails[223] when $V_{i(BAT)}$ drops below 1.6 V. When you see Date values corruptions it could highly possible that the battery power is low.

AI/ML Model Accuracy: Based on the battery power constraint, in terms of operating temperature region and perturbation of sensor data capture, the tradeoff on Model accuracy and battery operating temperature need to be analyzed and properly taken in the design. For instance, IoT devices that are designed for temperature extreme conditions, cold or high temperature regions, the data capture by the sensor exhibit perturbation or noise effect. The data capture and model inference should account variations of temperature and should handle data outliers. Additionally, the ML model needs to account the location temperature sensitivity as part of the model inference.

Storage Modeling

Connectivity: Connectivity mode influences battery power constraint. For no connectivity environments, it is imperative to have lower battery power consumption rates as compared to Wi-Fi & BLE powered connectivity environments, of course Wi-Fi connectivity enabled devices better served by electrical power. Other factors that influence workings of battery power is temperature sensitivity of a region or location.

---

[223] NXP Semiconductors N.V. 2015 - https://www.nxp.com/docs/en/data-sheet/LPC1759_58_56_54_52_51.pdf

Hardware Storage		Connectivity		
		No Connectivity	Bluetooth Low Energy (BLE)/ LoRa	Wi-Fi
Model Accuracy	Self-Contained & Updated only on Hardware Refresh	Storage footprint needs to be higher to support model with no connectivity.	Periodic offload of the data to Edge devices reduces the need to have higher storage capacities.	Availability of network could offload Application the data Cloud data servers.
	OTA Firmware	N/A	N/A	N/A
	Model Type: Eager Learner	Storage capacities are lower for eager learners compared to Lazy Learner based systems.	Storage capacities are lower for eager learners compared to Lazy Learner based systems.	Storage capacities are lower for eager learners compared to Lazy Learner based systems.
	Lazy Learner	The number of data points and the iteration counts will exert pressure on the storage footprint.	Though data offload via connectivity is available, the number of data points and the iteration counts will exert pressure on the storage footprint.	Though data offload via Wi-Fi is available, the number of data points and the iteration counts will exert pressure on the storage footprint.
	Model Invocation	Since inline model invocation, exerts lower pressure on storage.	Since inline model invocation, exerts lower pressure on storage.	Since inline model invocation, exerts lower pressure on storage.

AI/ML Model Accuracy: Storage needs are higher for lazy learner-based inferences as compared to eager learner.

AI Democratization—"Crossing the Chasm"

Artificial Intelligence offers unprecedented innovation and growth opportunities, provides strategic and competitive differentiation, empowers developments of new businesses and blue-ocean products, offers actionable insights to untapped-unrealized social good, and empowers humanity and makes the world a better place not only for current inhabitants but also for our future generations. The rate at which AI (and its many subfields) is moving, many new techniques are becoming available from various different subfields,[224] it is clear that AI is here to stay with unprecedented promise of future to offer [16]. However, many of AI algorithms are only made to run on very powerful research workstations without considering how they can be used on real-world hardware, embedded constrained hardware. Machine learning in embedded systems specifically target embedded system to gather data, process data, and apply mathematical rules to produce insights. The embedded systems typically consist of low memory, low Ram, limited power compared to regular computers. Increase in factors such as processing power[225] leads to higher accuracies—the cost to bear is battery life.

McKinsey[226] noted numerous use cases across many domains where AI could be applied for social good. For these AI-enabled interventions to be effectively applied, several barriers must be overcome [17]. These include the usual challenges of *data, computing*, and *talent availability*, as well as more basic challenges of *access, infrastructure*, and *financial resources* that are particularly acute in remote or economically challenged places and communities. To successfully disseminate AI to masses and enable successful democratization, we need to bring rural communities to digital revolution (people, technology and data together). By addressing the following challenges by the private sector, purpose built hardware manufacturers, AI developers, Cloud providers and local and national governments [18], can achieve AI for all, a true Fourth Industrial Revolution:[227]

---

[224] Real-World AI on constrained hardware - https://limo.libis.be/primo-explore/fulldisplay?docid=LIRIAS2372477&context=L&vid= Lirias&search_scope=Lirias&tab=default_tab&lang=en_US&fromSitemap=1

[225] Overclocked sensors - https://techcrunch.com/2016/11/21/overclocked-smartwatch-sensor-uses-vibrations-to-sense-gestures-objects-and-locations/

[226] The promise and challenge of the age of artificial intelligence - https://www.mckinsey.com/featured-insights/artificial-intelligence/the-promise-and-challenge-of-the-age-of-artificial-intelligence

[227] How AI can increase the reach of India's public welfare programmes and make them efficient - https://economictimes.indiatimes.com/tech/software/how-ai-can-increase-the-reach-of-indias-public-welfare-programmes-and-make-them-efficient/articleshow/66841527.cms?from=mdr

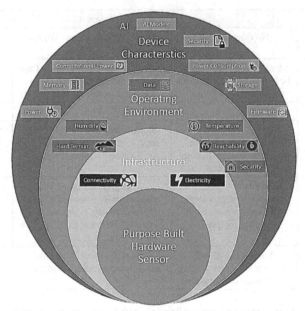

Figure 43: Democratization of AI for Small Devices (Class 0)

- Infrastructure Conditions (please see Figure 43)
- Operating Environment
- Device Characteristics

Please note: Infrastructure Conditions and Operating Environment constraints impose substantial design consideration of hardware and choice & pre-deployment of AI models. In some cases, AI models' accuracies need to be constrained to fit into hardware & firmware footprint.

Nevertheless, Customer Business Analytics, Service Level Agreement (SLA) and Tiered product offerings that are carefully carved to offer Software as a Service (SaaS) product SKUs with alacarte options to better serve customer on various product offerings per customer needs. Democratizing AI, we can address one of the perennial issues of the day—increasing the wealth gap (digital dystopia)[228] between rich and poor [19]. The chief reason for digital dystopia is lack of datasets that bring good to people on the bottom of the economic pyramid and that help people, generally senior citizens in rural areas, need social benefits and governance (see Table 1 for Use cases).

 "Large Scale Analytics, Machine Learning Algorithms, Artificial Intelligence Systems, Datasets & data assets that provide meaningful & actionable insights are national treasures and should be protected at any cost."

Infrastructure Issues

Lack of infrastructure [20] is deaccelerating the adoption of AI.[229] Infrastructure limitations such as availability of reliable electrical power (electricity), sustained and affordable network connectivity and availability of compute devices that operate in constrained environments are major impediments in successful dissemination and adoption of AI.

---

[228] Digital Dystopia - https://www.theguardian.com/technology/2019/oct/14/automating-poverty-algorithms-punish-poor
[229] The Long and Short of The Digital Revolution - https://www.imf.org/external/pubs/ft/fandd/2018/06/impact-of-digital-technology-on-economic-growth/muhleisen.htm

Table 1: Use Cases

Democratization of AI: Purpose Built Hardware (Small Device) Sensor Constraints Framework		
Type	Characteristics	Use Case
Operating Environment	Humidity Temperature Hard Terrain Reachability Security Lack of Data Availability	Sensor deployed in agriculture field
Infrastructure Conditions	Connectivity (intermittent or sustained) Electricity (Central Grid Power (Lack of))	Purpose built hardware deployed in rural areas where availability electricity & connectivity dismal
Device Characteristics	Computing Power Power Consumption Memory Storage Power (battery or electricity) AI Models (Software/Firmware) Firmware	Purpose built hardware with Hardware Economy vs. Device Form & Technical operating specification

Connectivity

Connectivity is essential for modern digital economy. We have entered a time when access to sustained broadband connectivity is a prerequisite for full participation in modern life. It is indispensable for education and health care, essential for business, the foundation for revolutionary techniques that will transform agriculture, and the key to participating in the cultural, social, and political life of the nation. As a new generation of solutions and capabilities built on artificial intelligence and the Internet of Things make their way into our day-to-day lives, broadband access will only become more important.

For farmers with broadband access, the ability to search for new customers, find buyers in more lucrative markets, and identify the most affordable sources of seeds, fertilizers, and farm equipment is boosting income. Broadband access is also essential for farmers to take advantage of a new generation of agricultural technologies built on sensors, data, and artificial intelligence that can help them conserve resources and increase yields. Having access to these new capabilities will improve rural prosperity and enable farmers to produce more food for a world that will add more than 2 billion people by 2050.[230]

Artificial Intelligence depends on large amounts of data being transferred, and the lack of phone reception and broadband makes this process impossible. Connecting our farmers to reliable, affordable, high-speed internet is the first step in supporting advancements in Artificial Intelligence in the agricultural industry.[231]

Yet, rural internet connectivity is still an issue. Rural America is still struggling for access to high-speed internet service.[232] Roughly, one-in-four rural residents [21] say access to high-speed internet is a major problem in their area.[233] In the developing world,[234] the situation is even dire—there is relatively sparse infrastructure in place to allow citizens of these areas to access the internet [22]. Moreover, even when there are connections available, many people in those regions simply cannot afford either the devices required or the account access.

[230] An update on connecting rural America - https://blogs.microsoft.com/uploads/prod/sites/5/2018/12/MSFT-Airband_InteractivePDF_Final_12.3.18.pdf
[231] Why rural internet is crucial to the advancement of AI - https://www.campbellpatterson.com/cpc-blog/2017/9/28/why-rural-internet-is-crucial-to-the-advancement-of-artificial-intelligence
[232] Rural America still struggling for access to high-speed internet service - https://www.foxnews.com/tech/southern-states-turn-to-electric-co-ops-to-speed-rural-internet
[233] About a quarter of rural Americans say access to high-speed internet is a major problem - https://www.pewresearch.org/fact-tank/2018/09/10/about-a-quarter-of-rural-americans-say-access-to-high-speed-internet-is-a-major-problem/
[234] Bringing the internet to the Developing World - https://cacm.acm.org/magazines/2018/7/229046-bringing-the-internet-to-the-developing-world/abstract

AI Model Design Considerations: Connectivity

Purpose built hardware deployed in resource-constrained areas where availability, electricity and connectivity dismal needs to adjust ML model design either incorporating complete model in the hardware or able to update model via Over the Air (OTA) firmware updates.

Electricity

Access to electricity is very stable across the world except in some pockets where electricity availability is sporadic due to seasonality. The hardware device may operate on electricity or powered by constrained source—"batteries". The purpose-built hardware or sensors operating under constrained environments needs to factor-in the AI model based on hardware—economy.

Operating Environment

Sensors are deployed worldwide with varying geolocations, territorial differences and environmental operation conditions. For example:

- Sensor deployed to measure watermarks in reservoirs deployed well below sea surface and very difficult to access
- Precision sensors that are deployed to measure health of cattle under constrained connectivity conditions and may not be reachable by human or AI agent easily
- Purpose built hardware deployed in agriculture settings subject to sturdy conditions with heavy noise, temperature and humidity variations, wear and tear and rough conditions.

AI Model Design Considerations: Operating Environmental Factors

The sturdiness of embedded systems and technical computational intensity and power needs are directly propositional to the environmental factors under which the real-time embedded systems operate. Some of the important environmental factors include:

Variations in Targeted Platform

Real world data is noisy and full of variations. Embedded ML systems under lab and real world have huge performance differences if variations in targeted platform not woven in the model design. During the training of the model, consider all real-world background conditions and account for perturbation.

Perturbations

Sensors and purpose-built hardware are subjected to varying electro-mechanical changes, ageing and calibration issues that could result in data perturbations. The ML model design and training must account all expected data and operational perturbations[235] to have a real-world AI model.

Thermal Characteristics

Let's consider the thermal effect on extremely constrained device (xCD) architecture for purpose built hardware—example would be agriculture sensors, dairy cow necklace, or automatic disease control dispensers for small healthcare settings. These purpose-built hardware build on top microcontroller architecture, in this case ARM Cortex-M3, for more detailed technical specification, please consider LPC1759/58/56/54/52/51

---

[235] Deep Learning: Perturbations and Diversity is All You Need - https://medium.com/intuitionmachine/deep-learning-perturbations-is-all-you-need-d630b6980587

are ARM Cortex-M3 technical specification.[236] Generally, the microcontrollers architecture-based applications featuring a high level of integration and low power consumption for the IoT or Edge based devices. The LPC 1759 operates at CPU frequencies of up to 100 MHz. The LPC1759 operates at CPU frequencies of up to 120 MHz. The ARM Cortex-M3 CPU incorporates a 3-stage pipeline and uses a Harward architecture with separate local instruction and data buses as well as a third bus for peripherals. The peripheral complement of the LPC1759/58/56/54/52/51 includes up to 512 kB of flash memory, up to 64 kB of data memory, Ethernet MAC, USB Device/Host/OTG interface, 8-channel general purpose DMA controller. Now let's look at the temperature effect on the operations of the MCU and its supported applications.

Extreme temperature and humidity will have performance and accuracy related issues with respect to purpose-built hardware. The average chip junction temperature, $T_j(°C)$, can be calculated using following formula:

$$T_j = T_{amb} + (P_D \times R_{th\,(j\text{-}a)})$$

where

T_j = Junction temperature

T_{amb} = Ambient temperature (°C) – range[237] –40°C to +85°C

$R_{th\,(j\text{-}a)}$ = Package junction-to-ambient thermal resistance (°C/W)

P_D = Sum of internal and I/O power dissipation

The internal power dissipation is the product of I_{DD} and V_{DD}. The I/O power dissipation of the I/O pins is often small, and many times can be negligible. However, it can be significant in some applications. The applications that operate with constrained resources in the environments that have huge swings in temperature and humidity. The direct negative affect of those perturbations will have adverse and performance impact on the extremely constrained devices (xCDs). In the following figure (please see Figure 44), the power dissipation above 20°C varies drastically for $I_{DD\,(REG)}$ and I_{BAT}, although one should see $I_{DD(REG)}$ supplied devices exhibit more battery drain.
where

- I_{DD} Regulator Supply Current (3.3 V)
- I_{BAT} Battery Supply Current
- V_{DD} Regulator Supply Voltage (3.3 V)

The extremely constrained devices (xCDs) design and AI/ML architecture must consider the thermal behavior to offset the effects of thermal on the device performance. As the developers of AI and ML for the extremely constrained devices (xCDs), it would be prudent and pragmatic to have the thermal behavior coefficients as part of the ship setting of the device. More on this, I will cover in Chapter 7—hierarchical cluster design.

 AI/ML & Environmental Consideration

Microcontroller operated purpose built hardware that operates in constrained environments and collects data under extreme environmental perturbations with Artificial Intelligence and Machine Learning algorithms deployed must have built-in error correction mechanism to offset data inaccuracies that show up when device battery power goes below 1.6 V. Generally, the Real Time Clock (RTC) typically fails[238] when $V_{i(BAT)}$ drops below 1.6 V.

---

[236] LPC 1759 Technical Specification - https://www.nxp.com/docs/en/data-sheet/LPC1759_58_56_54_52_51.pdf

[237] ARM Cortex – M3 MCU - https://www.nxp.com/docs/en/data-sheet/LPC1759_58_56_54_52_51.pdf

[238] NXP Semiconductors N.V. 2015 - https://www.nxp.com/docs/en/data-sheet/LPC1759_58_56_54_52_51.pdf

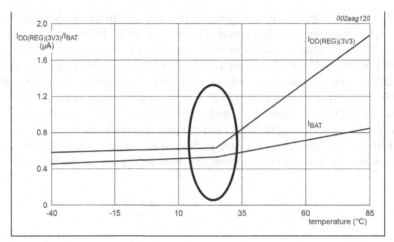

Figure 44: Deep Power-Down Mode: Typical Regulator Supply Current and Battery Supply Current Vs. Temperature

In the Figure 44, the data collected from agriculture fields—deployed extremely constrained Dairy Diagnostic device for Animal Husbandry. The sensor was deployed in 2016 and during the tail end of battery life (March 2017), the battery was below the operating threshold ($V_{i(bat)} < 1.6$ V). The voltage has generated Real Time Clock (RTC) error—you can see dates were wrong with June 30, 2065 (please see Table 2).

Table 2: Sensor Field Data

SensorTagID	Date	Time	Humidity
62CB2E1B-E964-4BCE-98BD-107D3D9F039E	12/29/2016	0:09	47.33
62CB2E1B-E964-4BCE-98BD-107D3D9F039E	12/29/2016	1:09	47.44
62CB2E1B-E964-4BCE-98BD-107D3D9F039E	12/29/2016	2:09	47.51
62CB2E1B-E964-4BCE-98BD-107D3D9F039E	12/29/2016	3:10	47.53
62CB2E1B-E964-4BCE-98BD-107D3D9F039E	12/29/2016	4:10	47.66
62CB2E1B-E964-4BCE-98BD-107D3D9F039E	12/29/2016	5:10	47.69
62CB2E1B-E964-4BCE-98BD-107D3D9F039E	12/29/2016	6:10	47.8
62CB2E1B-E964-4BCE-98BD-107D3D9F039E	12/29/2016	7:10	48.03
62CB2E1B-E964-4BCE-98BD-107D3D9F039E	12/29/2016	8:10	48.13
62CB2E1B-E964-4BCE-98BD-107D3D9F039E	1/20/2017	13:46	65.41
62CB2E1B-E964-4BCE-98BD-107D3D9F039E	1/20/2017	14:46	64.97
62CB2E1B-E964-4BCE-98BD-107D3D9F039E	2/6/2017	22:31	61.08
62CB2E1B-E964-4BCE-98BD-107D3D9F039E	6/30/2065	6:00	148
62CB2E1B-E964-4BCE-98BD-107D3D9F039E	6/30/2065	6:15	148
62CB2E1B-E964-4BCE-98BD-107D3D9F039E	6/30/2065	6:30	148
62CB2E1B-E964-4BCE-98BD-107D3D9F039E	6/30/2065	6:45	148
62CB2E1B-E964-4BCE-98BD-107D3D9F039E	6/30/2065	9:00	148
62CB2E1B-E964-4BCE-98BD-107D3D9F039E	6/30/2065	9:15	148
62CB2E1B-E964-4BCE-98BD-107D3D9F039E	6/30/2065	9:30	148
62CB2E1B-E964-4BCE-98BD-107D3D9F039E	6/30/2065	9:45	148

Data

Lack of data that could potentially help local businesses and societies are one of the most significant challenges in AI adoption. One limiting challenge is the availability of data for social good use cases, especially in rural communities. Lack of technologies, importantly, in the hands of the users exacerbate the issue. For instance, global penetration of internet in rural areas is very low[239] compared to suburban and urban areas and this is

---

[239] Digital gap between rural and nonrural America persists - https://www.pewresearch.org/fact-tank/2019/05/31/digital-gap-between-rural-and-nonrural-america-persists/

persisting a wider digital gap between rural and urban area. Lack of internet connectivity is causing inhibition of digital data services dissemination, resulting into sparse vital datasets capture for better governance purposes and preventing rural population to participate in the digital economy.[240]

> "Those who rule data will rule the entire world."
>
> Masayoshi Son[241] (phys.org)

Device Characteristics: Real-time, at the Edge, and a Reasonable Price Point

Real-time embedded systems offer huge advantages: gather business use case data, analyze in real-time, faster response, closed loop response. Nevertheless, price point is critical for success and adoption in the marketplace.

Most of the real-time embedded ML systems have one thing in common. They need to collect complex sensor streaming data—temperature, humidity, accelerometer, vibrations, sound, electrical and biometrical signals—in real-time to find specific signatures in data, outliers in data, anomaly detection, historical analysis to predict future pattern locally on device with the code that runs in firmware on a microcontroller with limited memory.[242] These embedded systems may have limited storage and be at mercy of intermittent network and connectivity. Many of the embedded ML systems are geographically deployed with attached limited power or functions on fixed set batteries. One most important point from device manufacturers is that the device should perform as indicated in technical specification with the useful lifetime of device must meet SLAs and should incur minimal service cost or truck roll numbers.

> **A new norm to build AI model-Interlace Sensor Data, albeit gather from Constrained device, with Survey Data**[243]
>
> Data is pristine! In my experience working on real agriculture fields across the world, I have discovered the ground truth that the World has very less agriculture data, especially for precision agriculture use cases. Perplexing observation to be noted, agriculture is one of the oldest professions and still lacks data. One way to address the deficiency is to collect Sensor data using precision sensors and interlace the data with the Survey or voice of the customer data [23]. The idea is to train models that link these large, but decontextualized, data with ground truth consumption or survey information, collected on small representative samples. This process of *developing proxies to link small, rich datasets with large, coarse ones can be viewed as a type of semi-supervised learning.*

> **Idea incubation**
>
> For successful flourishment of AI projects, idea incubation is very essential. As part of the incubation process, the idea and its potential use cases are discussed with the teams; both technical and business members. The goal of the idea incubation is to validate and evaluate the benefits presented by the idea to common people at large. Though commercials are essentials, the underpinning of a good AI project idea is based on the greater good and value it brings to humanity at large.

---

[240] A digital ASEAN for everyone - https://theaseanpost.com/article/digital-asean-everyone

[241] SoftBank CEO sees massive data, AI as key to future advances - https://phys.org/news/2017-07-softbank-ceo-massive-ai-key.html

[242] The Ultimate Guide to ML for Embedded Systems - https://reality.ai/ultimate-guide-to-machine-learning-for-embedded-systems/

[243] Tackling Climate Change with Machine Learning - https://arxiv.org/pdf/1906.05433.pdf

Chapter Summary:

After reading the chapter, you should fairly answer

- Design of custom-built hardware that operate under constrained compute environment
- The design view of building hardware, looking from three major hardware design characteristics: AI/ML Model Accuracy, Connectivity and Hardware
- Tradeoff modeling
- Environmental & Operating Constraints

References

1. Rob Thomas and Patrick McSharry. Big Data Revolution: What farmers, doctors and insurance agents teach us about discovering big data patterns. 1st Edition, Wiley, (March 2, 2015), ISBN-10: 9781118943717.
2. Rob van der Meulen. Edge computing promises near real-time insights and facilitates localized actions. October 3, 2018, URL:https://www.gartner.com/smarterwithgartner/what-edge-computing-means-for-infrastructure-and-operations-leaders/, Access Date: September 18, 2019.
3. David Cearley, Brian Burke, Mike Walker, Bob Gill and Thomas Bittman. Top 10 Strategic Technology Trends for 2018: Cloud to the Edge. 8 March 2018, URL: https://www.gartner.com/en/documents/3865403/top-10-strategic-technology-trends-for-2018-cloud-to-the0, Access Date: September 18, 2019.
4. Simone Cirani (Author), Gianluigi Ferrari (Author), Marco Picone (Author) and Luca Veltri (Author). Internet of Things: Architectures, Protocols and Standards. Wiley; 1 edn (November 5, 2018), ISBN-10: 1119359678.
5. Nagasai. Classification of IoT Devices. February 18, 2017 URL:https://www.cisoplatform.com/profiles/blogs/classification-of-iot-devices, Access Date : 22 November 2019.
6. Qusay F. Hassan, Atta ur Rehman Khan, Sajjad A. Madani (Editors). Internet of Things: Challenges, Advances, and Applications (Chapman & Hall/CRC Computer and Information Science Series) 1st Edition, Chapman and Hall/CRC; (January 12, 2018), ISBN-10: 9781498778510.
7. Øvrebekk, T. 2020. The Importance of Average Power Consumption to Battery Life. Blog.nordicsemi.com. URL: https://blog.nordicsemi.com/getconnected/the-importance-of-average-power-consumption-to-battery-life Accessed: 8 January 2020.
8. Silicon Laboratories. Maximizing Range and Battery Life in Low-Cost Wireless Networks. URL: "https://www.silabs.com/documents/public/white-papers/Maximize-Wireless-Network-Range-and-Battery-Life.pdf, Accessed: 8 January 2020.
9. Avishek Biswas. Energy-Efficient Smart Embedded Memory Design for IoT and AI. June 2018, URL: https://pdfs.semanticscholar.org/9edf/918044222ae684b7e3d7de434ddf7cbbec3f.pdf, Accessed: 8 January 2020.
10. megaAVR Data Sheet, Published: 2018 URL:http://ww1.microchip.com/downloads/en/DeviceDoc/ATmega48A-PA-88A-PA-168A-PA-328-P-DS-DS40002061A.pdf, Accessed: 8 January 2020.
11. Brian Bailey. AI Architectures Must Change. August 13, 2018 URL: https://semiengineering.com/ai-architectures-must-change/, Accessed: January 8, 2020.
12. Scott Thornton. What is DRAM (Dynamic Random Access Memory) vs SRAM. June 22, 2017 URL:https://www.microcontrollertips.com/dram-vs-sram/, Accessed: 8 January 2020.
13. Energizer. Battery Cross Sectional Drawing. June 2008, URL:https://data.energizer.com/PDFs/LithiumL91192_XSection.pdf, Accessed: 8 January 2020.
14. Jacob Millman. Integrated Electronics Analog and Digital. Published: International Edition, June, 1975 , ISBN-10: 0070854939 , Accessed: January 8, 2020.
15. Energizer. Alkaline Manganese Dioxide. June 2018. URL:https://data.energizer.com/pdfs/alkaline_appman.pdf, Accessed: January 8, 2020.
16. Van Ranst, W., Vennekens, J. (Supervisor), Goedemé, T. (Co supervisor). Real World Applications of Artificial Intelligence on Constrained Hardware. April 2019, URL:https://limo.libis.be/primo-explore/fulldisplay?docid=LIRIAS2372477&context=L&vid=Lirias&search_scope=Lirias&tab=default_tab&lang=en_US&fromSitemap=1 , Accessed: January 8, 2020.
17. James Manyika and Jacques Bughin. The Promise and Challenge of the Age of Artificial Intelligence. Oct. 2018, URL:https://www.mckinsey.com/featured-insights/artificial-intelligence/the-promise-and-challenge-of-the-age-of-artificial-intelligence,Accessed: January 8, 2020.
18. Sanjeev Sharma. How AI Can Increase the Reach of India's Public Welfare Programmes and Make Them Efficient. Nov. 28, 2018, URL:https://economictimes.indiatimes.com/tech/software/how-ai-can-increase-the-reach-of-indias-public-welfare-programmes-and-make-them-efficient/articleshow/66841527.cms?from=mdr , Accessed: 8 January 2020.

19. Ed Pilkington. Digital Dystopia: How Algorithms Punish the Poor. Oct. 14, 2019: URL:https://www.theguardian.com/technology/2019/oct/14/automating-poverty-algorithms-punish-poor, Accessed: January 8, 2020.

20. Martin Mühleisen. Finance & Development. June 2018, URL:https://www.imf.org/external/pubs/ft/fandd/2018/06/impact-of-digital-technology-on-economic-growth/muhleisen.htm, Accessed: January 8, 2020.

21. Monica Anderson. About a Quarter of Rural Americans Say Access to High-Speed Internet is a Major Problem. September 10, 2018, URL:https://www.pewresearch.org/fact-tank/2018/09/10/about-a-quarter-of-rural-americans-say-access-to-high-speed-internet-is-a-major-problem/ Accessed: 8 January 2020.

22. Keith Kirkpatrick. Bringing the Internet to the (Developing) World. Communications of the ACM, July 2018, Vol. 61 No. 7, Pages 20-21, URL: https://cacm.acm.org/magazines/2018/7/229046-bringing-the-internet-to-the-developing-world/abstract, Accessed: January 8, 2020.

23. Rolnick, David, Donti, Priya, L., Kaack, Lynn, H., Kochanski, Kelly, Lacoste, Alexandre, Sankaran, Kris, Slavin Ross, Andrew, Milojevic-Dupont, Nikola, Jaques, Natasha, Waldman-Brown, Anna, Luccioni, Alexandra, Maharaj, Tegan, Sherwin, Evan, D., Karthik Mukkavilli, S., Kording, Konrad, P., Gomes, Carla, Ng, Andrew, Y., Hassabis, Demis; Platt, John C.; Creutzig, Felix Chayes, Jennifer and Yoshua Bengio. Tackling Climate Change with Machine Learning. eprint arXiv:1906.05433, https://arxiv.org/pdf/1906.05433.pdf, June 2019 arXiv: arXiv:1906.05433 Bibcode: 2019arXiv190605433R.

## CHAPTER 5

Device Software and Hardware Engineering Tools

"It seems probable that once the machine thinking method had started, it would not take long to outstrip our feeble powers... They would be able to converse with each other to sharpen their wits. At some stage therefore, we should have to expect the machines to take control."

Alan Turing

This chapter introduces Machine Learning tools, Software Engineering tools, and Hardware Components. The chapter starts with Software Engineering Tools, Hardware Engineering Tools, and Libraries.

Software Engineering Tools

Software Engineering tools used for building Machine Learning models both on the Personal Computer and Cloud, deploying of Models into Firmware, Flashing Firmware to the device, and connecting the hardware to the Edge devices.

Machine Learning Tools

Anaconda

The open-source Anaconda Individual Edition[244] (formally Anaconda Distribution) is the easiest way to perform Python/R data science and machine learning on Linux, Windows, and Mac OS X. With over 19 million users worldwide, it is the industry standard for developing, testing, and training on a single machine, enabling individual data scientists to:

- Develop and train machine learning and deep learning models with scikit-learn, TensorFlow, and Theano
- Analyze data with scalability and performance with Dask, NumPy, pandas, and Numba
- Visualize results with Matplotlib, Bokeh, Datashader, and Holoviews.

Download: Anaconda Individual Edition can be downloaded from Anaconda Distribution. Please download for Python 3.7 version based Anaconda distribution.

URL: https://www.anaconda.com/distribution/

License: All rights reserved under the 3-clause BSD License - https://docs.anaconda.com/anaconda/eula/

Use in this book: We have Anaconda environment to build Python models. Additionally, used Jupyter Notebook and Spyder to build Machine Learning Models.

The following support toolsets from Anaconda Individual Edition (please see Figures 1,2):

---

[244] Anaconda - https://www.anaconda.com/distribution/

Figure 1: Anaconda Navigator

Figure 2: Anaconda tools

Jupyter Notebook

The Jupyter Notebook[245] is an open-source web application that allows you to create and share documents that contain live code, equations, visualizations and narrative text. Uses include data cleaning and transformation, numerical simulation, statistical modeling, data visualization, machine learning, and much more.

Download: Jupyter Notebook can be download from https://jupyter.org/install. You can refer Getting started with the classic Jupyter Notebook.

URL: https://jupyter.org/install

License: Project Jupyter is a non-profit, open-source project, born out of the IPython Project in 2014 as it evolved to support interactive data science and scientific computing across all programming languages. Jupyter will always be 100% open-source software, free for all to use and released under the liberal terms of the modified BSD license.

---

[245] Jupyter Notebook - https://jupyter.org/

Use in this book: Jupyter Notebook was used to create Supervised and Unsupervised Machine Learning Models both on personal desktop and Cloud software.

Spyder

Spyder[246] is a powerful scientific environment written in Python, for Python, and designed by and for scientists, engineers and data analysts. It offers a unique combination of the advanced editing, analysis, debugging, and profiling functionality of a comprehensive development tool with the data exploration, interactive execution, deep inspection, and beautiful visualization capabilities of a scientific package.

Beyond its many built-in features, its abilities can be extended even further via its plug in system and API. Furthermore, Spyder can also be used as a PyQt5 extension library, allowing developers to build upon its functionality and embed its components, such as the interactive console, in their own PyQt software.

Download: Spyder install was performed through Anaconda Distribution.

URL: https://www.anaconda.com/distribution/

License: Project Jupyter is a non-profit, open-source project, born out of the IPython Project in 2014 as it evolved to support interactive data science and scientific computing across all programming languages. Jupyter will always be 100% open-source software, free for all to use and released under the liberal terms of the modified BSD license.

Use in this book: Spyder was used to create Supervised and Unsupervised Machine Learning Models both on personal desktop.

Android Studio

Android Studio is the official integrated development environment for Google's Android operating system, built on JetBrains' IntelliJ IDEA software and designed specifically for Android development. It is available for download on Windows, macOS and Linux based operating systems.

Download: 3.6.1 for Windows 64-bit (749 MB) can be downloaded from following URL.

URL: https://developer.android.com/studio

License: https://developer.android.com/studio/terms

Use in this book: Mobile Edge software was built using Android Studio for Android Mobile deployments.

Google Colaboratory

Google Colaboratory or Co-Lab is a free Jupyter notebook environment that requires no setup and runs entirely in the cloud (please see Figure 3).

With Colaboratory you can write and execute code, save and share your analyses, and access powerful computing resources, all for free from your browser.

Download: Since Google Colaboratory is a Cloud based Jupyter environment, no need to install any specific Software on to the Desktop.

URL: https://colab.research.google.com/notebooks/welcome.ipynb#scrollTo=-Rh3-Vt9Nev9

License: https://research.google.com/colaboratory/faq.html

Use in this book: Google Co-Lab was used to build Machine Learning Models.

---

[246] Spyder - https://www.spyder-ide.org/

Figure 3: Google Co-Lab

Microsoft Azure Machine Learning

Azure Machine Learning empowers developers and data scientists with a wide range of productive experiences for building, training, and deploying Machine Learning models faster. Accelerates time to market and foster team collaboration with industry-leading MLOps-DevOps for machine learning. Innovates on a secure, trusted platform, designed for responsible AI (please see Figure 4).

Download: Since Microsoft Azure Machine Learning is a Cloud based ML platform, there is no need to install any specific software on to the desktop. You can sign up for a free (if available) subscription.

URL: https://azure.microsoft.com/en-us/services/machine-learning/#product-overview

License: https://research.google.com/colaboratory/faq.html

Use in this book: Microsoft Azure was used to build Machine Learning models.

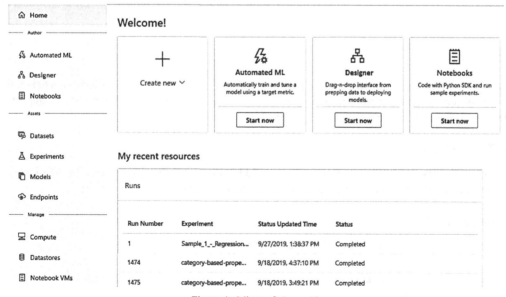

Figure 4: Microsoft Azure AI

Azure Databricks

Azure Databricks[247] is an Apache Spark-based analytics platform optimized for the Microsoft Azure cloud services platform (please see Figure 5). Designed with the founders of Apache Spark, Databricks is integrated with Azure to provide one-click setup, streamlined workflows, and an interactive workspace that enables collaboration between data scientists, data engineers, and business analysts.

Download: Since Microsoft Azure Databricks is a Cloud-based ML platform, there is no need to install any specific software on to the desktop. You can sign up for a free (if available) subscription.

URL: https://docs.microsoft.com/en-us/azure/azure-databricks/what-is-azure-databricks

License: https://research.google.com/colaboratory/faq.html

Use in this book: Azure Databricks was used to build Machine Learning models.

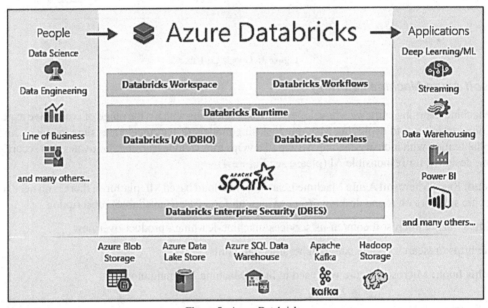

Figure 5: Azure Databricks

TensorFlow

TensorFlow[248] is an end-to-end open source platform for machine learning. It has a comprehensive, flexible ecosystem of tools, libraries and community resources that lets researchers push the state-of-the-art in ML and developers easily build and deploy ML powered applications.

Install TensorFlow for C

TensorFlow provides a C API that can be used to build bindings for other languages. The API is defined in c_api.h and designed for simplicity and uniformity rather than convenience.

Supported Platforms

TensorFlow for C is supported on the following systems:

---

[247] What is Azure Databricks? - https://docs.microsoft.com/en-us/azure/azure-databricks/what-is-azure-databricks
[248] TensorFlow - https://www.tensorflow.org/

- Linux, 64-bit, x86
- macOS X, Version 10.12.6 (Sierra) or higher
- Windows, 64-bit x86

We have used TensorFlow to build models for our machine learning models and deployed to the embedded hardware.

TensorFlow Lite for Microcontrollers

TensorFlow Lite for Microcontrollers[249] is an experimental port of TensorFlow Lite designed to run machine learning models on microcontrollers and other devices with only kilobytes of memory.

It doesn't require operating system support, any standard C or C++ libraries, or dynamic memory allocation. The core runtime fits in 16 KB on an Arm Cortex M3, and with enough operators to run a speech keyword detection model, takes up a total of 22 KB.

Hardware and Engineering Tools

Eclipse IDE for C/C++ Developers

The Eclipse[250] Foundation provides our global community of individuals and organizations with a mature, scalable, and business-friendly environment for open source software collaboration and innovation. The Foundation is home to the Eclipse IDE, Jakarta EE, and over 350 open source projects, including runtimes, tools, and frameworks for a wide range of technology domains such as the Internet of Things, automotive, geospatial, systems engineering, and many others.

Eclipse IDE for C/C++ (please see Figure 6) is used to build Firmware for boards ATmega and San Jose State Board for building of Machine Learning Modules into Device Firmware and the deployment/Flash of Firmware into the device.

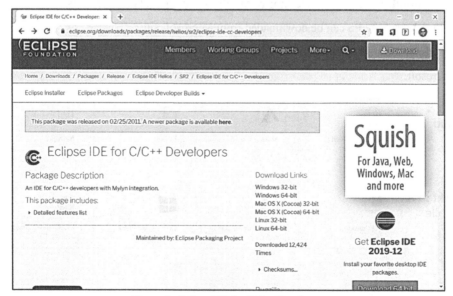

Figure 6: Eclipse IDE for C/C++ Developers

---

[249] TensorFlow for Microcontrollers - https://www.tensorflow.org/lite/microcontrollers
[250] Eclipse - https://www.eclipse.org/org/

Download: Download Eclipse IDE for C/C++ Developers.

URL: https://www.eclipse.org/downloads/packages/release/helios/sr2/eclipse-ide-cc-developers

License: Public License v1.0. Url: https://www.eclipse.org/legal/epl-v10.html

Use in this book: Eclipse for C/C++ developers is used to build firmware for iDispenser and Dairy products.

Scratch for Arduino

Scratch for Arduino (S4A) is a modified version of Scratch, ready to interact with Arduino boards. It was developed in 2010 by the Citilab Smalltalk Team and it has been used since by many people in a lot of different projects around the world. There is also a sensor report board similar to the PicoBoard one.

The main aim of the project is attracting people to the programming world. The goal is also to provide a high-level interface to Arduino programmers with functionalities such as interacting with a set of boards through user events.

Download: Download Scratch for Arduino.

URL: http://s4a.cat/

License: S4A is free software distributed under an MIT license.

Use in this book: We have used Scratch for Arduino for building purpose built hardware Sensor.

Microsoft Visual Studio 2019

Microsoft Visual Studio[251] 2019 (please see Figure 7) is a best-in class Integrated Development Environment for building Microsoft Web & Cloud, Desktop & Mobile, Gaming, and other toolsets.

Download: Free Visual Studio can be downloaded from Microsoft site.

URL: https://visualstudio.microsoft.com/free-developer-offers/

License: Microsoft.

Figure 7: Microsoft Visual Studio 2019

---

[251] Microsoft Visual Studio - https://visualstudio.microsoft.com/vs/

Use in this book: We have used Microsoft Visual Studio to build C/C++ code & Machine Learning Code (conversion of Python Models in C).

Cortex M3 Processor

We have used Cortex-m3 as ARM family Target Processor (please see Figure 8).

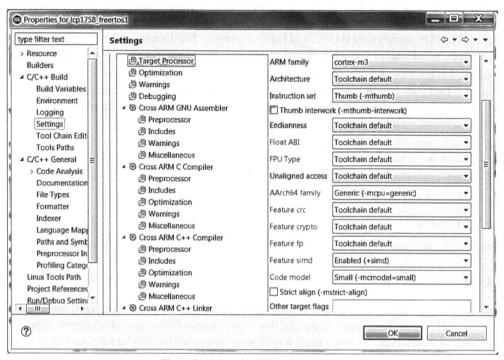

Figure 8: Eclipse GNU Compiler Settings

GNU Arm Embedded Toolchain

arm Developer

The tools we used to compile this project are: Eclipse IDE, arm-none-eabi GNU embedded tool chain, doxygen, graphviz and hyperload.

Pre-built GNU toolchain for Arm Cortex-M and Cortex-R processors

The GNU Arm Embedded toolchain contains integrated and validated packages featuring the Arm Embedded GCC compiler, libraries and other GNU tools necessary for bare-metal software development on devices based on the Arm Cortex-M and Cortex-R processors. The tool chains are available for cross-compilation on Microsoft Windows, Linux and Mac OS X host operating systems.

GNU C/C++ Compiler

You can find the sources to Arm Embedded GCC under svn://gcc.gnu.org/svn/gcc/branches/ARM/. All contributions are made to trunk and patches are cherry-picked on a need basis to the Arm Embedded branches.

URL: https://developer.arm.com/tools-and-software/open-source-software/developer-tools/gnu-toolchain/gnu-rm

License: GNU Public.

Use in this book: We have used GNU Arm Embedded tool to build the real-time embedded code.

Doxygen

Doxygen is the de facto standard tool for generating documentation from annotated C++ sources, but it also supports other popular programming languages such as C, Objective-C, C#, PHP, Java, Python, IDL (Corba, Microsoft, and UNO/OpenOffice flavors), Fortran, VHDL, Tcl, and to some extent D.

Doxygen can help you in three ways:

1. It can generate an on-line documentation browser (in HTML) and/or an off-line reference manual from a set of documented source files. There is also support for generating output in RTF (MS-Word), PostScript, hyperlinked PDF, compressed HTML, and Unix man pages. The documentation is extracted directly from the sources, which makes it much easier to keep the documentation consistent with the source code.

2. You can configure doxygen to extract the code structure from undocumented source files. This is very useful to quickly find your way in large source distributions. Doxygen can also visualize the relations between the various elements by means of include dependency graphs, inheritance diagrams, and collaboration diagrams, which are all generated automatically.

3. You can also use doxygen for creating normal documentation. Doxygen is developed under Mac OS X and Linux,but is set-up to be highly portable. As a result, it runs on most other Unix flavors as well. Furthermore, executables for Windows are available.

URL: http://www.doxygen.nl/

License: GNU Public.

Use in this book: We have used GNU Arm Embedded tool to build the real-time embedded code.

Hyperload

Hyperload (please see Figure 9) is a program based on Universal Asynchronous Receiver/Transmitter (UART) to program a CPU at a very high speed. Typically, this is 10–20 times faster than using other flash programming programs such as FlashMagic. This is a good way to write your own bootloader for your CPU.

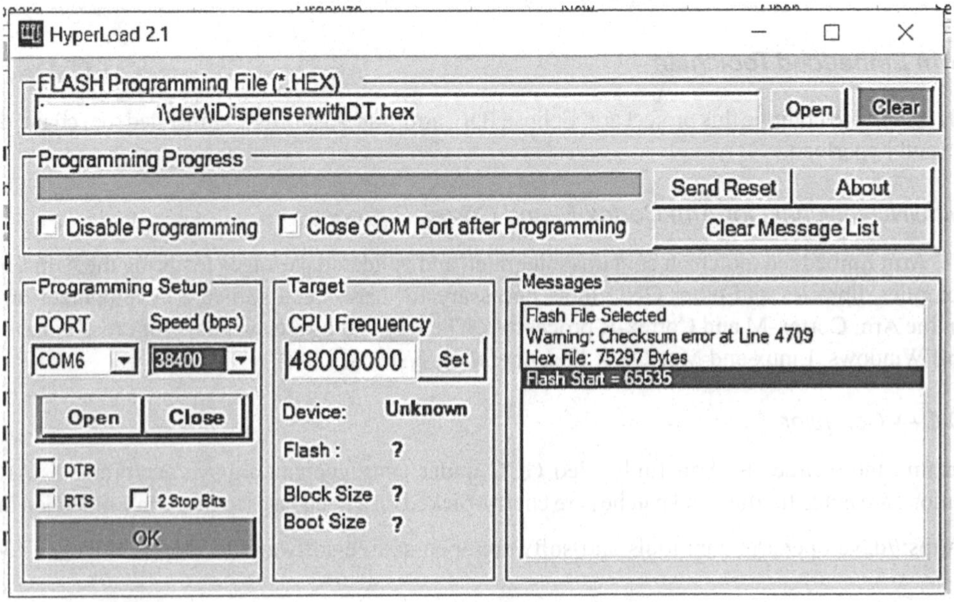

Figure 9: HyperLoader

Libraries

C Language – C11

The C Complier we have used is C11 – C11 is the current and latest standard of the C programming language and, as the name suggests, this standard was adopted in 2011. The formal document describing the C11 standard is called ISO/IEC 9899:2011.[252]

Here is the list of C languages: K&R C, ANSI C, C99, C11 and Embedded C

C Language	Description
Kernighan and Ritchie (K&R) C	C was developed by Dennis Ritchie between 1969 and 1973 at Bell Labs. K&R C not only acted as an informal standards specification for C but also added language features like the new data types long int and unsigned int and the compound assignment operator. A standardized I/O library was also proposed by K&R C.
ANSI C	The American National Standards Institute (ANSI) developed standards draft which was finally approved in 1989 – C89. ANSI C is dependent on the POSIX standard. POSIX is the Portable Operating System Interface, a family of standards specified by the IEEE Computer Society for maintaining compatibility between operating systems. The same standard proposed by ANSI was adopted by ISO officially as ISO/IEC 9899:1990 in 1990. This is the reason why ANSI C is also called as ISO C and C90.
C99	C99 is the informal name given to the ISO/IEC 9899:1999 standards specification for C that was adopted in 1999.
C11	C11 is the current and latest standard of the C programming language and, as the name suggests, this standard was adopted in 2011. The formal document describing the C11 standard is called ISO/IEC 9899:2011.
Embedded C	The Embedded C standard was proposed to customize the C language in such a way that it can cater to the needs of embedded system programmers.

GNU GCC Compiler

The GNU Compiler Collection[253] (please see Figure 10) includes front ends for C, C++, Objective-C, Fortran, Ada, Go, and D, as well as libraries for these languages (libstdc++,...). GCC was originally written as the compiler for the GNU operating system. The GNU system was developed to be 100% free software, free in the sense that it respects the user's freedom.

The Compiler Options

We have used GCC Option O$_s$—Optimize Size (see Figure 11) to build embedded modules—iDispenser Code.
The original version of this answer stated there were 7 options. GCC has since added -Og to bring the total to 8. From the main page:[254]

- -O (Same as -O1)
- -O0 (do no optimization, the default if no optimization level is specified)
- -O1 (optimize minimally)
- -O2 (optimize more)
- -O3 (optimize even more)
- -Ofast (optimize very aggressively to the point of breaking standard compliance)

---

[252] Different C Standards: The Story of C - https://opensourceforu.com/2017/04/different-c-standards-story-c/
[253] GCC, the GNU GCC Compiler Collection - https://gcc.gnu.org/
[254] Main Page - https://linux.die.net/man/1/gcc

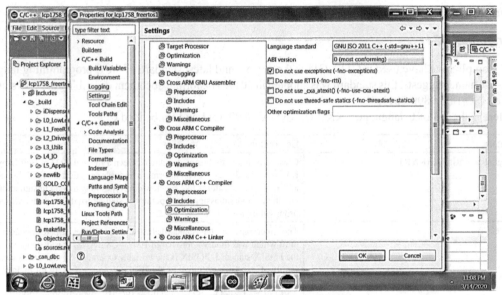

Figure 10: GNU Compiler Options - optimizations

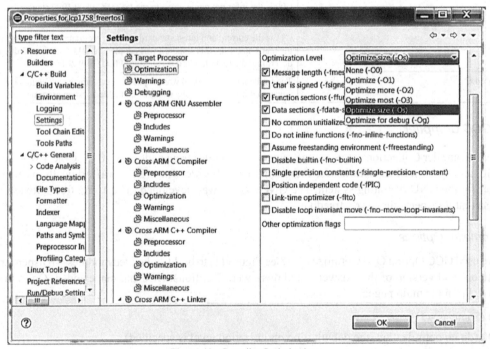

Figure 11: Compiler Optimizations

- -Og (Optimize debugging experience. -Og enables optimizations that do not interfere with debugging. It should be the optimization level of choice for the standard edit-compile-debug cycle, offering a reasonable level of optimization while maintaining fast compilation and a good debugging experience)

- -Os (Optimize for size. -Os enables all -O2 optimizations that do not typically increase code size. It also performs further optimizations designed to reduce code size. -Os disables the following optimization flags: -falign-functions -falign-jumps -falign-loops -falign-labels -freorder-blocks -freorder-blocks-and-partition -fprefetch-loop-arrays -ftree-vect-loop-version)

There may also be platform specific optimizations, as @pauldoo notes, OS X has -Oz. In summary, the compiler's optimization levels include.[255]

Option	Optimization Level	Execution Time	Code Size	Memory Usage	Compile Time
-O0	optimization for compilation time (default)	+	+	-	-
-O1 or -O	optimization for code size and execution time	-	-	+	+
-O2	optimization more for code size and execution time	--		+	++
-O3	optimization more for code size and execution time	---		+	+++
-Os	optimization for code size		--		++
-Ofast	O3 with fast none accurate math calculations	---		+	+++

+increase, ++increase more, +++increase even more, -reduce, --reduce more, ---reduce even more

Microsoft Visual Studio—C++ Compiler

A Visual Studio project[256] is a project based on the MSBuild build system. MSBuild is the native build system for Visual Studio and is generally the best build system to use for Windows-specific programs. MSBuild is tightly integrated with Visual Studio, but you can also use it from the command line (please see Figure 12).

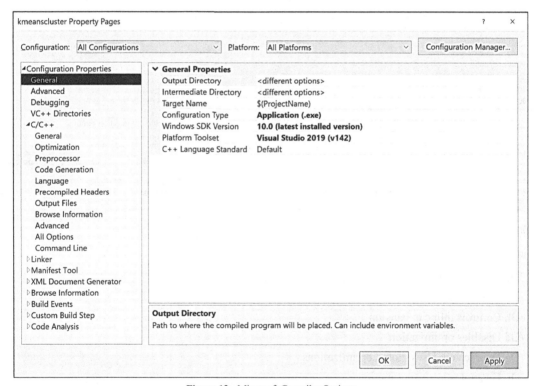

Figure 12: Microsoft Compiler Options

MSVC Compiler Options

C++ compiler and linker options are located under the C/C++ and Linker nodes in the left pane under Configuration Properties. These translate directly to command-line options that will be passed to the compiler.

[255] GCC Compiler Options - https://www.rapidtables.com/code/linux/gcc/gcc-o.html
[256] Visual Studio – C++ Projects - https://docs.microsoft.com/en-us/cpp/build/creating-and-managing-visual-cpp-projects?view=vs-2019

MSVC Compiler Options[257]

cl.exe is a tool that controls the Microsoft C++ (MSVC) C and C++ compilers and linker. cl.exe can be run only on operating systems that support Microsoft Visual Studio for Windows (please see Figure 13).

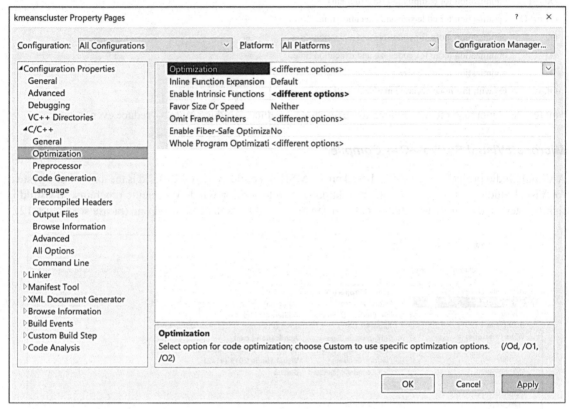

Figure 13: Microsoft Visual Studio C Compiler Optimization

Optimization

Option Purpose

- /O1 Creates small code
- /O2 Creates fast code
- /Ob Controls inline expansion
- /Od Disables optimization
- /Og Deprecated. Uses global optimizations
- /Oi Generates intrinsic functions
- /Os Favors small code
- /Ot Favors fast code
- /Ox A subset of /O2 that doesn't include /GF or /Gy
- /Oy Omits frame pointer (x86 only)

```
Custom
Disabled (/Od)
Maximum Optimization (Favor Size) (/O1)
Maximum Optimization (Favor Speed) (/O2)
Optimizations (Favor Speed) (/Ox)
```

After reading this chapter, you should be able understand the tools required and to practice tools described in the chapter. All the tools used in this chapter, you should able to download for free. Tools include:

- Anaconda
- Microsoft Visual Studio
- Jupyter Integrated Development Environment
- Eclipse for C/C++
- Android Integrated Development Environment

References

1. Anaconda. Anaconda Individual Edition. March 15. 2020. https://www.anaconda.com/, Access Date:September 18, 2019.
2. The Jupyter Notebook. Jupyter Development Environment. Jan 28, 2020, URL: https://jupyter.org/, Access Date: September 18, 2019.
3. Spyder. The Scientific Python Development Environment. Date: 2018. URL: https://www.spyder-ide.org/. Access Date: Jan. 8, 2020.
4. Android Developer Studio. Developer Platform. https://developer.android.com/studio, Access Date: Jan 8, 2020.
5. Google Colaboratory. Google Colaboratory Introduction. https://colab.research.google.com/notebooks/basic_features_overview.ipynb, Access Date: Jan 8, 2020.
6. Microsoft Azure. Azure Machine Learning. https://azure.microsoft.com/en-us/services/machine-learning/, March 15, 2020.
7. Microsoft. What is Azure Databricks? May 8, 2019. https://docs.microsoft.com/en-us/azure/azure-databricks/what-is-azure-databricks, Access Date: September 18, 2019.
8. TensorFlow. An End-To-End Open Source Machine Learning Platform. https://www.tensorflow.org/, Access Date: March 15, 2020.
9. Eclipse. Eclipse IDE for C/C++ Developers. https://www.eclipse.org/downloads/packages/release/helios/sr2/eclipse-ide-cc-developers, Access date : March 15, 2020.
10. S4A. Scratch For Arduino. http://s4a.cat/, Acess Date: March 15, 2020.
11. Microsoft. Visual Studio 2019. https://visualstudio.microsoft.com/vs/, Access Date: November 22, 2019.

SECTION-III

Model Development and Deployment

CHAPTER 6

Supervised Models

"Genius is one percent inspiration, ninety nine percent perspiration."

Thomas Alva Edison

The chapter starts with Supervised models design and apply Model Connectivity Hardware Design as part of the Machine Learning models. Supervised models discussed as part of the chapter includes Decision Trees, Random Forest, Adaptive Boosting, Extreme Gradient Boosting (XGB), Linear Regression Models, and Kalman Filter. Next, each of the Supervised models discussed as part of the chapter analyzed from constrained environment perspective and the goes in depth to optimize the Machine Learning model to be effectively functional in the constrained environment. The constrained environment modeling includes Hardware Connectivity Trade-offs, Connectivity Model Trade-offs, and Hardware Connectivity Trade-offs.

By following the procedures and the frameworks described as part of the chapter, the Machine Learning and Embedded Engineers can develop Artificial Intelligence models and analyze holistically to successfully deploy in resource constrained environments.

Decision Trees

During the late 1970s and early 1980s, J. Ross Quinlan, a researcher in Machine Learning, developed a decision tree algorithm known as ID3 (Iterative Dichotomiser). This work expanded on earlier work on concept learning systems, described by E. B. Hunt, J. Marin, and P. T. Stone. Quinlan later presented C4.5 (a successor of ID3), which became a benchmark to which newer supervised learning algorithms are often compared. In 1984, a group of statisticians (L. Breiman, J. Friedman, R. Olshen, and C. Stone) published the book *Classification and Regression Trees CART*), which described the generation of binary decision trees. ID3 and CART were invented independently of one another at around the same time yet follow a similar approach for learning decision trees from training tuples [1]. These two cornerstone algorithms spawned a flurry of work on decision tree induction [1].

ID3, C4.5, and CART adopt a greedy (i.e., nonbacktracking) approach in which decision trees are constructed in a top down recursive divide and conquer manner. Most algorithms for decision tree induction also follow such a top down approach, which starts with a training set of tuples and their associated class labels. The training set is recursively partitioned into smaller subsets as the tree is being built [1].

Decision trees are a simple, but powerful form of multiple variable analyses. They provide unique capabilities to supplement, complement, and substitute for [2]:

- traditional statistical forms of analysis (such as multiple linear regression)
- a variety of data mining tools and techniques (such as neural networks)
- recently developed multidimensional forms of reporting and analysis found in the field of business intelligence.

Decision trees are produced by algorithms that identify various ways of splitting a data set into branch-like segments. These segments form an inverted decision tree that originates with a root node at the top of the tree [2]. A decision tree algorithm is used to classify the attributes and select the outcome of the class attribute. To construct a decision tree, both class attribute and item attributes are required. A decision tree is a tree-like structure wherein the intermediate nodes represent attributes of the data, the leaf nodes represent the outcome of the data and the branches hold the attribute value. Decision trees are widely used in the classification process because no domain knowledge is needed to construct the decision tree. Figure 1 shows some simple decision trees [1].

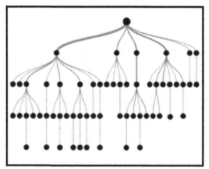

Figure 1: Decision Tree

An attribute selection measure is a heuristic for selecting the splitting criterion that "best" separates a given data partition, D, of class labeled training tuples into individual classes. If we were to split D into smaller partitions according to the outcomes of the splitting criterion, ideally each partition would be pure (i.e., all of the tuples that fall into a given partition would belong to the same class). Attribute selection measures are also known as splitting rules because they determine how the tuples at a given node are to be split [1]. The following two are applied to derive attribute selection:

- Information Gain
- Gain Ration

The primary first step in the decision tree algorithm is to identify the root node for the given set of data. Multiple methods exist to decide the root node of the decision tree. Information gain and Gini impurity are the primary methods used to identify the root node. The root node plays an important role in deciding which side of decision tree the data falls into. Like every classification method, decision trees are also constructed using the training data and tested with the test data.

Let node N represent or hold the tuples of partition D. The attribute with the highest information gain is chosen as the splitting attribute for node N. This attribute minimizes the information needed to classify the tuples in the resulting partitions and reflects the least randomness or "impurity" in these partitions. Such an approach minimizes the expected number of tests needed to classify a given tuple and guarantees that a simple (but not necessarily the simplest) tree is The expected information needed to classify a tuple in D is given by Information Gain.

Information Gain: Information gain is used to identify the root node and the branch nodes in the decision tree. Information gain is calculated using entropy and information. The purpose of the information gain calculation is to identify the node that has the least randomness or impurity. Entropy is calculated using the following formula

$$Info\ (D) = -\sum_{i=1}^{m} pi\ \log_2(pi) \tag{Eq. (1)}$$

Info(D) is also known as Entropy, a measure of disorder or impurity.
Information of the attribute is calculated using the following formula:

$$Info_A\ (D) = \sum_{j=1}^{v} \frac{|Dj|}{|D|} x\ info(Dj) \tag{Eq. (2)}$$

The information gain of an attribute is the difference between entropy and the information of that attribute. The attribute with the highest information gain is the root node, and the next level nodes are identified using the next highest information gain attributes.

$$Gain\ (A) = Info\ (D) - Info_A(D) \tag{Eq. (3)}$$

In the following section, decision tree technique is used to calculate temperature sensitivity coefficient for adaptive measuring purposes to detect influence of temperatures on electronic devices, a specific use case from a Diagnostic measuring sensor.

Managing Temperature Effects in Edge IoT Deployments and Adaptive Coefficients

Temperature has been known to cause changes in materials as far back as the early stone age some 800,000 years ago [3]. One of the most important temperature-related material discoveries was in 1833, when Michael Faraday discovered that silver sulfide (Ag_2S) was conductive at high temperatures and nearly insulating at low temperatures [3]. This was in direct contrast to the temperature dependence observed in materials, which become less conductive as temperature increased [3].

The findings of Faraday and other scientists have laid foundation for the impact of temperature in and on electronic systems. The main factor of the temperature on the electronic systems on the power dissipation and power density makes it among the most important of factors constraining electronic and nanoscale design [3].

Temperature Variations and Sensor Data Errors

There are two types of temperature variations that affect system performance: global temperature variations and local temperature variations [3]. Global temperature variations are caused by changes in ambient temperature or changes in cooling capacity. Increasing global chip temperatures will cause higher CPU clock periods, resulting in functional errors [3].

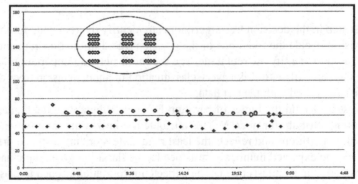

Figure 2: Sensor Data

The local temperature (also called intra-die temperature variations [3]), can result in communication errors between with a large temperature differential. Intra-die temperature exceeding 50°C have been reported [3] to causes micro-chip performance behavior due to temperature differential between a microprocessor core and an on-chip cache [3].

The following data captured by CLASS 10 Medical Diagnostics Wearable Sensor—"Cow Necklace". The Cow Necklace is a veterinary sensor for use in capturing a cow's vital signs, providing data to the farmer to monitor the cow's milk productivity, and improving its overall health and it is designed to work on extremely constrained environments and use the Internet of Things technologies. The sensors are deployed in fields of agriculture worldwide and the data captured is specifically from India.

The above figure (please see Figure 2) depicts Dairy Sensor data capture in the agriculture fields in Punjab and Andhra Pradesh, India. Dairy Sensor captures temperature, humidity and activity level details from dairy farms by connecting it to the neck of cattle. For some dates, the data humidity values are reported more than 140%. One reason could have been (a) corruption of value reading at the sensor due to variations in temperature and humidity due to geolocation, in this case Patiala, Punjab (please note we did not observe the same behavior in Vizag—a coastal town in Andhra Pradesh), or (b) could be due to voltage corruption. In order to weed out temperature and humidity variations, we have developed adaptive engineering data design coefficient that factors in geolocation thermal and extreme temperature changes to balance Machine Learning Algorithm computations. Second, the data collection of the sensor records show the dates are incorrect, a factor reveals the microcontroller battery running low.

Design of Adaptive System to Autocorrect

Understanding and managing temperature effects in Edge IoT devices deployment is crucial for the successful working of the devices that are distributed across different geolocations with varying temperature and humidity conditions.

In order to improve the device performance against thermal and environmental perturbations, combination supervised learning algorithms, such as decision trees, are used so as to make the device an adaptive edge to factor-in the environmental fluctuations. For instance, we have collected the following temperature data and applied Machine Learning to devise edge algorithms that fine tune in real-time by adjusting Machine Learning calculations and algorithm coefficients to improve the algorithms accuracies (please see Table 1).

Table 1: Adaptive Coefficients Sensor Table Data

Outlook	Temp (°F)	Humidity (%)	Windy	Adaptive Coefficient
Sunny	75	70	True	Low Value
Sunny	80	90	True	High Value
Sunny	85	85	False	High Value
Sunny	72	95	False	High Value
Sunny	69	70	False	Low Value
Overcast	72	90	true	Low Value
Overcast	83	78	False	Low Value
Overcast	64	65	True	Low Value
Overcast	81	75	False	Low Value
Rain	71	80	True	High Value
Rain	65	70	True	High Value
Rain	75	80	False	Low Value
Rain	68	80	False	Low Value
Rain	70	96	False	Low Value

Model Development (on Paper)

For the given dataset, the class label "Adaptive Coefficient" has two distinct outcomes (High Value, Low Value). Therefore, there are two distinct classes ($m = 2$)

C1 = Low Value   C2 = High Value

9 tuples of class **Low Value**

5 tuples of class **High Value**

Information Gain (D) = $-\Sigma$ (Probability of 'Low Value * Log_2 (Probability of 'Low Value) + Probability of 'High Value * Log_2 (Probability of High Value)) (please see Eqs. 1, 2 & 3)

Information Gain (D) = $-\Sigma$ ((5/14) * Log2 (5/14) + (9/14) * Log2 (9/14))

Information gain (D) = 0.9402 bits

Expected information required for each attribute:

1. For Outlook attribute

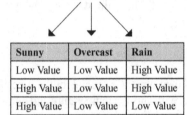

Outlook

Sunny	Overcast	Rain
Low Value	Low Value	High Value
High Value	Low Value	High Value
High Value	Low Value	Low Value
High Value	Low Value	Low Value
Low Value	-	Low Value

2. For temperature attribute
   For continuous valued attributes, there will be many possible split points.

- Evaluate information gain for every possible split point for the attribute
- Choose the best split point
- Information gain for the best split point becomes information gain for that attribute

Sort the values in the increasing order

64	65	68	69	70	71	72	72	75	75	80	81	83	85
Low	High	Low	Low	Low	High	High	Low	Low	Low	High	Low	Low	High

Consider a split at the value 71.5

Example: Temperature < 71.5 & Temperature > = 71.5

Temperature

< 71.5	> = 71.5
Low Value	High Value
High Value	Low Value
Low Value	Low Value
Low Value	Low Value
Low Value	High Value
High Value	Low Value
-	Low Value
	High Value

Information Gain (Temperature):

Info Temperature (D) = 8/14 * (–(5/8 * Log2(5/8)) – (3/8 * Log2(3/8))) + 6/14 * (–(4/6 * Log2(4/6)) – (2/6 * Log2(2/6)))

= 0.9388 bits

Gain (A) = 0.9402 – 0.9388

Gain (A) = **0.0014 bits**

1. For attribute Humidity

Consider split at value 79.5

Humidity

< 79.5	> = 71.5
Low Value	High Value
Low Value	High Value
Low Value	High Value
Low Value	Low Value
Low Value	High Value
High Value	Low Value
	Low Value
	Low Value

Information Gain (Humidity):

Info Humidity (D) = 10/14 * (–(6/10 * Log2(6/10)) – (4/10 * Log2(4/10))) + 4/14 * (–(3/4 * Log2(3/4)) – (1/4 * Log2(1/4)))

= 0.8963

Gain (A) = 0.9402 – 0.8963

Gain (A) = **0.0439 bits**

For attribute **Windy**

Windy

True	False
Low	High
Low	High
High	Low
High	Low
Low	Low
Low	Play
	Low
	Low

Information Gain (Windy):

Info Wind (D) = 8/14 * (–(6/8 * Log2(6/8)) – (2/8 * Log2(2/8))) + 6/14 * (–(3/6 * Log2(3/6)) – (3/6 * Log2(3/6)))

= 0.8921

Gain (A) = 0.9402 – 0.8921

Gain (A) = **0.0479 bits**

Therefore, Outlook is selected as root node based on Gain values (please see Figure 3)

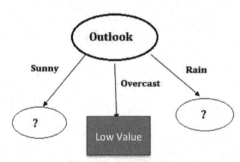

Figure 3: Outlook node

Once the root node is known, the next step is to select which attributes become internal nodes (see Figure 4).

Info(D) = –25log2(25) –35log2(35) = 0.971 bits

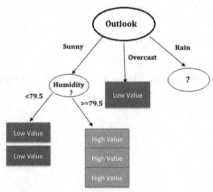

Figure 4: Humidity Node

Once the root node is known, the step is to select with attributes become internal nodes (see Figure 5).

Info (D) = –25log2(25) –35log2(35) = 0.971 bits

Infowindy(D) = 25 × (−12log2(12) + −12log2(12)) + 35 × (−32log2(23) + −31log2(13))

0.4 + 0.551

0.951 bits (see Figure 5)

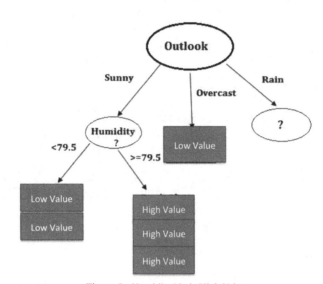

Figure 5: Humidity Node High Value

Temperature

Information Gain (Temperature) = 0.6490 bits

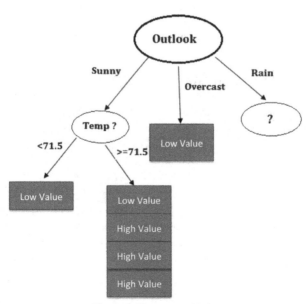

Figure 6: Temperature Node

Gain(Humidity) = Info(D) – InfoHumidity(D) = 0.971 – 0 = 0.971 bits (see Figure 6)

Gain(Windy) = Info(D) – Infowindy(D) = 0.971 – 0.951 = 0.02 bits

Gain(Temperature) = Info(D) – InfoTemperature(D) = 0.971 – 0.6490 = 0.322 bits

Therefore, Humidity is selected as internal node (see Figures 6 and 7).

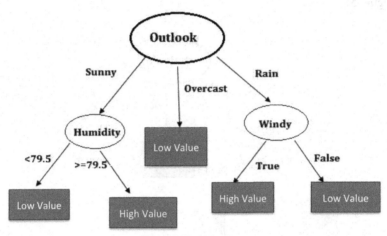

Figure 7: Decision Tree

Model Development (Python)

For the above given dataset, the following python code generates the decision tree (please see Table 2 and Figure 8).
We have used scikit Learn Decision Tree[258] to classifier to develop following model:

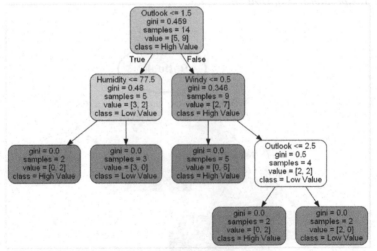

Figure 8: Decision Tree (Python model)

---

[258] Scikit Decision Tree - https://scikit-learn.org/stable/modules/generated/sklearn.tree.DecisionTreeClassifier.html

Table 2: Decision Tree Python Code

```
# -*- coding: utf-8 -*-
"""
Created on Sat Dec 8 18 : 33 : 27 2018
@author: cvuppalapati
"""

from sklearn.tree import DecisionTreeClassifier, export_graphviz
from subprocess import call
import numpy as np
"""
```

Outlook	Temp(oF)	Humidity(%)	Windy	Adaptive Coefficient
Sunny	75	70	True	Low Value
Sunny	80	90	True	High Value
Sunny	85	85	False	High Value
Sunny	72	95	False	High Value
Sunny	69	70	False	Low Value
overcast	72	90	true	Low Value
overcast	83	78	False	Low Value
overcast	64	65	True	Low Value
overcast	81	75	False	Low Value
Rain	71	80	True	High Value
Rain	65	70	True	High Value
Rain	75	80	False	Low Value
Rain	68	80	False	Low Value
Rain	70	96	False	Low Value

Codification

Outlook Attributes

Tuples :

Sunny 1

Overcast 2

Rain 3

Windy Attribute

Tuples

True 1

False 0

Class Variable - adaptive coefficent

Low Value 0

High Value 1

Outlook	Temp(oF)	Humidity(%)	Windy	Adaptive Coefficient
1	75	70	1	0
1	80	90	1	1
1	85	85	0	1
1	72	95	0	1
1	69	70	0	0
2	72	90	1	0
2	83	78	0	0
2	64	65	1	0
2	81	75	0	0
3	71	80	1 1	
3	65	70 1	1	
3	75	80	0	0
3	68	80	0	0
3	70	96 0	0	

```
"""
```

Table 2 contd. ...

...Table 2 contd.

```
X = np.array([[1, 75, 70, 1],
         [1, 80, 90, 1],
         [1, 85, 85, 0],
         [1, 72, 95, 0],
         [1, 69, 70, 0],
         [2, 72, 90, 1],
         [2, 83, 78, 0],
         [2, 64, 65, 1],
         [2, 81, 75, 0],
         [3, 71, 80, 1],
         [3, 65, 70, 1],
         [3, 75, 80, 0],
         [3, 68, 80, 0],
         [3, 70, 96, 0]])
         y = ['Low Value', 'High Value', 'High Value', 'High Value', 'Low Value', 'Low Value', 'Low Value', 'Low Value',
'Low Value', 'High Value', 'High Value', 'Low Value', 'Low Value', 'Low Value']
         dtree = DecisionTreeClassifier()
         dtree.fit(X, y)
         dot_data = export_graphviz(dtree,
                 out_file = 'edgedecisiontree.dot',
                 feature_names = ['Outlook', 'Temperature', 'Humidity', 'Windy'],
                 class_names = ['Low Value', 'High Value'],
                 filled = True,
                 rounded = True)
         call(['dot', '-Tpng', 'edgedecisiontree.dot', '-o', 'edgedecisiontree.png'])
```

Model Development (Citizen Data Scientist) Using Machine Learning User Interface

Data Sciences and Machine Learning algorithms can be developed on powerful desktops or the central Cloud compute architectures. We have used Microsoft Azure[259] Cloud Machine Learning to validate the model that we have developed before deploying into the hardware. Please see Figure 9, that initiatives the ML development phase by creating dataset (see Table 1) either upload from local file or via cloud storage:

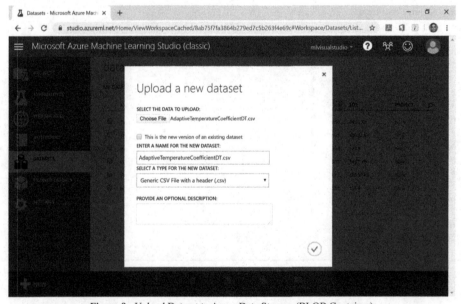

Figure 9: Upload Dataset to Azure Data Storage (BLOB Container)

[259] Azure Machine Learning - https://docs.microsoft.com/en-us/azure/machine-learning/studio-module-reference/

Next step, create Dataset (please see Figure 10):

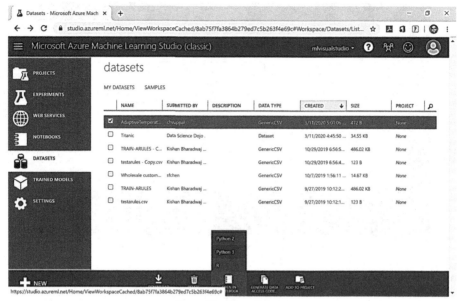

Figure 10: Create Dataset

Verify the data:[260] Please verify data by looking data statistics (please see Figure 11)

Figure 11: Verify Data

If you want to view the cleaned dataset, click the left output port of the Clean Missing Data module and select Visualize. Notice that the normalized-losses column is no longer included, and there are no missing values.

Start the model build process: For building model, using Azure ML User Interface, drag the model constructs on to Canvas (see Figure 12).

---

[260] Microsoft Azure - https://docs.microsoft.com/en-us/azure/machine-learning/studio/create-experiment

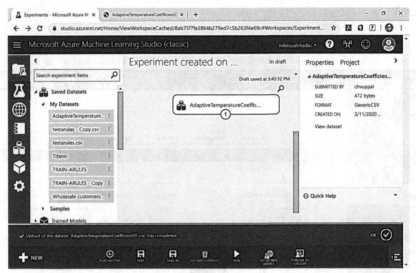

Figure 12: Build Model (UI Elements Drag and Drop)

Run the experiment: Once the model is constructed completely, i.e., Selection of data sources, selection of attributes, Decision Tree Classifier, Split Data, Prepare model and Validate the model, click the Run button on the experiment[261] (Please see Figure 13).

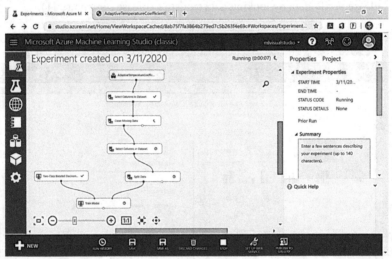

Figure 13: Run Model

Model Prediction: the following figure (Figure 14) provides actual vs. predicted values.

Upon verifying the accuracies, the next step is to deploy the model: (please see Figure 15)

View Data: the dataset can be viewed by connecting through Jupyter Notebook (please see Figure 16)

```
from azureml import Workspace
ws = Workspace()
ds = ws.datasets['AdaptiveTemperatureCoefficientDT.csv']
frame = ds.to_dataframe()
```

---

[261] Create Microsoft Azure Experiment - https://docs.microsoft.com/en-us/azure/machine-learning/studio/create-experiment

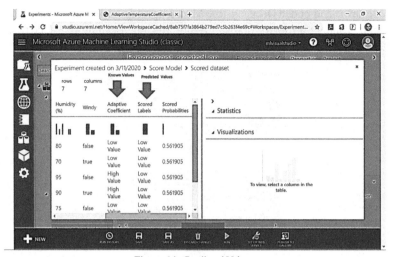

Figure 14: Predicted Values

Figure 15: Model Deployment

Figure 16: Data to Jupyter Notebook

At this point, we have hand-coded Decision Tree Model, developed the model using Python on the desktop, validated the model by running through Microsoft Azure Machine Leaning Interface (Step 1 is complete as shown in Figure 17). The next step is to deploy the model into hardware (Step 2 in the Figure 17).

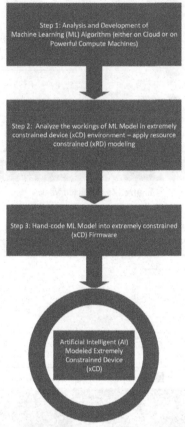

Figure 17: xRC AI-ML Process

Constrained Environment Considerations

Decision Rule

The above code is converted into decision rules within the edge processor. The values for the Humidity, Windy, and Temperature are used to dynamically compute the adaptive edge coefficient (see Figure 7 and Table 3).

Build the Model

In order to Flash the model into hardware, first integrate the model as part of the Firmware of the device and second flash the device. Using Build Eclipse of C/C++ IDE, build the project (please see Figure 18). The following is the build summary (see Figure 19):

Table 3: Edge Processor Decision Rules

```
/*
Codification
Outlook Attributes
Tuples:
Sunny 1
Overcast 2
Rain 3
Windy Attribute
Tuples
True 1
False 0
*/
double computeAdaptiveEdgeCoefficent(intOutlook, boolWindy)
{
        If(Outlook == 1)
        {
                if (Humidity < 79.5)
                {
                        adaptiveEdgeCoefficent = 0.3
                }
                else {
                        adaptiveEdgeCoefficent = 0.8
                }
        } Else if (Outlook == 2)
        {
                adaptiveEdgeCoefficent = 0.3
        }
        elseif (outlook == 3) {
                if (Windy == True)
                {
                        adaptiveEdgeCoefficent = 0.8
                }
                else {
                        adaptiveEdgeCoefficent = 0.3
                }
        }
        else {
                // Do nothing
        }
}
```

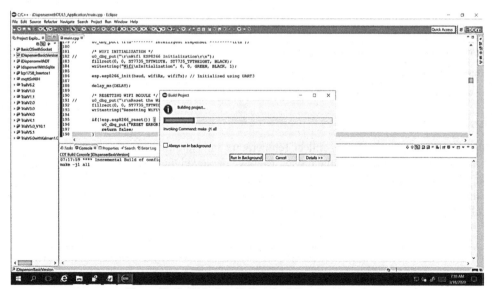

Figure 18: Build Project (Eclipse for C/C++ Developers)

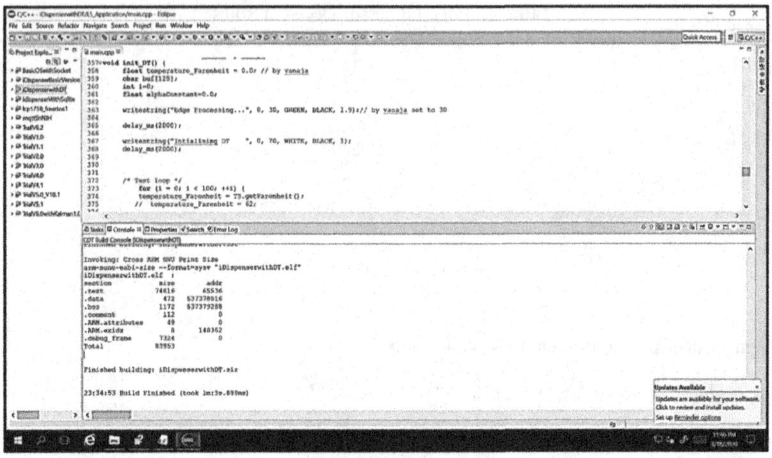

Figure 19: Build Output

Next, Flash the Firmware into the device. The cross-ARM GNU provides the program & application Flash summary that is useful to understand the footprint of the firmware. Please see following table (all numbers are in Bytes):

Invoking: Cross ARM GNU Print Size
arm-none-eabi-size --format=sysv "iDispenserwithDT.elf"
iDispenserwithDT.elf :

Section	Size	Addr
.text	74816	65536
.data	472	537378816
.bss	1172	537379288
.comment	112	0
.ARM.attributes	49	0
.ARM.exidx	8	140352
.debug_frame	7324	0

Please note:

'text' is written in FLASH and has code and constant data.

'data' is used for initialized data that written in RAM.

The 'bss' contains all the uninitialized data and written in RAM.

Inference

The final step is to start the Hardware, collect Temperature and Humidity Sensor data and see the predictions. In order to see runtime prediction of the model, we have deployed the ML algorithm into custom built edge processor that is used to develop both Cow necklace and Intelligence Dispenser used om healthcare settings. The run-time inference can be seen in Figure 20: We have tested the algorithm in San Ramon, California. The device collected Temperature 80°F—edge coefficient computed 0.05.

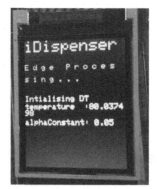

Figure 20: Model Deployment

Pre-Run Compute & Memory Statistics

During the running of the device, following are the Device Memory statistics before run:

Memory Details	Used (in Bytes)
Global Used	1644 Bytes
Malloc Used	7740 Bytes
Malloc Available	0 Bytes
System Available Memory	56152 Bytes
Next Heap Pointer	0x10001E3C
Last Stack Back Pointer	0x10001DDC
Last Stack Back Pointer Size	96
Number of Stack calls	44

Post Run - Compute & Memory Statistics

The following contains the memory information after the run:

Memory Details	Used (in Bytes)
Global Used	1644 Bytes
Malloc Used	7740 Bytes
Malloc Available	0 Bytes
System Available Memory	56152 Bytes
Next Heap Pointer	0x10001E3C
Last Stack Back Pointer	0x10001DDC
Last Stack Back Pointer Size	96
Number of Stack calls	44

As we can see from above two tables, there is no computational differences before and after the run (please see Figure 21). This leads to the conclusion that the Decision Trees have time complexity of O(1) and will not be expensive to run on extremely constrained devices (xCDs).

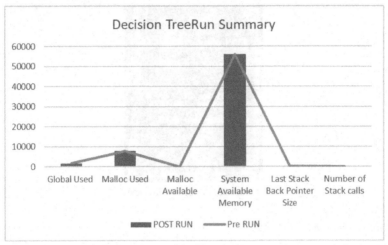

Figure 21: Decision Tree Run Summary

Memory Details	Differentials (in Bytes)
Global Used	0 Bytes
Malloc Used	0 Bytes
Malloc Available	0 Bytes
System Available Memory	0 Bytes
Next Heap Pointer	0 Bytes
Last Stack Back Pointer	0 Bytes
Last Stack Back Pointer Size	0 Bytes
Number of Stack calls	0 Bytes

In the next section, we will apply xCD framework and model Decision Tree deployment in Constrained environment.

 Decision Model is a callable or in-line code that could be embedded as part of constrained device code. For constraint environments, there would not be huge data movements between Processor to Memory space. And hence, no penalty with respect to Energy and Computation power consumption.

Hardware Economy—Model Accuracy tradeoff

The purpose of the section is to analyze the decision tree deployment in extremely constrained devices (xCD) or Tiny Compute hardware.

Modeling for no Connectivity

Modeling for no connectivity influences model upgradability, Storage and Computational Power. Due to no connectivity, the Decision Tree model could only updated during new device refresh cycle. Additionally, the lack of connectivity inhibits model upgrade via Over the air (OTA) firmware update.

Engineering Metrics

- Space Complexity: constant space
- Time Complexity: Constant Time (O (1)) as the algorithm only evaluates If-Else conditions.

Connectivity No Connectivity	Hardware 🔒 Email: 🔓 No Constraint			
	Memory	Power Source	Storage	Computational Power
Self-Contained & Updated only on Hardware Refresh	Space complexity and time complexity are low – O (1)	Time complexity is same but increased storage space could exert higher power consumption.	Application data storage could increase due to no connectivity.	Space complexity and time complexity are low – O (1)
OTA Firmware	N/A	N/A	N/A	N/A
Model Type: Eager Learner	Decision Tree is family of eager learner algorithm and could have less influence memory due to no connectivity – O (1).	Decision path evaluation calls are if-else conditions – O (1)	No influence on storage	Linear Time complexity.
Lazy Learner	Not applicable as decision tree is if-else condition evaluator	Not applicable as decision tree is if-else condition evaluator	Not applicable as decision tree is if-else condition evaluator	Not applicable as decision tree is if-else condition evaluator
Model Invocation	Inline model invocation O (1)	Inline model invocation O (1) does not demand huge power.	Not applicable	Inline model invocation requires less computational power compared to stack or heap based.

(left vertical label: ML Model Accuracy)

Overall design consideration: ☞ Operating Decision Tree Machine Learning algorithm in a constrained or Tiny device with "No Connectivity", the engineering consideration is storage of the device needs to be well defined and should be sufficient for the Service Level Agreement (SLA) duration. If the time window of SLA is half-yearly or yearly refresh, the high device storage needs could impact the battery power. As far as the model computational and performance goes, these should not exert huge pressure on the device footprint.

Modeling Low Power Bluetooth Connectivity

Bluetooth Low Energy connectivity availability enables to offload Data storage to Edge devices. ☞ Only tax is extra power consumption due to BLE power.

Connectivity Bluetooth Low Energy (BLE)	Hardware 🔒 Constraint 🔓 No Constraint			
	Memory	Power Source	Storage	Computational Power
Self-Contained & Updated only on Hardware Refresh	🔒 DTs are O (1) linear Time complex;	🔒 Continuous Bluetooth advertisements could increase power needs.	🔓 Data could be offloaded to external devices.	🔓 The model data and storage are offloaded to Edge devices (Smartphone) and thus could reduce hardware footprint.
OTA Firmware	N/A	N/A	N/A	N/A
Model Type: Eager Learner	🔒 DTs are eager learner with linear time complex O (1)	🔒 Bluetooth advertisements need higher power.	🔓 Periodic offloading of Sensor data to Edge devices lowers the storage needs.	🔓 Lower Data Size and Code size reduce the needs for more computational power.
Lazy Learner	N/A	N/A	N/A	N/A
Model Invocation	🔓 Linear time complex O (1) and Inline if-else model invocation.	🔒 Though Linear time complex, continuous Bluetooth advertisements could increase power needs	🔓 Periodic offloading of Sensor data to Edge devices lowers the storage needs.	🔓 Linear time complex & Inline model invocation

(Left vertical label: ML Model Accuracy)

Overall design consideration: Operating Decision Tree ML algorithms in an extremely constrained devices (xCD) or Tiny device with "Bluetooth Low Energy (BLE) Connectivity" will increase battery power consumption. Model performance & invocation should be linear as time complexity is O (1). ☞ Only caution to be noted is sustainability of Battery for radio advertisements—that is turn on radio & BLE advertisement on a need basis. For instance, Cow necklace sensors are equipped with BLE button to TURN ON Bluetooth connectivity on need basis (See Figure 22). This mechanical feature is vital to reduce the cost of service of the sensor especially rural and hard to reach deployments areas.

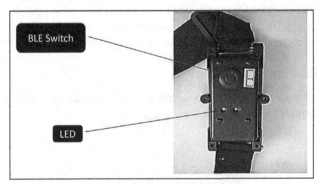

Figure 22: Sensor BLE Switch

Modeling Wi-Fi Connectivity

Wi-Fi connectivity enables to offload Data storage to the Edge devices or call to the central Cloud Server. Additionally, enables to upload Firmware via OTA. The major limitation is many of the constrained devices are not suited for Wi-Fi as it drains huge power from the battery.

Connectivity Wi-Fi		Hardware 🔒 Constraint ⭘ No Constraint			
		Memory	Power Source	Storage	Computational Power
ML Model Accuracy	Self-Contained & Updated only on Hardware Refresh	🔒 DTs are O (1) linear Time complex;	🔒 On board Wi-Fi demands more power. For battery power devices, Wi-Fi option is less favorable.	🔒 Data could be offloaded to external devices or central Cloud.	🔒 The model data and storage are offloaded to Edge devices (Smartphone) and thus could reduce hardware footprint.
	OTA Firmware	🔒 OTA allows to deploy the latest and greatest optimized model.	🔒 Onboard Wi-Fi demands more power.	🔒 Data could be offloaded to external devices or to the central Cloud.	🔒 OTA allows to deploy the latest and greatest optimized model.
	Model Type: Eager Learner	🔒 DTs are eager learner with linear time complex O (1)	🔒 Bluetooth advertisements need higher power.	🔒 Periodic offloading of Sensor data to Edge devices lowers the storage needs.	🔒 Lower Data Size and Code size reduce the needs for more computational power.
	Lazy Learner	N/A	N/A	N/A	N/A
	Model Invocation	🔒 Linear time complex O (1) and Inline if-else model invocation.	🔒 Though Linear time complex, continuous Bluetooth advertisements could increase power needs	🔒 Periodic offloading of Sensor data to Edge devices lowers the storage needs.	🔒 Linear time complex & Inline model invocation

Overall design consideration: Operating Decision Tree Machine Learning algorithms in a constrained or Tiny device with "Wi-Fi Connectivity" increases battery power consumption and may not be a suitable option due to power consumption. Next, the Model performance & invocation are linear as time complexity is O (1) and thus won't exert any penalty on battery performance.

Connectivity—Model Accuracy tradeoff

Modeling Memory

Since the model is a decision rule evaluation, it does not require to be memory resident[262] nor heap[263] stored computation. Plus, doesn't require historical data for inference. These model characteristics make non-memory constraint.

☞ That is, the memory (lack of Data movement or access) does not influence power or computational hardware economy.

---

[262] Memory resident - https://www.trendmicro.com/vinfo/us/security/definition/memory-resident
[263] Stack vs. Heap allocation - https://www.geeksforgeeks.org/stack-vs-heap-memory-allocation/

Hardware Memory	Connectivity 🔒 Constraint / No Constraint		
	No Connectivity	Bluetooth Low Energy (BLE)/LoRa	Wi-Fi
Self-Contained & Updated only on Hardware Refresh	🔒	🔒 Since model update is cycled on Hardware refresh (SLA half-yearly or yearly), there would be staleness of the model. However, **memory constraint has no influence on the Accuracy.**	🔒
OTA Firmware	🔒 Since the model gets updated OTA as part of Firmware update, no staleness of model. Memory constraint has no influence on the model accuracy.	🔒	🔒
Model Type: Eager Learner	🔒	🔒	🔒
Lazy Learner	N/A	N/A	N/A
Model Invocation	🔒 In-line model invocation and not memory intense—no memory constraint on the model.	🔒	🔒

(Left axis label: Model Accuracy)

Design Considerations: Since model update is cycled on Hardware refresh (SLA half-yearly or yearly), there would be staleness of the model. However, memory constraint has no influence on the Accuracy. Model is decision rule and will not require data movement between processor & memory space. In-line model invocation and not memory intense—no memory constraint on the model.

Modeling Processing Power

The Model deployed as decision rule and invoked as an in-line function and does not require extra computation power.

Hardware Processing Power	Connectivity 🔒 Constraint / No Constraint			
	No Connectivity	BLE	Wi-Fi	Design Notes
Hardware Refresh	🔒	🔒	🔒	Since model update is cycled on Hardware refresh (SLA half-yearly or yearly), there could be staleness in Model accuracy. This has no influence on the Power.
OTA Firmware Update	🔒	🔒	🔒	Processing Power is not a constraint with respect to Model Accuracy.
Model Type: Eager Learner	🔒	🔒	🔒	Processing Power is not a constraint with respect to Model Accuracy.
Lazy Learner	N/A	N/A	N/A	
Model Invocation	🔒	🔒	🔒	Processing Power is not a constraint with respect to Model Accuracy.

(Left axis label: Model Accuracy)

Design Considerations: Since model update is cycled on Hardware refresh (SLA half-yearly or yearly), there could be staleness in Model accuracy. This has no influence on the Power. Processing Power is not a constraint with respect to Model Accuracy.

Modeling Storage

The Model deployed as an in-line function and does not require huge storage. Nonetheless, as hardware device collects sensor data, the data needs to be stored on device local storage (offload to Edge Gateway or Mobile device) before uploads to Central Storage or Corporate Cloud storage.

Hardware Storage		Connectivity 🔒 Constraint ⚪ No Constraint			
		No Connectivity	BLE	Wi-Fi	Design Notes
Model Accuracy	Hardware Refresh	⚪	⚪	⚪	Since model update is cycled on Hardware refresh (SLA half-yearly or yearly), the data storage needs to be large to store on the device. A flat file-based storage. For BLE and Wi-Fi modes, the storage could be offloaded to gateway or mobile device & Circular storage data pattern could help to concise storage.
	OTA Firmware Update	⚪	⚪	🔒	Since the model gets updated OTA as part of Firmware update, no staleness of model. Storage could be a constraint under "No Connectivity" but not a constraint for BLE and Wi-Fi.
	Model Type: Eager Learner	⚪	⚪	⚪	Model is linear equation and will not require data movement between processor & memory space. Storage could be a constraint under "No Connectivity" but not a constraint for BLE and Wi-Fi.
	Lazy Learner	N/A	N/A	N/A	
	Model Invocation	⚪	⚪	⚪	In-line model invocation and not memory intense – no memory constraint on the model. Storage could be constraint under "No Connectivity" but not a constraint for BLE and Wi-Fi.

Design Considerations: Since model update is cycled on Hardware refresh (SLA half-yearly or yearly), there could be ☛ staleness in Model accuracy. This has no influence on the Power. Processing Power is not a constraint with respect to Model Accuracy.

Modeling Environmental Perturbations

Model deployed under geographical conditions with huge temperature or humidity variations, ☛ the influence of temperature on the battery could lead into accuracy consideration (thermal characteristics). The Humidity accumulation is also a factor to be considered in the model calculation

Hardware Environmental Perturbations		Connectivity 🔒 Constraint ⚪ No Constraint			
		No Connectivity	BLE	Wi-Fi	Design Notes
Model Accuracy	Hardware Refresh	⚪	⚪	🔒	Model should a factor in geographical sensitivity as part of the calculations. The model coefficients need to tame environmental perturbations.
	OTA Firmware Update	⚪	⚪	⚪	The OTA could fine-tune sensitive model coefficients to overcome environmental perturbations.
	Model Type: Eager Learner	N/A	N/A	N/A	No influence of environmental perturbations on data movement.
	Lazy Learner	N/A	N/A	N/A	No influence of environmental perturbations on data movement.
	Model Invocation	N/A	N/A	N/A	No influence of environmental perturbations on data movement.

Connectivity to Hardware Tradeoff

Connectivity to Hardware Tradeoff (See Figure 23) models Model Accuracy attributes trade-off:

Model constraints include refresh cycle and Model invocation modes. Nonetheless, Data Movement and Learner Type is not a constraint for the decision rule.

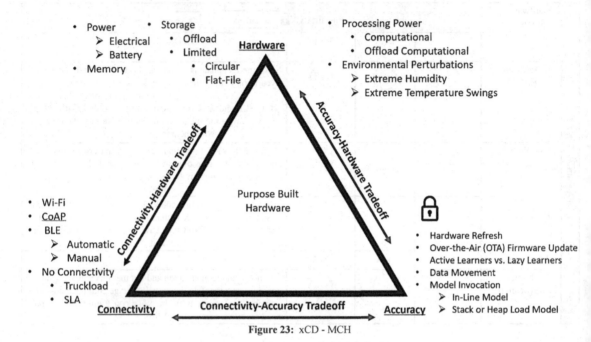

Figure 23: xCD - MCH

Here are the Connectivity considerations:

- Wi-Fi
- CoAP
- Bluetooth Low Energy (Manual or Automatic Connection with the device)
- No Connectivity at all

The Hardware parameters include:

- Power
- Memory
- Storage
- Processing power
- Environmental perturbations

Modeling: Hardware Refresh

In this mode, the Model gets refreshed whenever hardware refresh takes place. Generally, this could be half-yearly or yearly, depending upon the SLA signed with the customer. With this constraint, there is staleness on the model accuracy. That is, adaptive learning is only deferred when a new model gets deployed. The model is a decision rule function that is invoked as part of in-line code. So, no special requirements for memory, power, and computational processing. And hence, no constraints.

Design Considerations: Since model update is cycled on Hardware refresh (SLA half-yearly or yearly), there could be staleness in Model accuracy. This has no influence on the Power. Processing Power is not a constraint with respect to Model Accuracy.

Modeling Over-The-Air (OTA) Firmware Update

In this mode, Model is only getting refreshed over the air as part of firmware update. Except for no connectivity mode, the model is the latest and learned with the most recent collected data. That is, adaptive learning is current and gets deployed as part of OTA. The model is a linear function that is invoked as part of in-line code. So there is no special requirements for memory, power, and computational processing. And hence, no constraints.

Model Accuracy OTA		Hardware 🔒 Constraint ⬤ No Constraint					
		Power	Memory	Storage	Processing	Environmental	Design Notes
Connectivity	No Connectivity	⬤	⬤	🔒	⬤	🔒	Storage needs are higher & inaccuracies due to environmental perturbations.
	BLE	⬤	⬤	🔒	⬤	🔒	With the model is updated vis OTA, the latest learnings will have higher accuracies.
	Wi-Fi	⬤	⬤	⬤	⬤	⬤	

Modeling Active Learner vs. Lazy Learner

The model is an active learner with no special requirements to be in-memory, memory resident or data movements. The model is a linear function that is invoked as part of in-line code. So, there is no special requirement for memory, power, and computational processing. And hence, no constraints. No restrictions except in No connectivity mode with or without OTA firmware model.

Model Accuracy Learner		Hardware 🔒 Constraint ⬤ No Constraint					
		Power	Memory	Storage	Processing	Environmental	Design Notes
Connectivity	No Connectivity	⬤	⬤	🔒	⬤	🔒	Storage needs are higher & inaccuracies due to environmental perturbations.
	BLE	⬤	⬤	⬤	⬤	⬤	With the model is updated via OTA, the latest learnings will have higher accuracies.
	Wi-Fi	⬤	⬤	⬤	⬤	⬤	

Modeling Model Invocation

The Model is invoked as an in-line code, not a heap or memory resident call. Except storage constraint as part of No Connectivity mode, the model has no special restrictions.

Model Accuracy Invocation		Hardware 🔒 Constraint ⬤ No Constraint					
		Power	Memory	Storage	Processing	Environmental	Design Notes
Connectivity	No Connectivity	⬤	⬤	🔒	⬤	🔒	Storage needs are higher & inaccuracies due to environmental perturbations.
	BLE	⬤	⬤	⬤	⬤	⬤	With the model is updated via OTA, the latest learnings will have higher accuracies.
	Wi-Fi	⬤	⬤	⬤	⬤	⬤	

XGBoost

Boosting machine learning is one of the advanced and sophisticated machine learning (ML) technique that can be used to such complex data driven real-world problems.[264] The idea of boosting came out of the idea of whether a weak learner can be modified to become better [4].A weak hypothesis or weak learner is defined as one whose performance is at least slightly better than random chance. Three types of boosting include: adaptive boosting, gradient boosting, XGBoost.

Adaptive boosting: AdaBoost, short for "Adaptive Boosting", is the first practical boosting algorithm proposed by Freund and Schapire in 1996. It focuses on classification problems and aims to convert a set of weak classifiers into a strong one.[265] The weak learners in AdaBoost are decision trees with a single split, called decision stumps for their shortness [4]. AdaBoost works by weighing the observations, putting more weight on difficult to classify instances and less on those already handled well. New weak learners are added sequentially that focus their training on the more difficult patterns.

The final equation for classification can be represented as

$$F(x) = \text{sign} \left(\sum_{m=1}^{M} \theta m \, fm(x) \right)$$ Eq. (4)

Where

- f_m stands for the m weak classifier and Θ is the corresponding weight.

It is exactly the weighted combination of M weak classifiers.

AdaBoost Algorithm

Given a dataset containing n points, where

$$x_i \in \mathbb{R}^d, y_i \in \{-1, 1\}.$$

Data point x_n belongs to sample space R with output -1 denotes the negative class and 1 represents the positive class.
Initialize the weight for each data point as:

$$w(x_i, y_i) = \frac{1}{n}, \, i = 1,\ldots, n.$$

For iteration $m = 1, \ldots, M$

(1) Fit weak classifier to the data set and select the one with the lowest weighted classification error:

$$\epsilon_m = E_{w_m} [1_{y \neq f}(x)]$$

(2) Calculate the weight for the m weak classifier:

$$\theta_m = \frac{1}{2} \ln \left(\frac{1 - \epsilon_m}{\epsilon_m} \right).$$

For any classifier with accuracy higher than 50%, the weight is positive. The more accurate the classifier, the larger the weight. While for the classifier with less than 50% accuracy, the weight is negative. It means that we combine its prediction by flipping the sign.

(3) Update the weight for each data point as:

---

[264] A Complete Guide to XGBoost Model in Python using scikit-learn - https://hackernoon.com/want-a-complete-guide-for-xgboost-model-in-python-using-scikit-learn-sc11f31bq

[265] Boosting Algorithm - https://towardsdatascience.com/boosting-algorithm-adaboost-b6737a9ee60c

$$w_{m+1}(x_i, y_i) = \frac{w_m(x_i, y_i)\, exp[-\theta_m y_i f_m(x_i)]}{Z_m}$$

where Z_m is a normalization factor that ensures the sum of all instance weights is equal to 1.

If a misclassified case is from a positive weighted classifier, the "exp" term in the numerator would be always larger than 1 ($y*f$ is always -1, is positive). Thus, misclassified cases would be updated with larger weights after an iteration. The same logic applies to the negative weighted classifiers. The only difference is that the original correct classifications would become misclassifications after flipping the sign.

(4) After M iteration we can get the final prediction by summing up the weighted prediction of each classifier.

Adaptive boosting starts by assigning equal weight edge to all of your data points and you draw out a decision stump for a unique input feature, so the next step is the results that you get from the first decision stump which are analyzed.

Generalization of AdaBoost as Gradient Boosting

Gradient Boosting algorithm employs gradient descent algorithm to minimize errors in sequential models. Gradient boosting involves three elements:

1. A loss function to be optimized
2. A weak learner to make predictions
3. An additive model to add weak learners to minimize the loss function.

XGBoost

XGBoost is one of the most used and best decision evaluation-based algorithm.[266] XGBoost is a decision-tree-based ensemble Machine Learning algorithm that uses a gradient boosting framework.

In prediction problems involving unstructured data (images, text, etc.) artificial neural networks tend to outperform all other algorithms or frameworks. However, when it comes to small-to-medium structured/tabular data, decision tree-based algorithms are considered best-in-class right now. Please see the chart below for the evolution of tree-based algorithms over the years.

Install

The XGB can be installed by pip command (please see Figure 24):

Pip install xgboost

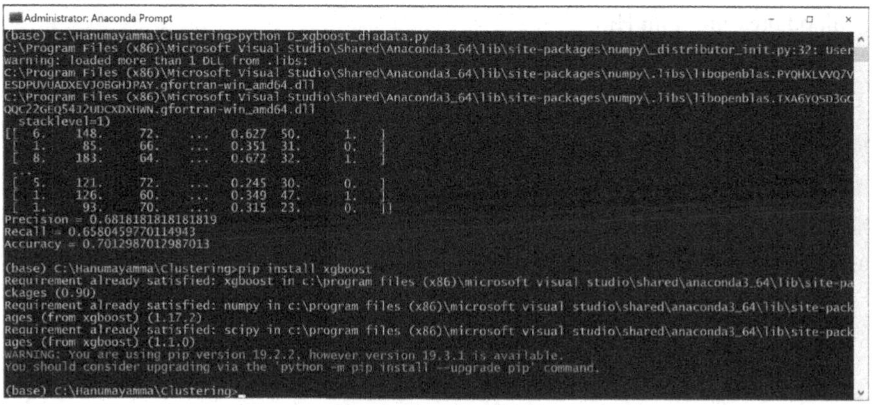

Figure 24: XGB Install

---

[266] XGBoost Algorithm - https://towardsdatascience.com/https-medium-com-vishalmorde-xgboost-algorithm-long-she-may-rein-edd9f99be63d

Coding XGBoost

In the following section, let's develop XGBoost model and deploy it in real-time embedded system.
Dataset:

Predicts the onset of diabetes based on diagnostic measures.[267] Please see Figure 25.

Dataset location:

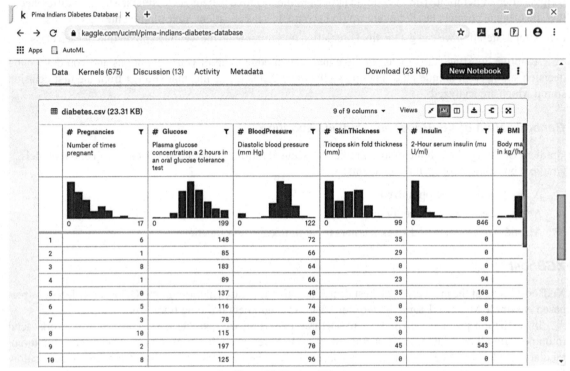

Figure 25: Dataset View

https://www.kaggle.com/kumargh/pimaindiansdiabetescsv

Attributes:[268] Please see Figure 26

1. Number of times pregnant
2. Plasma glucose concentration 2 hours in an oral glucose tolerance test
3. Diastolic blood pressure (mm Hg)
4. Triceps skin fold thickness (mm)
5. 2-Hour serum insulin (mu U/ml)
6. Body mass index (weight in kg/(height in m)^2)
7. Diabetes pedigree function
8. Age (years)
9. Class variable (0 or 1)

---

[267] Diabetes dataset - https://www.kaggle.com/uciml/pima-indians-diabetes-database
[268] https://raw.githubusercontent.com/jbrownlee/Datasets/master/pima-indians-diabetes.names

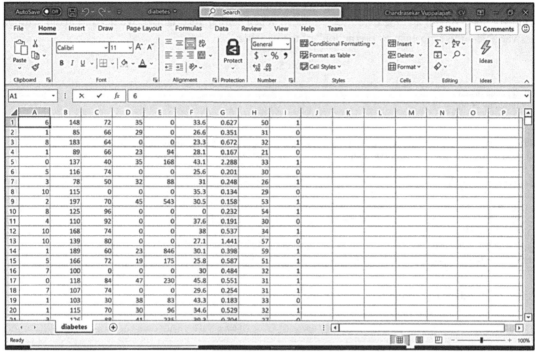

Figure 26: Dataset View

Code[269]

The following Google Co-Lab Jupyter Notebook that prepares and builds the XGB Model (please see Figure 27).

Import required libraries that include Pandas, NUmpy, XGBoost and re. One some systems, XGBoost needs to be installed by running

```
pip install xgboost
```

Figure 27: XGB Code

[269] XGBoost - https://machinelearningmastery.com/develop-first-xgboost-model-python-scikit-learn/

Next step, load train and test datasets. Set up columns—please note, for the training dataset, we have provided Label column and the data (See Table 5). When setting Test dataset, we need to avoid setting the class variable—this is to ensure "prediction" of the variable.

Table 5: XGB Code

```
import pandas as pd
import numpy as np
import xgboost as xgb
import re
train = pd.read_csv("https://storageconstts.blob.core.windows.net/tsdata/pima-indians-diabetes_xgboost_train.csv")
test  = pd.read_csv("https://storageconstts.blob.core.windows.net/tsdata/pima-indians-diabetes_xgboost_test.csv")
X_y_train = xgb.DMatrix(data=train[['Numberoftimespregna
nt', 'PlasmaGlucoseConcentration', 'Diastolicbloodpressure', 'TricepsSkinFoldThickness', '2Hourseruminsulin','BodyMassIndex'
,'Diabetes','Age']], label= train['Class'])
X_test   = xgb.DMatrix(data=test[['Numberoftimespregna
nt', 'PlasmaGlucoseConcentration', 'Diastolicbloodpressure', 'TricepsSkinFoldThickness', '2Hourseruminsulin','BodyMassIndex'
,'Diabetes','Age']])
train[['Numberoftimespregnant', 'PlasmaGlucoseConcentration', 'Diastolicbloodpressure', 'TricepsSkinFoldThickness',
'2Hourseruminsulin','BodyMassIndex','Diabetes','Age']].head()
```

Display the head dataset of the train data frame (see Figure 28).

	Numberoftimespregnant	PlasmaGlucoseConcentration	Diastolicbloodpressure	TricepsSkinFoldThickness	2Hourseruminsulin	BodyMassIndex	Diabetes	Age
0	6	148	72	35	0	33.6	0.627	50
1	1	85	66	29	0	26.6	0.351	31
2	8	183	64	0	0	23.3	0.672	32
3	1	89	66	23	94	28.1	0.167	21
4	0	137	40	35	168	43.1	2.288	33

Figure 28: Jupyter Code

Setup training configuration—most important class variable and objective with valuation metric (See Table 6).
Visualize the tree by calling to Graphviz.

Table 6: Code Run

```
params = {
        ‹base_score›: np.mean(train[‹Class›]),
        ‹eta›:  0.1,
        ‹max_depth›: 3,
        ‹gamma› :3,
        ‹objective›  :›reg:linear›,
        ‹eval_metric› :›mae›
    }
model = xgb.train(params=params,
        dtrain=X_y_train,
        num_boost_round=3)
xgb.to_graphviz(booster = model, num_trees=0)
```

The Output

The Figure 29 contains XGB Model decision structure:

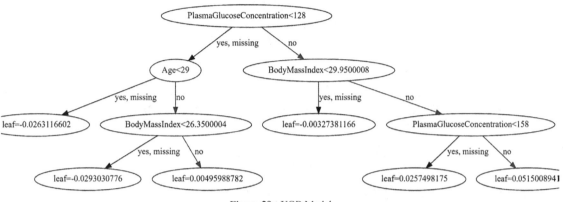

Figure 29: XGB Model

```
xgb.to_graphviz(booster = model, num_trees=1)
```

In order to generate single tree XGM, please use above command—see Figure 30 for output.

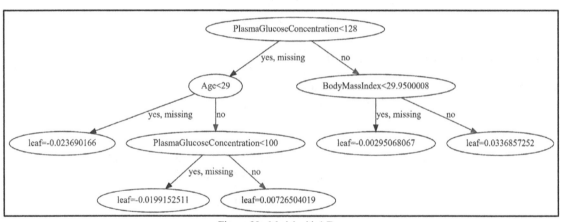

Figure 30: Model with 1 Tree

```
xgb.to_graphviz(booster = model, num_trees=2)
```

To create model with two trees, use above command. The output is as per Figure 31

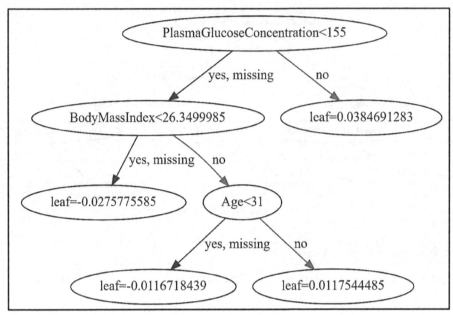

Figure 31: XGB With Two Trees

Constrained Environment Considerations

Constrained Environment considerations for the XGBoost all most all similar to that of considerations for Decision Tree as XGB generates decision paths.

Decision Paths—Simplified Rules:

Code in embedded System - Convert model into callable if-else-tree:[270] (please see Table 7)
The plot_tree() function generates graphical representation of the model.[271]

Table 7: XGB Decision Rules

['0:[PlasmaGlucoseConcentration<128] yes=1,no=2,missing=1\n\t1:[Age<29] yes=3,no=4,missing=3\n\t\t3:leaf=-0.0263116602\n\t\t4:[BodyMassIndex<26.3500004] yes=9,no=10,missing=9\n\t\t\t9:leaf=-0.0293030776\n\t\t\t10:leaf=0.00495988782\n\t2:[BodyMassIndex<29.9500008] yes=5,no=6,missing=5\n\t\t5:leaf=-0.00327381166\n\t\t6:[PlasmaGlucoseConcentration<158] yes=13,no=14,missing=13\n\t\t\t13:leaf=0.0257498175\n\t\t\t14:leaf=0.0515008941\n',
  '0:[PlasmaGlucoseConcentration<128] yes=1,no=2,missing=1\n\t1:[Age<29] yes=3,no=4,missing=3\n\t\t3:leaf=-0.023690166\n\t\t4:[PlasmaGlucoseConcentration<100] yes=9,no=10,missing=9\n\t\t\t9:leaf=-0.0199152511\n\t\t\t10:leaf=0.00726504019\n\t2:[BodyMassIndex<29.9500008] yes=5,no=6,missing=5\n\t\t5:leaf=-0.00295068067\n\t\t6:leaf=0.0336857252\n',
  '0:[PlasmaGlucoseConcentration<155] yes=1,no=2,missing=1\n\t1:[BodyMassIndex<26.3499985] yes=3,no=4,missing=3\n\t\t3:leaf=-0.0275775585\n\t\t4:[Age<31] yes=9,no=10,missing=9\n\t\t\t9:leaf=-0.0116718439\n\t\t\t10:leaf=0.0117544485\n\t2:leaf=0.0384691283\n']

Time Complexity

Let d be the maximum depth of the tree and K be total number of trees. For the exact greedy algorithm, the time complexity of original sparse aware algorithm is $O(K\,d\,\|x\|_0\,\log n)$. Here we use $\|x\|_0$ to denote number of

---

[270] https://towardsdatascience.com/xgboost-deployment-made-easy-6e11f4b3f817
[271] How to Visualize Gradient Boosting Decision Trees With XGBoost in Python - https://machinelearningmastery.com/visualize-gradient-boosting-decision-trees-xgboost-python/

non-missing entries in the training data. On the other hand, tree boosting on the block structure only cost $O(K$ $d\|x\|_0 + \|x\|_0 \log n)$. Here $O(\|x\|_0 \log n)$ is the one-time preprocessing cost that can be amortized. This analysis shows that the block structure helps to save an additional log n factor, which is significant when kljcis large.

Putting it in other words, the original sparse greedy algorithm doesn't use block storage. Thus, to find the optimal split at each node, you needed to re-sort the data on each column. This ends up incurring a time complexity at each layer that is very crudely approximated by $O(\|x\|_0 \log n)$: basically, say you have $\|x\|_{0i}$ nonzero entries for each feature $1 \le I \le m$; then at each layer you're sorting lists, each of length at most n, whose lengths sum to $\sum_{i=1}^{m} \|x\|0i = \|x\|_0$, which can't take more than $O(\|x\|_0 \log n)$ time. Multiplying by K trees and d layers per tree gives you the original $O(Kd\|x\|_0 \log n)$ time complexity.[272]

For the approximate algorithm, the time complexity of the original algorithm with binary search is $O(Kd$ $\|x\|_0 \log q)$. Here q is the number of proposal candidates in the dataset. The most time-consuming part of the tree learning algorithm is getting the data in sorted order.[273] This makes the time complexity of learning each tree $O(K d \|x\|_0 \log q)$[5] [6].

Inference

We used XGBoost models for Epsilon (400 K samples, 2000 features) dataset. For each model we limit number of trees used for evaluation to 8000. Thus, comparison gives only some insights of how fast the models can be applied. The model predictions on Intel Xeon E5-2660 CPU with 128GB RAM.[274] Please see Figure 32.

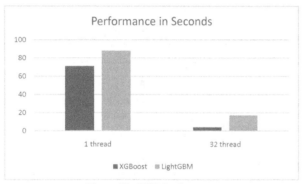

Figure 32: XGB Performance

Output

The output of XGB is shown in Figure 33.

Hardware Economy—Model Accuracy trade-off

The purpose of this section is to analyze the decision tree deployment in constrained or Tiny Compute hardware.

Modeling No Connectivity

Modeling for no connectivity influences model upgradability, storage and power source. Since, there is no connectivity environment, the XGB model could only update during new device refresh cycle. Additionally, the lack of connectivity inhibits model upgrade via OTA firmware update.

---

[272] How to Tune the Number and Size of Decision Trees with XGBoost in Python - https://machinelearningmastery.com/tune-number-size-decision-trees-xgboost-python/

[273] XGBoost: Reliable Large-Scale Tree Bossing System - http://learningsys.org/papers/LearningSys_2015_paper_32.pdf

[274] Best in class inference and a ton of speedups - https://catboost.ai/news/best-in-class-inference-and-a-ton-of-speedups

Figure 33: XGB Output

Connectivity No Connectivity	Hardware 🔒 Constraint ⭕ No Constraint			
	Memory	Power Source	Storage	Computational Power
Self-Contained & Updated only on Hardware Refresh ⭕	⭕ Memory at inference level is independent of time complexity- dependent upon Number of trees (K), depth (d), Number of features (q) – **O (K d ||x||0 log q)**	🔒 Time complexity is proportional to number of trees (K) and depth (d) and increased storage space could exert higher power consumption.	🔒 Application data storage could increase due to no connectivity.	🔒 High computational power for XGB models with high number of trees & depth - O (K d ||x||0 log q)
OTA Firmware	N/A	N/A	N/A	N/A
Model Type: Eager Learner	⭕ XGB is family of eager learner algorithm and could have less influence on memory due to no connectivity.	🔒 Decision path evaluation calls are if-else conditions – **O (K d ||x||0 log q)**	🔒 No influence on storage	🔒 High computational power for high number of trees & depth - O (K d ||x||0 log q).
Lazy Learner	Not applicable as decision tree is if-else condition evaluator	Not applicable as decision tree is if-else condition evaluator	Not applicable as decision tree is if-else condition evaluator	Not applicable as decision tree is if-else condition evaluator
Model Invocation	⭕ Inline model invocation Number of trees (K), depth (d), Number of features (q) – **O (K d ||x||0 log q)**	🔒 Inline model invocation O (1) does demand huge power for deeper (high d) & spread out (high K) trees.	⭕ Not applicable	🔒 Inline model invocation requires more computational power for deeper (high d) & spread out (high K) trees. **O (K d ||x||0 log q)**

(Left side vertical header: ML Model Accuracy)

Overall design consideration: Operating XGB Machine Learning algorithm in an constrained devices (xCDs) or Tiny device with "No Connectivity", Storage of the device needs to well defined and should be sufficient for the SLA time of the device. If the SLA is half-yearly or yearly refresh, the device storage capacity could impact battery power. As far as the model computational performance goes, XGB Machine Learning Models with deeper (d) and spread out (K) trees could impact battery performance.

> ⚛ For Boosting Tree ML models such as XGBoost, the impact of battery power consumption is based on the size of the model, i.e., in terms of spread and depth, and frequency of calls to the model invocation.

Modeling Low Power Bluetooth Connectivity

Bluetooth Low Energy connectivity availability enables to offload Data storage to Edge devices. Only tax is extra power consumption due to BLE power. Additionally, computational power levy penalties on battery power source by XGB ML models that have deeper depths (d) and expanded tree structure (K).

Connectivity Bluetooth Low Energy (BLE)		Hardware 🔒 Constraint ⭕ No Constraint			
		Memory	Power Source	Storage	Computational Power
ML Model Accuracy	Self-Contained & Updated only on Hardware Refresh ⭕	Memory at inference level is independent of time complexity- dependent upon Number of trees (K), depth (d), Number of features (q) – $O\,(K\,d\,\|x\|0\,\log q)$	🔒 Continuous Bluetooth advertisements and time complexity could increase power needs.	⭕ Data could be offloaded to external devices.	🔒 High computational power for XGB models with high number of trees & depth - $O\,(K\,d\,\|x\|0\,\log q)$
	OTA Firmware	N/A	N/A	N/A	N/A
	Model Type: Eager Learner ⭕	XGB is a family of eager learner algorithm and could have less influence on memory due to no connectivity.	🔒 *Decision paths evaluation calls are if-else conditions* – $O\,(K\,d\,\|x\|0\,\log q)$	No influence on storage	🔒 High computational power for high number of trees & depth - O (K d $\|x\|0 \log q$).
	Lazy Learner	N/A	N/A	N/A	N/A
	Model Invocation ⭕	Time complex $O\,(K\,d\,\|x\|0\,\log q)$ and Inline if-else model invocation.	🔒 Time complexity-dependent upon Number of trees (K), depth (d), Number of features (q) – $O\,(K\,d\,\|x\|0\,\log q)$, continuous Bluetooth advertisements could increase power needs	⭕ Periodic offloading of Sensor data to Edge devices lowers the storage needs.	🔒 Time complexity-dependent upon Number of trees (K), depth (d), Number of features (q) – $O\,(K\,d\,\|x\|0\,\log q)$,

Overall design consideration: Operating XGB based Machine Learning algorithms in *a* constrained or Tiny device with "Bluetooth Low Energy (BLE) Connectivity" increases battery power consumption. Model performance and invocation are logarithmic time complexity with $O\,(K\,d\,\|x\|0\,\log q)$ with if-else invocation.

> ⚛ For Boosting Tree ML models such as XGBoost, the impact of BLE connectivity & model invocation, i.e., the size of the model, in terms of spread and depth, and frequency of calls will have considerable battery power implications.

Modeling Wi-Fi Connectivity

Wi-Fi connectivity enables to offload data storage to the Edge devices or call to the central Cloud Server. Additionally, enables to download Firmware via OTA. The major limitation is many of the constrained devices are not suited for Wi-Fi as it drains huge power from the battery.

Connectivity Wi-Fi		Hardware Constraint No Constraint			
		Memory	Power Source	Storage	Computational Power
ML Model Accuracy	Self-Contained & Updated only on Hardware Refresh	Memory at inference level is independent of time complexity- dependent upon Number of trees (K), depth (d), Number of features (q) – **O (K d ||x||0 log q)**	On board Wi-Fi demands more power. For battery power devices, Wi-Fi option is less favorable.	Data could be offloaded to external devices or central Cloud.	Time complexity- dependent upon Number of trees (K), depth (d), Number of features (q) – **O (K d ||x||0 log q)**
	OTA Firmware	OTA allows to deploy the latest and greatest optimized model.	Onboard Wi-Fi demands more power & Time complexity- dependent upon Number of trees (K), depth (d), Number of features (q) – **O (K d ||x||0 log q)**	Data could be offloaded to external devices or to the central Cloud.	OTA allows to deploy the latest and greatest optimized model. However, Time complexity demands high computational power.
	Model Type: Eager Learner	XGB MLs are eager learner with time complexity- dependent upon Number of trees (K), depth (d), Number of features (q) – **O (K d ||x||0 log q)**	Both Wi-Fi and Time Complexity make higher power consumption	Periodic offloading of Sensor data to Edge devices lowers the storage needs.	Time complexity- dependent upon Number of trees (K), depth (d), Number of features (q) – **O (K d ||x||0 log q)**
	Lazy Learner	N/A	N/A	N/A	N/A
	Model Invocation	Time complex **O (K d ||x||0 log q)** and Inline if-else model invocation.	Time complexity- dependent upon Number of trees (K), depth (d), Number of features (q) – **O (K d ||x||0 log q)**, continuous Bluetooth advertisements could increase power needs	Periodic offloading of Sensor data to Edge devices lowers the storage needs.	Time complexity- dependent upon Number of trees (K), depth (d), Number of features (q) – **O (K d ||x||0 log q)**,

Overall design consideration: Operating XGB based Machine Learning algorithms in a constrained or Tiny device with "Wi-Fi Connectivity" increases battery power consumption. Model performance & invocation is logarithmic time complexity is $O\ (K\ d\ \|x\|0\ \log\ q)$ with if-else invocation.

Connectivity—Model Accuracy tradeoff

Modeling Memory

Since the model is a decision rule and does not require to be memory resident[275] nor heap[276] computation and doesn't require historical data for accuracy inference, the hardware memory is not a constraint for the Model accuracy. That is, the memory (Data movement or access) does not influence power or computational hardware economy.

Hardware Memory		Connectivity 🔒 Constraint ⚪ No Constraint		
		No Connectivity	Bluetooth Low Energy (BLE)/LoRa	Wi-Fi
Model Accuracy	Self-Contained & Updated only on Hardware Refresh	⚪	⚪	⚪
			Since model update is cycled on Hardware refresh (SLA half-yearly or yearly), there would be staleness of the model. However, **memory constraint has no influence on the Accuracy.**	
	OTA Firmware	⚪	⚪	⚪
		Since the model gets updated OTA as part of Firmware update, no staleness of model. Memory constraint has no influence on the model accuracy.		
	Model Type: Eager Learner	⚪	⚪	⚪
	Lazy Learner	N/A	N/A	N/A
	Model Invocation	⚪	⚪	⚪
		In-line model invocation and not memory intense—no memory constraint on the model.		

Design Considerations: Since model update is cycled on Hardware refresh (SLA half-yearly or yearly), there would be staleness of the model. However, memory constraint has no influence on the Accuracy. Model is decision rule and will not require data movement between processor & memory space. In-line model invocation and not memory intense—no memory constraint on the model.

Modeling Processing Power

The Model deployed as decision rule and invoked as an in-line function and does not require extra-computation power.

---

[275] Memory resident - https://www.trendmicro.com/vinfo/us/security/definition/memory-resident
[276] Stack vs. Heap allocation - https://www.geeksforgeeks.org/stack-vs-heap-memory-allocation/

Hardware Processing Power	Connectivity 🔒 Constraint ⭘ No Constraint			
	No Connectivity	BLE	Wi-Fi	Design Notes
Model Accuracy — Hardware Refresh	🔒	🔒	🔒	Since model update is cycled on Hardware refresh (SLA half-yearly or yearly), there could be staleness in Model accuracy. Additionally, due to logarithmic time complexity, the power consumption is higher.
OTA Firmware Update	🔒	🔒	🔒	Logarithmic Time complexity- dependent upon Number of trees (K), depth (d), Number of features (q) – O (K d ‖x‖0 log q) – increases processing power needs.
Model Type: Eager Learner	🔒	🔒	🔒	Logarithmic Time complexity- dependent upon Number of trees (K), depth (d), Number of features (q) – O (K d ‖x‖0 log q) – increases processing power needs.
Lazy Learner	N/A	N/A	N/A	
Model Invocation	🔒	🔒	🔒	Logarithmic Time complexity- dependent upon Number of trees (K), depth (d), Number of features (q) – O (K d ‖x‖0 log q) – increases processing power needs.

Design Considerations: Processing power, given logarithmic time complexity, is higher as the model time complexity logarithmic and increases at logarithmic scale based on number of feature, depth of Tree and breadth of tree.

Modeling Storage

The Model deployed as an in-line function and does not require huge storage. Nonetheless, as hardware device collects sensor data, the data needs to be stored on device local storage (offload to Edge Gateway or Mobile device) before uploads to Central Storage or Corporate Cloud storage.

Hardware **Storage**	Connectivity 🔒 Constraint ⭘ No Constraint			
	No Connectivity	BLE	Wi-Fi	Design Notes
Model Accuracy — Hardware Refresh	🔒	⭘	⭘	Since model update is cycled on Hardware refresh (SLA half-yearly or yearly), the data storage needs to be large to store on the device. For BLE and Wi-Fi connectivity modes, the storage could be offloaded to gateway or mobile device and could help to concise storage.
OTA Firmware Update	🔒	⭘	⭘	Since the model gets updated OTA as part of Firmware update, no staleness of model. Storage could be constraint under "No Connectivity" but not a constraint for BLE and Wi-Fi.
Model Type: Eager Learner	🔒	⭘	⭘	Model is a Logarithmic Time complexity- dependent upon Number of trees (K), depth (d), Number of features (q) – O (K d ‖x‖0 log q) and will not require data movement between processor & memory space. Storage could be constraint under "No Connectivity" but not a constraint for BLE and Wi-Fi.
Lazy Learner	N/A	N/A	N/A	
Model Invocation	🔒	⭘	⭘	In-line model invocation and not memory intense—no memory constraint on the model. Storage could be constraint under "No Connectivity" but not a constraint for BLE and Wi-Fi.

Design Considerations: Since model update is cycled on hardware refresh (SLA half-yearly or yearly), there could be staleness in Model accuracy.

Modeling Environmental Perturbations

Model deployed under geographical conditions with huge temperature or humidity variations, the influence of temperature on the battery could lead into accuracy consideration. The Humidity accumulation is also a factor to be considered in the model calculation.

Hardware Environmental Perturbations	Connectivity 🔒 Constraint ⚪ No Constraint			
	No Connectivity	BLE	Wi-Fi	Design Notes
Model Accuracy — Hardware Refresh	🔒 Staleness of the model is a constraint in taming (adaptive) environmental perturbations.	⚪	⚪	Model should factor in geographical sensitivity as part of the calculations. The model coefficients need to tame environmental perturbations. Given model is a logarithmic time complex, processing and computational power requirement are higher.
OTA Firmware Update	⚪	⚪	⚪	The OTA could fin-tune sensitive model coefficients to overcome environmental perturbations.
Model Type: Eager Learner	🔒	🔒	🔒	Model exhibits Logarithmic Time complexity- dependent upon Number of trees (K), depth (d), Number of features (q) – O (K d ‖x‖0 log q) – this could lead into more computational as model lead to traverse more conditions.
Lazy Learner	N/A	N/A	N/A	No influence of environmental perturbations on data movement.
Model Invocation	🔒	🔒	🔒	Model exhibits Logarithmic Time complexity- dependent upon Number of trees (K), depth (d), Number of features (q) – O (K d ‖x‖0 log q) – this could lead into more computational as model lead to traverse more conditions.

Design Considerations: Environmental perturbations increase the complexity and depth of XGBoost based decision models. Given the conditions, the code complexity is higher for the model. As a result, more design constraints could be a factor for XGBoost models.

 XGBoost models computational and power processing are higher under environmental perturbations.

Connectivity to Hardware trade-off

Modeling Hardware Refresh

In this mode, the Model is only getting refreshed whenever hardware refresh takes place. Generally, this could be half-yearly or yearly, depending upon the SLA signed with customer. With this constraint, there is staleness on the model accuracy. That is, adaptive learning is only deferred when a new model gets deployed. The model is a decision rule function that is invoked as part of in-line code. So, no special requirements for memory, power, and computational processing. And hence, no constraints.

Model Accuracy Model Refresh		Hardware 🔒 Constraint ◯ No Constraint					
		Power	Memory	Storage	Processing	Environmental	Design Notes
Connectivity	No Connectivity	🔒	◯	🔒	🔒	🔒	Model exhibits Logarithmic Time complexity-dependent upon Number of trees (K), depth (d), Number of features (q) – O (K d ‖x‖0 log q) – except for memory, K, d, q constraints all other hardware design characteristics.
	BLE	🔒	◯	◯	🔒	🔒	Model exhibits Logarithmic Time complexity-dependent upon Number of trees (K), depth (d), Number of features (q) –
	Wi-Fi	🔒	◯	◯	🔒	🔒	O (K d ‖x‖0 log q) – except for memory and storage, K, d, q constraints all other hardware design characteristics.

Design Considerations: Since model update is cycled on Hardware refresh (SLA half-yearly or yearly), there could be staleness in Model accuracy. ☞ This has influence on Power, Processing Power and Environmental perturbations.

Modeling Over-The-Air (OTA) Firmware Update

In this mode, the Model is only getting refreshed over the air as part of firmware update. Except for no connectivity mode, the model is the latest and learned with the most recent collected data. That is, adaptive learning is current and gets deployed as part of OTA. The model is a logarithmic function that is invoked as part of in-line code. So, extra care required to handle power and computational processing.

Model Accuracy OTA		Hardware 🔒 Constraint ◯ No Constraint					
		Power	Memory	Storage	Processing	Environmental	Design Notes
Connectivity	No Connectivity	🔒	◯	🔒	🔒	🔒	Storage needs are higher & inaccuracies due to environmental perturbations.
	BLE	🔒	◯	◯	🔒	🔒	With the model is updated vis OTA, the latest learnings will have higher accuracies.
	Wi-Fi	🔒	◯	◯	🔒	🔒	

Design Considerations: Firmware updates over the air enables deployment of the latest and greatest model. This would reduce the need to have higher hardware capacities.

 Over the air Firmware updates enable the model to be equipped with most recent learnings.

Modeling Active Learner vs. Lazy Learner

The model is active learner with no special requirements to be in-memory, memory resident or data movements. The model is a Logarithmic Time complexity—dependent upon Number of trees (K), depth (d), Number of features (q) – $O (K d ‖x‖0 \log q)$ that is invoked as part of the in-line code. So, extra processing care needs to take care of power and computational processing.

Model Accuracy Learner		Hardware 🔒 Constraint ○ No Constraint					
		Power	Memory	Storage	Processing	Environmental	Design Notes
Connectivity	No Connectivity	🔒	○	🔒	🔒	🔒	Storage needs are higher & inaccuracies due to environmental perturbations.
	BLE	🔒	○	○	🔒	🔒	With the model is updated via OTA, the latest learnings will have higher accuracies.
	Wi-Fi	🔒	○	○	🔒	🔒	

Design Considerations: Model Learner type is not subject to change for different connectivity modes and hardware modes.

Modeling Model Invocation

The model is invoked as an in-line code, and not a heap or memory resident call. Except storage constraint as part of No Connectivity mode, the model has no special restrictions.

Model Accuracy Invocation		Hardware 🔒 Constraint ○ No Constraint					
		Power	Memory	Storage	Processing	Environmental	Design Notes
Connectivity	No Connectivity	🔒	○	🔒	🔒	🔒	Storage needs are higher & inaccuracies due to environmental perturbations.
	BLE	🔒	○	○	🔒	🔒	With the model is updated via OTA, the latest learnings will have higher accuracies.
	Wi-Fi	🔒	○	○	🔒	🔒	

Design Considerations: Model invocation is not subject to change for different connectivity modes and hardware modes.

Random Forests

Random Forests are ensembles of decision trees. Multiple decision trees are trained and aggregated to form a model that is more performant than any of the individual trees. This general idea is the purpose of ensemble learning.

There are many types of ensemble methods. Random Forests are an instance of bootstrap aggregating, also called bagging, where models are trained on randomly drawn subsets of the training set. Random Forests yield information about the importance of each feature for the classification or regression task.samples of the training dataset are taken with replacement, but the trees are constructed in a way that reduces the correlation between individual classifiers.[277] Specifically, rather than greedily choosing the best split point in the construction of the tree, only a random subset of features is considered for each split.

Random Forest Classifier

A Random Forest is a meta estimator that fits a number of decision tree classifiers on various sub-samples of the dataset and uses averaging to improve the predictive accuracy and control over-fitting. The sub-sample

---

[277] Ensemble Machine Learning Algorithms in Python with scikit-learn - https://machinelearningmastery.com/ensemble-machine-learning-algorithms-python-scikit-learn/

size is always the same as the original input sample size but the samples are drawn with replacement if bootstrap=True (default).

The example below (see Table 8) provides an example of Random Forest for classification with 100 trees and split points chosen from a random selection of 3 features.

Table 8: Random Forest Mode Code

```
# Random Forest Classification
import pandas
from sklearn import model_selection
from sklearn.ensemble import RandomForestClassifier
url = "https://storageconstts.blob.core.windows.net/tsdata/pima-indians-diabetes.csv"
names = ['preg', 'plas', 'pres', 'skin', 'test', 'mass', 'pedi', 'age', 'class']
dataframe = pandas.read_csv(url, names=names)
array = dataframe.values
X = array[:,0:8]
Y = array[:,8]
seed = 7
num_trees = 100
max_features = 3
kfold = model_selection.KFold(n_splits=10, random_state=seed)
model = RandomForestClassifier(n_estimators=num_trees, max_features=max_features)
#results = model_selection.cross_val_score(model, X, Y, cv=kfold)
# Train
results= model.fit(X,Y)
# Extract single tree
estimator = model.estimators_[5]
```

Output

The output of Random Forest is presented in Table 9.

Table 9: Random Forest Output

```
/usr/local/lib/python3.6/dist-packages/sklearn/model_selection/_split.py:296: FutureWarning: Setting a random_state has no effect
since shuffle is False. This will raise an error in 0.24. You should leave random_state to its default (None), or set shuffle=True.
 FutureWarning
0.7720779220779221
```

Print Tree: The following code contains Radom Forest model Tree generator (Table 10) please see Figure 34 for the developed Tree:

Table 10: Random Forest Print Tree

```
from sklearn.tree import export_graphviz
# Export as dot file
export_graphviz(estimator, out_file='tree.dot',
         rounded = True, proportion = False,
         precision = 2, filled = True)
# Convert to png using system command (requires Graphviz)
from subprocess import call
call(['dot', '-Tpng', 'tree.dot', '-o', 'tree.png', '-Gdpi=600'])
# Display in jupyter notebook
from IPython.display import Image
Image(filename = 'tree.png')
```

Code snapshot: the code for Random Forest is as presented in Figure 35

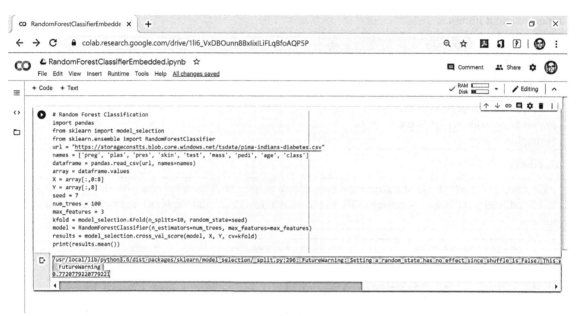

Figure 34: Random Forest Code IDE

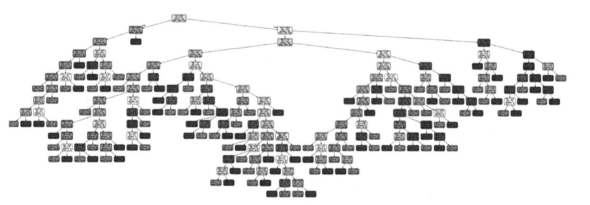

Figure 35: Random Forest Tree

Random Forest Regressor

A Random Forest[278] is a meta estimator that fits a number of classifying decision trees on various sub-samples of the dataset and uses averaging to improve the predictive accuracy and control over-fitting. The sub-sample size is always the same as the original input sample size but the samples are drawn with replacement if bootstrap=True (default).

The example below provides an example of Random Forest for classification with 100 trees and split points chosen from a random selection of 3 features.

---

[278] Random Forest Classifier - https://scikit-learn.org/stable/modules/generated/sklearn.ensemble.RandomForestClassifier.html?highlight=randomforestclassifier#sklearn.ensemble.RandomForestClassifier

Import required libraries:

```
import numpy as np
import pandas as pd
import sklearn as sk
import sklearn.datasets as skd
import sklearn.ensemble as ske
import matplotlib.pyplot as plt
%matplotlib inline
```

Load dataset:

```
train = pd.read_csv("https://storageconstts.blob.core.windows.net/tsdata/xgboost_dataset_train.csv")
test  = pd.read_csv("https://storageconstts.blob.core.windows.net/tsdata/xgboost_dataset_test.csv")
train
```

	PassengerId	Survived	Pclass	Name	Sex	Age	SibSp	Parch	Ticket	Fare	Cabin	Embarked
0	1	0	3	Braund, Mr. Owen Harris	male	22.0	1	0	A/5 21171	7.2500	NaN	S
1	2	1	1	Cumings, Mrs. John Bradley (Florence Briggs Th...	female	38.0	1	0	PC 17599	71.2833	C85	C
2	3	1	3	Heikkinen, Miss. Laina	female	26.0	0	0	STON/O2. 3101282	7.9250	NaN	S
3	4	1	1	Futrelle, Mrs. Jacques Heath (Lily May Peel)	female	35.0	1	0	113803	53.1000	C123	S
4	5	0	3	Allen, Mr. William Henry	male	35.0	0	0	373450	8.0500	NaN	S
...	...	...	...	...	...	...	...	...	...	...	...	...
886	887	0	2	Montvila, Rev. Juozas	male	27.0	0	0	211536	13.0000	NaN	S
887	888	1	1	Graham, Miss. Margaret Edith	female	19.0	0	0	112053	30.0000	B42	S
888	889	0	3	Johnston, Miss. Catherine Helen "Carrie"	female	NaN	1	2	W./C. 6607	23.4500	NaN	S
889	890	1	1	Behr, Mr. Karl Howell	male	26.0	0	0	111369	30.0000	C148	C
890	891	0	3	Dooley, Mr. Patrick	male	32.0	0	0	370376	7.7500	NaN	Q

891 rows × 12 columns

Load train labels:

```
data=train[['Pclass', 'Age', 'Fare', 'SibSp', 'Parch']]
label= train['Survived']
```

	Pclass	Age	Fare	SibSp	Parch
1	1	38.0	71.2833	1	0
3	1	35.0	53.1000	1	0
6	1	54.0	51.8625	0	0
10	3	4.0	16.7000	1	1
11	1	58.0	26.5500	0	0
...	...	...	...	...	...
871	1	47.0	52.5542	1	1
872	1	33.0	5.0000	0	0
879	1	56.0	83.1583	0	1
887	1	19.0	30.0000	0	0
889	1	26.0	30.0000	0	0

183 rows × 5 columns

Model the Random Forest Regressor:

```
reg = ske.RandomForestRegressor()
reg.fit(data, label)
fig, ax = plt.subplots(1, 1, figsize=(8, 3))
labels = data['Age'][fet_ind]
pd.Series(fet_imp, index=labels).plot('bar', ax=ax)
ax.set_title('Features importance')
```

The following feature importance:

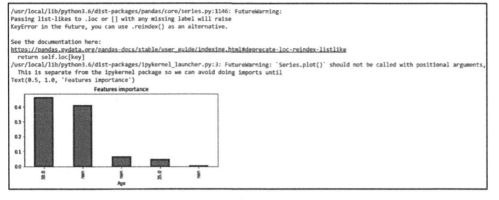

Convert model to equation & image:

```
from sklearn import tree
tree.export_graphviz(reg.estimators_[0],
            'tree.dot')
from sklearn import tree
tree.export_graphviz(reg.estimators_[0],'tree.dot')
import pydot
(graph,) = pydot.graph_from_dot_file('tree.dot')
graph.write_png('RandomForrest.png')
```

Random Forest Model Image

The following Random Forest Regressor (Figure 36) is the output of the model:

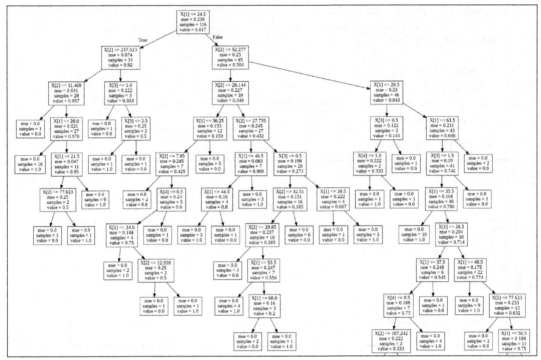

Figure 36: Random Forest Model Regressor Image

Model Generated Code

The Model Code is as per the following table (please see Table 10):

Table 10: Random Forest Graph Code

```
'digraph Tree {\nnode [shape=box] ;\n0 [label="X[1] <= 24.5\\nmse = 0.236\\nsamples = 116\\nvalue = 0.617"] ;\n1 [label="X[2]
<= 237.523\\nmse = 0.074\\nsamples = 31\\nvalue = 0.92"] ;\n0 -> 1 [labeldistance=2.5, labelangle=45, headlabel="True"] ;\n2
[label="X[2] <= 11.469\\nmse = 0.041\\nsamples = 28\\nvalue = 0.957"] ;\n1 -> 2 ;\n3 [label="mse = 0.0\\nsamples = 1\\nvalue
= 0.0"] ;\n2 -> 3 ;\n4 [label="X[1] <= 20.0\\nmse = 0.021\\nsamples = 27\\nvalue = 0.978"] ;\n2 -> 4 ;\n5 [label="mse = 0.0\\
nsamples = 16\\nvalue = 1.0"] ;\n4 -> 5 ;\n6 [label="X[1] <= 21.5\\nmse = 0.047\\nsamples = 11\\nvalue = 0.95"] ;\n4 -> 6 ;\n7
[label="X[2] <= 77.623\\nmse = 0.25\\nsamples = 2\\nvalue = 0.5"] ;\n6 -> 7 ;\n8 [label="mse = 0.0\\nsamples = 1\\nvalue = 0.0"]
;\n7 -> 8 ;\n9 [label="mse = 0.0\\nsamples = 1\\nvalue = 1.0"] ;\n7 -> 9 ;\n10 [label="mse = 0.0\\nsamples = 9\\nvalue = 1.0"] ;\n6
-> 10 ;\n11 [label="X[3] <= 1.0\\nmse = 0.222\\nsamples = 3\\nvalue = 0.333"] ;\n1 -> 11 ;\n12 [label="mse = 0.0\\nsamples = 1\\
nvalue = 0.0"] ;\n11 -> 12 ;\n13 [label="X[3] <= 2.5\\nmse = 0.25\\nsamples = 2\\nvalue = 0.5"] ;\n11 -> 13 ;\n14 [label="mse =
0.0\\nsamples = 1\\nvalue = 1.0"] ;\n13 -> 14 ;\n15 [label="mse = 0.0\\nsamples = 1\\nvalue = 0.0"] ;\n13 -> 15 ;\n16 [label="X[2]
<= 52.277\\nmse = 0.25\\nsamples = 85\\nvalue = 0.504"] ;\n0 -> 16 [labeldistance=2.5, labelangle=-45, headlabel="False"] ;\n17
[label="X[2] <= 26.144\\nmse = 0.227\\nsamples = 39\\nvalue = 0.349"] ;\n16 -> 17 ;\n18 [label="X[1] <= 36.25\\nmse = 0.133\\
nsamples = 12\\nvalue = 0.158"] ;\n17 -> 18 ;\n19 [label="X[2] <= 7.85\\nmse = 0.245\\nsamples = 7\\nvalue = 0.429"] ;\n18 -> 19
;\n20 [label="mse = 0.0\\nsamples = 2\\nvalue = 0.0"] ;\n19 -> 20 ;\n21 [label="X[4] <= 0.5\\nmse = 0.24\\nsamples = 5\\nvalue
= 0.6"] ;\n19 -> 21 ;\n22 [label="X[1] <= 34.0\\nmse = 0.188\\nsamples = 4\\nvalue = 0.75"] ;\n21 -> 22 ;\n23 [label="mse = 0.0\\
nsamples = 2\\nvalue = 1.0"] ;\n22 -> 23 ;\n24 [label="X[2] <= 12.938\\nmse = 0.25\\nsamples = 2\\nvalue = 0.5"] ;\n22 -> 24 ;\
n25 [label="mse = 0.0\\nsamples = 1\\nvalue = 0.0"] ;\n24 -> 25 ;\n26 [label="mse = 0.0\\nsamples = 1\\nvalue = 1.0"] ;\n24 ->
26 ;\n27 [label="mse = 0.0\\nsamples = 1\\nvalue = 0.0"] ;\n21 -> 27 ;\n28 [label="mse = 0.0\\nsamples = 5\\nvalue = 0.0"] ;\n18
-> 28 ;\n29 [label="X[2] <= 27.735\\nmse = 0.245\\nsamples = 27\\nvalue = 0.432"] ;\n17 -> 29 ;\n30 [label="X[1] <= 46.5\\nmse
= 0.083\\nsamples = 7\\nvalue = 0.909"] ;\n29 -> 30 ;\n31 [label="X[1] <= 44.5\\nmse = 0.16\\nsamples = 4\\nvalue = 0.8"] ;\n30
-> 31 ;\n32 [label="mse = 0.0\\nsamples = 3\\nvalue = 1.0"] ;\n31 -> 32 ;\n33 [label="mse = 0.0\\nsamples = 1\\nvalue = 0.0"] ;\
n31 -> 33 ;\n34 [label="mse = 0.0\\nsamples = 3\\nvalue = 1.0"] ;\n30 -> 34 ;\n35 [label="X[3] <= 0.5\\nmse = 0.198\\nsamples
= 20\\nvalue = 0.273"] ;\n29 -> 35 ;\n36 [label="X[2] <= 32.51\\nmse = 0.151\\nsamples = 16\\nvalue = 0.185"] ;\n35 -> 36 ;\n37
[label="X[2] <= 29.85\\nmse = 0.237\\nsamples = 10\\nvalue = 0.385"] ;\n36 -> 37 ;\n38 [label="mse = 0.0\\nsamples = 3\\nvalue
= 0.0"] ;\n37 -> 38 ;\n39 [label="X[1] <= 53.5\\nmse = 0.247\\nsamples = 7\\nvalue = 0.556"] ;\n37 -> 39 ;\n40 [label="mse = 0.0\\
nsamples = 4\\nvalue = 1.0"] ;\n39 -> 40 ;\n41 [label="X[1] <= 68.0\\nmse = 0.16\\nsamples = 3\\nvalue = 0.2"] ;\n39 -> 41 ;\n42
```

Table 10 contd. ...

...Table 10 contd.

[label="mse = 0.0\\nsamples = 2\\nvalue = 0.0"] ;\n41 -> 42 ;\n43 [label="mse = 0.0\\nsamples = 1\\nvalue = 1.0"] ;\n41 -> 43 ;\n44 [label="mse = 0.0\\nsamples = 6\\nvalue = 0.0"] ;\n36 -> 44 ;\n45 [label="X[1] <= 39.5\\nmse = 0.222\\nsamples = 4\\nvalue = 0.667"] ;\n35 -> 45 ;\n46 [label="mse = 0.0\\nsamples = 1\\nvalue = 0.0"] ;\n45 -> 46 ;\n47 [label="mse = 0.0\\nsamples = 3\\nvalue = 1.0"] ;\n45 -> 47 ;\n48 [label="X[1] <= 29.5\\nmse = 0.23\\nsamples = 46\\nvalue = 0.643"] ;\n16 -> 48 ;\n49 [label="X[3] <= 0.5\\nmse = 0.122\\nsamples = 3\\nvalue = 0.143"] ;\n48 -> 49 ;\n50 [label="X[4] <= 1.0\\nmse = 0.222\\nsamples = 2\\nvalue = 0.333"] ;\n49 -> 50 ;\n51 [label="mse = 0.0\\nsamples = 1\\nvalue = 1.0"] ;\n50 -> 51 ;\n52 [label="mse = 0.0\\nsamples = 1\\nvalue = 0.0"] ;\n50 -> 52 ;\n53 [label="mse = 0.0\\nsamples = 1\\nvalue = 0.0"] ;\n49 -> 53 ;\n54 [label="X[1] <= 63.5\\nmse = 0.211\\nsamples = 43\\nvalue = 0.698"] ;\n48 -> 54 ;\n55 [label="X[3] <= 1.5\\nmse = 0.19\\nsamples = 41\\nvalue = 0.746"] ;\n54 -> 55 ;\n56 [label="X[1] <= 35.5\\nmse = 0.168\\nsamples = 40\\nvalue = 0.786"] ;\n55 -> 56 ;\n57 [label="mse = 0.0\\nsamples = 10\\nvalue = 1.0"] ;\n56 -> 57 ;\n58 [label="X[1] <= 38.5\\nmse = 0.204\\nsamples = 30\\nvalue = 0.714"] ;\n56 -> 58 ;\n59 [label="X[1] <= 37.5\\nmse = 0.248\\nsamples = 8\\nvalue = 0.545"] ;\n58 -> 59 ;\n60 [label="X[4] <= 0.5\\nmse = 0.188\\nsamples = 7\\nvalue = 0.75"] ;\n59 -> 60 ;\n61 [label="X[2] <= 107.242\\nmse = 0.222\\nsamples = 3\\nvalue = 0.333"] ;\n60 -> 61 ;\n62 [label="mse = 0.0\\nsamples = 2\\nvalue = 0.0"] ;\n61 -> 62 ;\n63 [label="mse = 0.0\\nsamples = 1\\nvalue = 1.0"] ;\n61 -> 63 ;\n64 [label="mse = 0.0\\nsamples = 4\\nvalue = 1.0"] ;\n60 -> 64 ;\n65 [label="mse = 0.0\\nsamples = 1\\nvalue = 0.0"] ;\n59 -> 65 ;\n66 [label="X[1] <= 49.5\\nmse = 0.175\\nsamples = 22\\nvalue = 0.774"] ;\n58 -> 66 ;\n67 [label="mse = 0.0\\nsamples = 9\\nvalue = 1.0"] ;\n66 -> 67 ;\n68 [label="X[2] <= 77.623\\nmse = 0.233\\nsamples = 13\\nvalue = 0.632"] ;\n66 -> 68 ;\n69 [label="mse = 0.0\\nsamples = 2\\nvalue = 0.0"] ;\n68 -> 69 ;\n70 [label="X[1] <= 50.5\\nmse = 0.188\\nsamples = 11\\nvalue = 0.75"] ;\n68 -> 70 ;\n71 [label="X[3] <= 0.5\\nmse = 0.222\\nsamples = 2\\nvalue = 0.333"] ;\n70 -> 71 ;\n72 [label="mse = 0.0\\nsamples = 1\\nvalue = 1.0"] ;\n71 -> 72 ;\n73 [label="mse = 0.0\\nsamples = 1\\nvalue = 0.0"] ;\n71 -> 73 ;\n74 [label="X[4] <= 1.5\\nmse = 0.13\\nsamples = 9\\nvalue = 0.846"] ;\n70 -> 74 ;\n75 [label="X[4] <= 0.5\\nmse = 0.076\\nsamples = 8\\nvalue = 0.917"] ;\n74 -> 75 ;\n76 [label="mse = 0.0\\nsamples = 4\\nvalue = 1.0"] ;\n75 -> 76 ;\n77 [label="X[2] <= 81.404\\nmse = 0.16\\nsamples = 4\\nvalue = 0.8"] ;\n75 -> 77 ;\n78 [label="mse = 0.0\\nsamples = 1\\nvalue = 0.0"] ;\n77 -> 78 ;\n79 [label="mse = 0.0\\nsamples = 3\\nvalue = 1.0"] ;\n77 -> 79 ;\n80 [label="mse = 0.0\\nsamples = 1\\nvalue = 0.0"] ;\n74 -> 80 ;\n81 [label="mse = 0.0\\nsamples = 1\\nvalue = 0.0"] ;\n55 -> 81 ;\n82 [label="mse = 0.0\\nsamples = 2\\nvalue = 0.0"] ;\n54 -> 82 ;\n}'

Model Equation

☛ First, the model equation is not possible for random forest. This is because the nature of random forest algorithm *inherently leads to destruction of any simple mathematical representation.*

Second, Random Forest has no coefficient of importance like that of regression model. The regression coefficients have an important role when dealing with parametric models, as they describe the parameters. In the case of linear regression, for example, the parameter $\mu = X\beta$, where β is the vector of regression coefficients. For trees, in the other hand, the "μ" (quoted, because there is no official μ in a non-parametric model) is just a constant, such as the mean for the regions formed by the explanatory variables. Then, the variables have nothing to do with the prediction, besides creating the regions. Random forest works by building decision trees & then aggregating them & hence the Beta values like regression have no counterpart in random forest. Though you do get the 'Variable Importance /Gini Index' values for the forest,

 The model equation is not possible for random forest. This is because the nature of random forest algorithm inherently leads to destruction of any simple mathematical representation.

Naïve Bayesian

Bayesian classifiers are statistical classifiers. They can predict class membership probabilities, such as the probability that a given tuple belongs to a particular class Bayesian.

Bayesian classification is based on Bayes' theorem. Studies comparing classification algorithms have found a simple Bayesian classifier known as the naïve Bayesian classifier to be comparable in performance with decision tree and selected neural network classifiers. Bayesian classifiers have also exhibited high accuracy and speed when applied to large databases [1].

Naïve Bayesian classifiers assume that the effect of an attribute value on a given class is independent of the values of the other attributes. This assumption is called class conditional independence . It is made to simplify the computations involved and, in this is considered "naïve" [1].

Before defining Bayesian theorem, I would like to introduce two more probability concepts that help in defining Bayesian theorem:

Marginal Probability: The probability of an event irrespective of the outcomes of other random variables, e.g., P(A). Recall that marginal probability is the probability of an event, irrespective of other random variables. if the variable is dependent upon other variables, then the marginal probability is the probability of the event summed over all outcomes for the dependent variables, called the sum rule.

Joint Probability: Probability of two (or more) simultaneous events, e.g., P(A and B) or P(A, B). The joint probability is the probability of two (or more) simultaneous events, often described in terms of events A and B from two dependent random variables, e.g., X and Y. The joint probability is often summarized as just the outcomes, e.g., A and B.

Conditional Probability: Probability of one (or more) event given the occurrence of another event, e.g., P(A given B) or P(A | B) or probability of A given B. Conditional probability in terms of joint probability:

$$P(A, B) = P(A \mid B) * P(B)$$
$$P(A \mid B) = P(A, B)/P(B)$$

Eq. (5)

The conditional probability can be calculated using the joint probability.

Another Way of Calculating Conditional Probability: Specifically, one conditional probability can be calculated using the other conditional probability; for example:

$$P(A|B) = P(B|A) * P(A)/P(B)$$

The reverse is also true, for example:

$$P(B|A) = P(A|B) * P(B)/P(A)$$

Naming the terms in the Bayesian theorem:

Mathematically, Bayes' theorem gives the relationship between the probabilities of A and B, P(A) and P(B), and the conditional probabilities of A given B and B given A, P(A|B) and P(B|A). In its most common form, it is:

$$P(A|B) = \frac{P(A|B) * P(B)}{P(A)}$$

Eq. (6)

First, in general, the result P(A|B) is referred to as the Posterior Probability and P(A) is referred to as Prior Probability [7].

- P(A|B): Posterior probability
- P(A): Prior probability

Sometimes P(B|A) is referred to as the likelihood and P(B) is referred to as the evidence

- P(B|A): Likelihood.
- P(B): Evidence.

This allows Bayes theorem to be restated as [7]:

Posterior = Likelihood * Prior/Evidence

Bayesian for Multivariate

X is given as:

$$X = (x_1, x_2, x_3, ..., x_n)$$

Where x_1, x_2, x_3, and x_n represent features of a dataset. The class variable, Y, is (please see Eq. 6).

$$P(y|\,x_1, x_2, x_3, ..., x_n) = \frac{P(x1|y) * P(x2|y) * P(x3|y)...P(xn|y)}{P(x1) * P(x2)*P(x3)...\& \,(xn)}$$

Eq. (7)

The denominator in the equation remains same and does not change:

$$P(y|\,x_1, x_2, x_3, ..., x_n)\; \alpha \; P(y) \prod_{i=1}^{n} P(xi|y)$$

$$y = \text{argmax}_y\, P(y) \prod_{i=1}^{n} P(xi|y)$$

Eq. (8)

Using the above function (Eq. 8), we can obtain the following types of Naïve Bayesian Classifier—that is defined by distribution function:

Multinomial Naïve Bayes

The term Multinomial Naive Bayes simply lets us know that each is a multinomial distribution, rather than some other distribution. This works well for data which can easily be turned into counts, such as word counts in text. The multinomial Naive Bayes classifier[279] is suitable for classification with discrete features (e.g., word counts for text classification). The multinomial distribution normally requires integer feature counts. However, in practice, fractional counts such as tf-idf may also work. Please see Figure 37 for Naïve Bayesian algorithm[280] to extract data from a document.

class `sklearn.naive_bayes.`**`MultinomialNB`**`(alpha=1.0, fit_prior=True, class_prior=None)`

```
TRAINMULTINOMIALNB(C, D)
1   V ← EXTRACTVOCABULARY(D)
2   N ← COUNTDOCS(D)
3   for each c ∈ C
4   do N_c ← COUNTDOCSINCLASS(D, c)
5       prior[c] ← N_c/N
6       text_c ← CONCATENATETEXTOFALLDOCSINCLASS(D, c)
7       for each t ∈ V
8       do T_ct ← COUNTTOKENSOFTERM(text_c, t)
9       for each t ∈ V
10      do condprob[t][c] ← (T_ct+1)/(∑_t'(T_ct'+1))
11  return V, prior, condprob

APPLYMULTINOMIALNB(C, V, prior, condprob, d)
1   W ← EXTRACTTOKENSFROMDOC(V, d)
2   for each c ∈ C
3   do score[c] ← log prior[c]
4       for each t ∈ W
5       do score[c] += log condprob[t][c]
6   return arg max_{c∈C} score[c]
```

Figure 37: Naive Bayes algorithm (multinomial model): Training and testing.

---

[279] MultinomialNB - https://scikit-learn.org/stable/modules/generated/sklearn.naive_bayes.MultinomialNB.html
[280] Naïve Bayesian Text Classification - https://nlp.stanford.edu/IR-book/html/htmledition/naive-bayes-text-classification-1.html

Bernoulli Naive Bayes

Naive Bayes classifier for multivariate Bernoulli models or multivariate Bernoulli model.[281]

Like MultinomialNB, this classifier is suitable for discrete data. The difference is that while MultinomialNB works with occurrence counts, BernoulliNB is designed for binary/Boolean features [8]. This is similar to the multinomial NaiveBayesian, but the predictors are Boolean variables. The parameters that we use to predict the class variable take up only values yes or no, for example if a word occurs in the text or not. Please see Figure 38 for Bernoulli algorithm[282] [8].

```
TRAINBERNOULLINB(C, D)
1   V ← EXTRACTVOCABULARY(D)
2   N ← COUNTDOCS(D)
3   for each c ∈ C
4   do Nc ← COUNTDOCSINCLASS(D, c)
5       prior[c] ← Nc/N
6       for each t ∈ V
7       do Nct ← COUNTDOCSINCLASSCONTAININGTERM(D, c, t)
8           condprob[t][c] ← (Nct + 1)/(Nc + 2)
9   return V, prior, condprob

APPLYBERNOULLINB(C, V, prior, condprob, d)
1   Vd ← EXTRACTTERMSFROMDOC(V, d)
2   for each c ∈ C
3   do score[c] ← log prior[c]
4       for each t ∈ V
5       do if t ∈ Vd
6           then score[c] += log condprob[t][c]
7           else score[c] += log(1 − condprob[t][c])
8   return arg max_{c∈C} score[c]
```

Figure 38: NB Algorithm (Bernoulli) [8]

Gaussian Naive Bayes

Gaussian NB implements the Gaussian Naive Bayes algorithm[283] for classification. The likelihood of the features assumed to be Gaussian is based on the following (see Eq. 9):
Where

- μ is mean
- σ is standard deviation of the normal distribution (please see Figure 39)

The distribution equation is as follows:

$$P(x_i \mid y) = \frac{1}{\sqrt{2\pi\sigma_y^2}} exp\left(-\frac{(x_i - \mu_y)^2}{2\sigma_y^2} \right)$$

Eq. (9)

The parameters σy and μy are estimated using maximum likelihood.

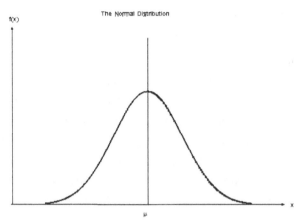

Figure 39: Normal Distribution

Device Health Prognostics

Naïve Bayesian is a very useful Machine Learning algorithm for devices that are operating in constrained environments with no connectivity or intermittent connectivity issues. The Bayesian algorithm is used to identify device prognostics, device calibration, and device intra-die [3] outliers.

Data Issues

The following field dataset (please see Field Data Issues) contains device intra-die Sensor readings issues. The embedded system was deployed in fields to collect the temperature and humidity data for five months. The hourly data collected from October 30, 2016 to January 18, 2017. The Sensors are installed in Vizag and Punjab, India.

We have observed the reporting of humidity values greater than 100% during the five months. For instance, the below figure has the hourly temperature plot grouped by time of the day for the period of October 30, 2016 to January 18, 2017. The x-axis 0:00 represents mid-night 12:00 hour and value 9.36 data at morning 9:36 am. For some dates, the data humidity values are reported more than 140%. This could have been (a) corruption of value reading at the sensor due to variations in temperature and humidity (intra-die issue), (b) residual humidity that could be as part of the Sensor case, or (c) could be due to voltage corruption. During periods of low VCC, the Flash program can be corrupted because the supply voltage is too low for the CPU and the Flash to operate properly. Irrespective of source corruption, the goal of the Naïve Bayesian machine learning is to detect the corruption and auto-adjust or inform data corruptions.

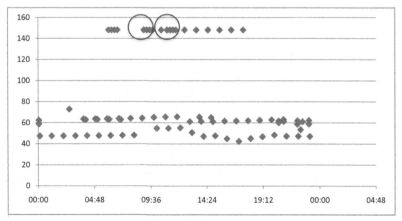

Figure: Field Data Issues

Device Malfunction Observations

- Device battery was low in first three months
- Device battery was full in greater than 6 months
- Sensor reading normal for extreme temperatures during high battery mode for first three months
- Sensor reading normal for extreme humidity during high battery mode for three to six months
- Sensor reading normal for extreme humidity event during high battery mode for first three months
- Sensor reading normal for extreme humidity event during high battery mode for three to six months
- Sensor reading was outlier during normal temperature during low battery mode – six months and greater
- Sensor reading was normal for extreme temperature for low battery mode six months or greater
- Sensor reading was normal for extreme humidity for low battery mode six months or greater.

Data Models

Device Prognostics refers to the prediction of device health. The prognostics is performed on devices large, small, and tiny. The advantage of employing device prognostics is to identify useful life of a device. Through application of device prognostics, the device manufacturer can initiate a service call, a truck roll, or sending a field agent to the device install base to correct issues of the device in a proactive manner. This will help the customer to get just-in service and prevents any business outages, thus ensuring the continuity of customer care.

The device prognostics plays an even important role when analyzing the health of extremely constrained devices as these devices are distributed geographically and serviceability is extremely expensive [9]. In the following dataset, the device call records of extremely constrained device data have been captured to perform device prognostics using Naïve Bayesian (Table 12).

Table 12: Device Malfunction Dataset

Weather	Sensor Readings	Battery Charge	Days Deployed	Device Malfunction
ExtremeTemperature	Normal	Full	3 months – 6 months	Yes
Normal	Outlier	Medium	First Three Months	Yes
Normal	Outlier	Low	Greater than 6 months	Yes
ExtremeTemperature	Normal	Low	First Three Months	Yes
ExtremeTemperature	Outlier	Low	Greater than 6 months	No
ExtremeTemperature	Normal	Medium	First Three Months	Yes
ExtremeHumidity	Normal	Medium	First Three Months	Yes
ExtremeHumidity	Normal	Low	Greater than 6 months	Yes
ExtremeHumidity	Outlier	Full	First Three Months	No
ExtremeHumidity	Outlier	Low	Greater than 6 months	No
Normal	Outlier	Low	First Three Months	Yes
Normal	Outlier	Low	3 months – 6 months	No
ExtremeHumidity	Outlier	Low	Greater than 6 months	No
ExtremeTemperature	Outlier	Full	First Three Months	No
Normal	Normal	Medium	3 months – 6 months	No
Normal	Normal	Low	3 months – 6 months	Yes
ExtremeHumidity	Normal	Full	First Three Months	Yes
Normal	Normal	Medium	3 months – 6 months	Yes

Figure 40: Tiny IoT PCB Board

The device that collected data consists of the following (please see Figure 40):

Electrical board circuit: The Sensor consists of Temperature and Humidity Sensors and casing of the device [9] [10]. The sensor is from Hanumayamma Innovations and Technologies, Inc.

Data Features & Attributes: The above table contains device prognostics data table that will be shipped as part of the device to aid device prognostics in the extreme constrained devices.

1. Weather
    a. Has three tuple values: Normal Temperature, High Temperature, and High Humidity. Generally, other Climate Events include: Rise of tide, Flooding, Rise of noise and Wind chill warning. For the specific use case, the rise of temperature and humidity play an important role.
2. Sensor Readings: this feature or attribute has two tuple values – Normal and Outlier. Normal Temperature or Humidity values are within range of ambient temperatures. Outlier values are extreme temperature or humidity values. For example, 140% of Humidity value is consider extreme and outlier. The Outlier values could be due to battery or data corruption issues.
3. Battery Charge: The amount of unused battery charge. It has three tuple values: Low (0% – 20%), Medium (20% – 60%) and Full (60% to 100%).
4. Days Deployed: This attribute is a numeric compute value. It indicates the number of days the Sensor is being deployed in the field. It has three tuple values: First Three Months (0 to 100 days), 3 months – 6 months, and greater than 6 months.

The purpose of Naïve Bayesian model is to predict device malfunction upon receiving the next temperature and humidity value.

Since there are 30% more "Device Malfunction = No" (10 examples) as compared to examples of "Device Malfunction = Yes" (7 examples), it is reasonable to believe that a new unobserved case is almost 30% as likely to have class of "No" as compared to "Yes". In the Bayesian paradigm, this belief, based on previous experience, is known as the prior probability.

Since there are 18 available data records, 7 of which are No and 11 are Yes, our prior probabilities for class membership are as follows:

- Prior Probability P (DeviceMalfunction = No) = 7/18
- Prior Probability P(DeviceMalfunction = Yes) = 11/18

Likelihood

Let new sensor values (temperature and humidity) read by the device. Assuming that the device buckets the value either one of the two values: Normal or Outlier. Based on the temperature and humidity, the Weather value is derived—either Normal, Extreme Temperature or Extreme Humidity.[284]

---

[284] Naïve Bayesian - . https://dzone.com/articles/naive-bayes-tutorial-naive-bayes-classifier-in-pyt

Let X = (Weather = Normal, Sensor Readings = Outlier, Battery Charge = Full, Days Deployed = First Three Months), then we have to compute the conditional probabilities that are given as underlined text in the following table:

- First, we will create a frequency table using each attribute of the dataset.

Frequency Table		DeviceMalfunction	
		No	Yes
Weather	Normal	3	5
	ExtremeTemperature	3	1
	ExtremeHumidity	4	2

Frequency Table		DeviceMalfunction	
		No	Yes
SensorReadings	Normal	2	5
	Outlier	8	3

Frequency Table		DeviceMalfunction	
		No	Yes
Battery Charge	Full	4	2
	Low	4	4
	Medium	1	2

Frequency Table		DeviceMalfunction	
		No	Yes
DaysDeployed	First Three Months	4	5
	Greater than 6 months	4	1
	3 months - 6 months	2	2

- For each frequency table, we will generate a likelihood table.

Likelihood Table		DeviceMalfunction		
		No	Yes	
Weather	Normal	3/10 $P(x\|c) = P(Normal\|No)$	5/8 $P(x\|c) = P(Normal\|Yes)$	8/18 $P(x) = P(Normal)$
	ExtremeTemperature	3/10 $P(x\|c) = P(ExtTemp\|No)$	1/8 $P(x\|c) = P(ExtTemp\|Yes)$	4/18 $P(x) = P(ExtTemp)$
	ExtremeHumidity	4/10 $P(x\|c) = P(ExtHumd\|No)$	2/8 $P(x\|c) = P(ExtHumd\|Yes)$	6/18 $P(x) = P(ExtHumd)$
		10/18 $P(c) = P(No)$	8/18 $P(c) = P(Yes)$	

$P(x|c) = P$ (Normal|Yes) = 5/8

$P(x) = P(\text{Normal}) = 8/18$

$P(c) = P(\text{Yes}) = 8/18$

Likelihood of Device Malfunction 'Yes' given weather is 'Normal' is

$P(c|x) = P(Yes|Normal) = P(Normal|Yes)* P(Yes)/P(Normal) = (5/8 \times 8/18)/8/18 = 5/18 = 0.27$ (see Eq. 6)

Similarly, the likelihood of Device Malfunction 'No' given weather is 'Normal' is

$P(c|x) = P(No|Normal) = P(Normal|No)* P(No)/P(Normal) = (3/10) \times (10/18)/(8/18) = 3/8 = 0.375$

Likelihood of Device Malfunction 'Yes' given weather is 'ExtremeTemperature':

P(c|x) = P(Yes| ExtremeTemperature) = P(ExtremeTemperature|Yes)* P(Yes)/P(ExtremeTemperature) = (1/8 x 8/18)/4/18 = 1/4 = 0.25

Similarly, the likelihood of Malfunction 'No' given weather is 'ExtremeTemperature'is

P(c|x) = P(No|ExtremeTemperature) = P(ExtremeTemperature|No)* P(No)/P(ExtremeTemperature) = (3/10) X (10/18)/(4/18) = 3/4 = 0.75

Likelihood of 'Yes' given 'ExtremeHumid' is

P(c|x) = P(Yes|ExtremeHumid) = P(ExtremeHumid|Yes)* P(Yes)/P(ExtremeHumid) = (2/8 x 8/18)/6/18 = 1/3 = 0.33

Similarly, the likelihood of 'No' given 'ExtremeHumid' is

P(c|x) = P(No|ExtremeHumid) = P(ExtremeHumid|No)* P(No)/P(ExtremeHumid) = (4/10) X (10/18)/ (6/18) = 2/3 = 0.166

Likelihood Table		DeviceMalfunction		
		No	Yes	
SensorReadings	Normal	2/10 P(x|c) = P(Normal|No)	5/8 P(x|c) = P(Normal|Yes)	7/18 P(x) = P(Normal)
	Outlier	8/10 P(x|c) = P(outlier|No)	3/8 P(x|c) = P(outlier|Yes)	11/18 P(x) = P(outlier)
		10/18 P(c) = P(No)	8/18 P(c) = P(Yes)	

P(x|c) = P (Normal|Yes) = 5/8

P(x) = P(Normal) = 7/18

P(c) = P(Yes) = 8/18

Likelihood of Device Malfunction 'Yes' given 'Normal' Sensor Reading is:

P(c|x) = P(Yes|Normal) = P(Normal|Yes)* P(Yes)/P(Normal) = (5/8) x (8/18)/(7/18) = 5/7 = 0.714

Similarly, the likelihood of Device Malfunction 'No' given 'Normal' Sensor Reading is:

P(c|x) = P(No|Normal) = P(Normal|No)* P(No)/P(Normal) = (2/10) x (10/18) / (7/18) = 0.285

Likelihood of Device Malfunction 'Yes' given 'Outlier' Sensor Reading is:

P(c|x) = P(Yes|Outlier) = P(Outlier|Yes)* P(Yes)/P(Outlier) = (3/8) x (8/18)/(11/18) = 3/11 = 0.27

Similarly, the likelihood of Device Malfunction 'No' given 'Outlier' Sensor Reading is:

P(c|x) = P(No|Outlier) = P(Outlier|No)* P(No)/P(Outlier) = (8/10) x (10/18)/(8/10) = 0.8

Likelihood Table		DeviceMalfunction		
		No	Yes	
Battery Charge	Full	4/10 P(x|c) = P(Full|No)	2/8 P(x|c) = P(Full|Yes)	6/18 P(x) = P(Full)
	Low	4/10 P(x|c) = P(Low|No)	4 / 8 P(x|c) = P(Low|Yes)	8/18 P(x) = P(Low)
	Medium	1 /10 P(x|c) = P(Medium|No)	2 / 8 P(x|c) = P(Medium|Yes)	3/18 P(x) = P(Medium)
		9/18 P(c) = P(No)	8/18 P(c) = P(Yes)	

P(x|c) = P (Full|Yes) = 2/8

P(x) = P(Full) = 6/18

P(c) = P(Yes) = 8/18

Likelihood of 'Yes' given 'Battery Change = Full' is:

P(c|x) = P(Yes|Full) = P(Full|Yes)* P(Yes)/P(Full) = (2/8) x (6/18)/(8/18) = 3/16 = 0.1875

Similarly, the likelihood of 'No' given 'Battery Change = Full' is:

P(c|x) = P(No|Full) = P(Full|No)* P(No)/P(Full) = (4/10) x (9/18)/(6/18) = 6/10 = 0.6

Likelihood of Device Malfunction 'Yes' given Low 'Battery Change' is:

P(c|x) = P(Yes|Low) = P(Low|Yes)* P(Yes)/P(Low) = (4/8) x (8/18)/(8/18) = 1/2= 0.5

Similarly, the likelihood of Device Malfunction 'No' given Low 'Battery Change' is:

P(c|x) = P(No|Low) = P(Low|No)* P(No)/P(Low) = (4/10) x (9/18)/(8/18) = 36/80 = 0.45

Likelihood of Device Malfunction 'Yes' given Medium 'Battery Change' is:

P(c|x) = P(Yes|Medium) = P(Medium|Yes)* P(Yes)/P(Medium) = (2/8) x (8/18)/(3/18) = 2/3 = 0.66

Similarly, the likelihood of Device Malfunction 'No' given Medium 'Battery Change' is:

P(c|x) = P(No|Medium) = P(Medium|No)* P(No)/P(Medium) = (1/10) x (9/18)/(3/18) = 3/10 = 0.3

Frequency Table		DeviceMalfunction		
		No	Yes	
DaysDeployed	First Three Months	4/10 P(x|c) = P(First Three Months|No)	5/8 P(x|c) = P(First Three Months|Yes)	9/18 P(x) = P(First Three Months)
	Greater than 6 months	4/10 P(x|c) = P(greater than 6 Months|No)	1/8 P(x|c) = P(greater than 6 Months|Yes)	5/18 P(x) = P(greater thanSix Months)
	3 months - 6 months	2/10 P(x|c) = P(3 Months – 6 Months|No)	2/8 P(x|c) = P(3 Months – 6 Months|Yes)	4/18 P(x) = P(3 Months – 6 Months)
		10/18 P(c) = P(No)	8/18 P(c) = P(Yes)	

P(x|c) = P (First Three Months|Yes) = 5/8

P(x) = P (First Three Months) = 9/18

P(c) = P (Yes) = 8/18

Likelihood of Device Malfunction 'Yes' given 'Days Deployed = First Three Months' is:

P(c|x) = P(Yes|First Three Months) = P(First Three Months|Yes)* P(Yes)/P(First Three Months) = (5/8) x (9/18)/(8/18) = 3/16= 0.7

Similarly, the likelihood of Device Malfunction 'No' given 'Days Deployed = First Three Months' is:

P(c|x) = P(No|First Three Months) = P(First Three Months|No)* P(No)/P(First Three Months) = (4/10) x (10/18)/(9/18) = 45/64 = 0.44

Likelihood of Device Malfunction 'Yes' given 'Days Deployed = greater than 6 Months' is:

P(c|x) = P(Yes|greater than 6 Months) = P(greater than 6 Months|Yes)* P(Yes)/P(greater than 6 Months) = (1/8) x (8/18)/(5/18) = 1/5 = 0.2

Similarly, the likelihood of Device Malfunction 'No' given 'Days Deployed = greater than 6 Months' is:

P(c|x) = P(No|greater than 6 Months) = P(greater than 6 Months|No)* P(No)/P(greater than 6 Months) = (4/10) x (10/18)/(5/18) = 4/5= 0.8

Likelihood of Device Malfunction 'Yes' given 'Days Deployed = 3 months—6 Months' is:

P(c|x) = P(Yes|3 months – 6 Months) = P(3 months – 6 Months|Yes)* P(Yes)/P(3 months – 6 Months) = (2/8) x (8/18)/(4/18) = 1/2 = 0.5

Similarly, the likelihood of Device Malfunction 'No' given 'Days Deployed = 3 months – 6 Months' is:

P(c|x) = P(No|3 months – 6 Months) = P(3 months – 6 Months|No)* P(No)/P(3 months – 6 Months) = (2/10) x (10/18)/(4/18) = 1/2 = 0.5

Suppose Sensor reads following values :

Weather = Normal

Sensor Reading = Normal

Battery = Full Charged

Days – First Three Months

Is device malfunctioned =?

So, with the data, we have to predict whether "Sensor Malfunctioned or not?"

Likelihood of Malfunctioned 'Yes' = P(Weather = Normal|Yes) * P(Sensor Reading = Normal|Yes)* P(Battery = Full Charge|Yes) * P(Days Deployed = First Three Months|Yes) * P(Yes) = 5/8 x 5/8 x 2/8 x 5/8 x 11/18 = 0.037 or 3.7% (based on Eq. 6)

Likelihood of Malfunctioned No = P(Weather = Normal|No)*P(Sensor Reading= Normal|No)* P(Battery = Full Charge|No)* P(Days Deployed = First Three Months|No)* P(No) = 3/10 x 2/10 x 4/10 x 4/10 x 7/18 = **0.0037 or 0.37%**
The likelihood of device malfunction is 3.7%.

Naive Bayesian Model in Python

The following Python code creates a Naïve Bayesian Model using Python code. Here are the steps to create Naïve Bayesian Model:

1. Concert arrays Weather, Sensor Readings, Battery Charges, and Days Deployed into Label encoder.
2. Create Naïve Bayesian Label Encoder and encode labels.
3. Create Naïve Bayesian Gaussian class object.
4. Prepare the model.
5. Predict the model.

```
# -*- coding: utf-8 -*-
"""
Created on Fri Mar 20 15:21:04 2020
@author: CHVUPPAL
"""
# Naive Bayesian for Device Health Prognastics
# Import LabelEncoder
from sklearn import preprocessing
import matplotlib as mlp
#creating labelEncoder
le = preprocessing.LabelEncoder()
Weather=['ExtremeTemperature','Normal','Normal','ExtremeTemperature','ExtremeTemperature','Ext
remeTemperature','ExtremeHumidity','ExtremeHumidity','ExtremeHumidity',
'ExtremeHumidity','Normal','Normal','ExtremeHumidity','ExtremeTemperature','Normal','Normal',
'ExtremeHumidity','Normal']
SensorReadings=['Normal','Outlier','Outlier','Normal','Outlier','Normal','Normal','Normal','Outlier
','Outlier','Outlier','Outlier','Outlier','Outlier',
 'Normal','Normal','Normal','Normal']
BatteryCharge =['Full','Medium','Low','Low','Low','Medium','Medium','Low','Full','Low','Low',
'Low','Low','Full',
 'Medium','Low','Full','Medium']
DaysDeployed = ['3 months - 6 months','First Three Months','Greater than 6 months','First Three Months',
 'Greater than 6 months','First Three Months','First Three Months','Greater than 6 months',
 'First Three Months','Greater than 6 months','First Three Months','3 months - 6 months','Greater than
6 months',
 'First Three Months','3 months - 6 months','3 months - 6 months','First Three Months','3 months - 6
months']
DeviceMalfunction=['Yes','Yes','Yes','Yes','No','Yes','Yes','Yes','No','No','Yes','No','No','No','N
o','Yes','Yes','Yes']
# Converting string labels into numbers.
weather_encoded=le.fit_transform(Weather)
print("weather:",weather_encoded)
# Converting string labels into numbers
temp_encoded_SensorReadings=le.fit_transform(SensorReadings)
temp_encoded_BatteryCharge=le.fit_transform(BatteryCharge)
temp_encoded_DaysDeployed=le.fit_transform(DaysDeployed)
label=le.fit_transform(DeviceMalfunction)
print ("SensorReadings:",temp_encoded_SensorReadings)
print ("BatteryCharge:",temp_encoded_BatteryCharge)
print ("DaysDeployed:",temp_encoded_DaysDeployed)
print ("DeviceMalfunction:",label)
#https://stackoverflow.com/questions/19777612/python-range-and-zip-object-type
#https://www.datacamp.com/community/tutorials/naive-bayes-scikit-learn
#Combinig weather and temp into single listof tuples
features=zip(weather_encoded,temp_encoded_SensorReadings,temp_encoded_BatteryCharge,temp_
encoded_DaysDeployed)
print(list(features))
#Import Gaussian Naive Bayes model
from sklearn.naive_bayes import GaussianNB
```

```
#Create a Gaussian Classifier
model = GaussianNB()
# Train the model using the training sets
model.fit(list(zip(weather_encoded,temp_encoded_SensorReadings,temp_encoded_BatteryCharge,temp_
encoded_DaysDeployed)),label)
#Predict Output
predicted= model.predict([[1,0,0,0]]) # 1:ExtremeTemperature, 0:Normal, 0:Full, 0:3 months - 6 months
print ("Predicted Value:", predicted) # 1:Yes, 0:No
predicted= model.predict([[1,1,1,0]]) # 1:ExtremeTemperature, 1:Outlier, 1:Low, 2:Greater than 6 months
print ("Predicted Value:", predicted) # 1:Yes, 0:No
```

Output:

The output of Naïve Bayesian (please see Figure 41):

Figure 41: Naive Bayesian Output

Pre-Run Compute & Memory Statistics

Please find Device Memory statistics before run

Memory Details	Used (in Bytes)
Global Used	1644 Bytes
Malloc Used	7740 Bytes
Malloc Available	0 Bytes
System Available Memory	56152 Bytes
Next Heap Pointer	0 x 10001E3C
Last Stack Back Pointer	0 x 10001DDC
Last Stack Back Pointer Size	96
Number of Stack calls	44

Please see Naïve Bayesian Run capture (see Figure 42):

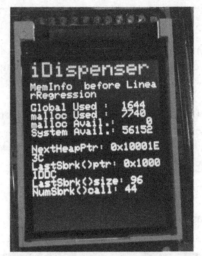

Figure 42: Linear Regression Run Capture

Post Run - Compute & Memory Statistics

Please find Device Memory statistics after run:

Memory Details	Used (in Bytes)
Global Used	1644 Bytes
Malloc Used	7740 Bytes
Malloc Available	0 Bytes
System Available Memory	56152 Bytes
Next Heap Pointer	0x10001E3C
Last Stack Back Pointer	0x10001DDC
Last Stack Back Pointer Size	96
Number of Stack calls	44

Please see Naïve Bayesian Run capture (see Figure 43):

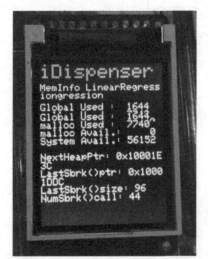

Figure 43: Naive Bayesian Post Run Screen Capture

As after the run (please see Figures 42 and 43) we can see from the above two tables, there is no computational differences before and after the run (Please see Figure 44).

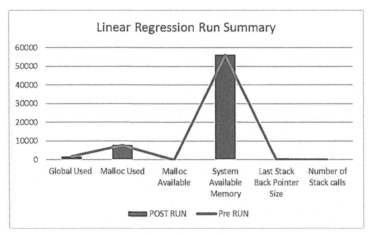

Figure 44: Naive Bayesian Run Summary

☞ This leads to the conclusion that the Naïve Bayesian models with complexity of O (1) will not be expensive to run on extremely constrained devices (xCDs). Put it plainly, the impact of Naïve Bayesian inference on the ream-time embedded xCDs is very minimal and the algorithm is one of the chosen algorithms for xCDs.

Memory Details	Differentials (in Bytes)
Global Used	0 Bytes
Malloc Used	0 Bytes
Malloc Available	0 Bytes
System Available Memory	0 Bytes
Next Heap Pointer	0 Bytes
Last Stack Back Pointer	0 Bytes
Last Stack Back Pointer Size	0 Bytes
Number of Stack calls	0 Bytes

Constrained Environment Considerations

Naïve Bayesian C Code & ML deployment using device Firmware

The device constants that are constructed as part of the ML Hand code Model section, are included as C constants and will be built into the Firmware and deployed into the custom built hardware.

ML Mode C Constants

The device constants shown as part of the table, are constructed based on the device data collected (please see Table 13:

Table 13: Naive Bayesian Device Constants

```
// Probability of Device Mal Function Yes
// Given Normal Weather
double Prob_devmalFunc_Y_Given_Weather_Normal = 0.27;
// Probability of Device Malfunction No
// Given Normal Weather
double Prob_devmalFunc_N_Given_Weather_Normal = 0.375;
// Probability of Device Malfunction Yes
// Given Weather is Extreme Temperature
double Prob_devmalFunc_Y_Given_Weather_ExtremeTemp = 0.375;
// Probability of Device Malfunction No
// Given Weather is Extreme Temperature
double Prob_devmalFunc_N_Given_Weather_ExtremeTemp = 0.75;
// Probability of Device Malfunction Yes
// Given Weather is Extreme Humid
double Prob_devmalFunc_Y_Given_Weather_ExtremeHumid = 0.33;
// Probability of Device Malfunction No
// Given Weather is Extreme Humid
double Prob_devmalFunc_N_Given_Weather_ExtremeHumid = 0.166;
// Probability of Device Malfunction Yes
// Given Sensor Reading is Normal
double Prob_devmalFunc_Y_Given_SensorRead_Normal = 0.71;
// Probability of Device Malfunction No
// Given Sensor Reading is Normal
double Prob_devmalFunc_N_Given_SensorRead_Normal = 0.27;
// Probability of Device Malfunction Yes
// Given Sensor Reading is Outlier
double Prob_devmalFunc_Y_Given_SensorRead_Outlier = 0.71;
// Probability of Device Malfunction No
// Given Sensor Reading is Outlier
double Prob_devmalFunc_N_Given_SensorRead_Outlier = 0.27;
// Probability of Device Malfunction Yes
// Given Battery Charge is Full
double Prob_devmalFunc_Y_Given_BatteryCharge_Full = 0.1875;
// Probability of Device Malfunction No
// Given Battery Charge is Full
double Prob_devmalFunc_N_Given_BatteryCharge_Full = 0.6;
// Probability of Device Malfunction Yes
// Given Battery Charge is Low
double Prob_devmalFunc_Y_Given_BatteryCharge_Low = 0.5;
// Probability of Device Malfunction No
// Given Battery Charge is Low
double Prob_devmalFunc_N_Given_BatteryCharge_Low = 0.45;
// Probability of Device Malfunction Yes
// Given Battery Charge is Medium
double Prob_devmalFunc_Y_Given_BatteryCharge_Medium = 0.66;
// Probability of Device Malfunction No
// Given Battery Charge is Medium
double Prob_devmalFunc_N_Given_BatteryCharge_Medium = 0.3;
// Probability of Device Malfunction Yes
// Given Days Deployed is First Three Months
double Prob_devmalFunc_Y_Given_Daysdeployede_FirstThreeMonths = 0.7;
// Probability of Device Malfunction Yes
// Given Days Deployed is First Three Months
double Prob_devmalFunc_N_Given_Daysdeployede_FirstThreeMonths = 0.44;
// Probability of Device Malfunction Yes
// Given Days Deployed is greater than 6 Months
double Prob_devmalFunc_Y_Given_Daysdeployed_GreaterThanSixMonths = 0.2;
// Probability of Device Malfunction Yes
// Given Days Deployed is greater than 6 Months
double Prob_devmalFunc_N_Given_Daysdeployed_GreaterThanSixMonths = 0.8;
```

Table 13 contd. ...

...Table 13 contd.

```
// Probability of Device Malfunction Yes
// Given Days Deployed is 3 months - 6 Months
double Prob_devmalFunc_Y_Given_Daysdeployed_ThreeMonthsToSixMonths = 0.2;
// Probability of Device Malfunction No
// Given Days Deployed is 3 months - 6 Months
double Prob_devmalFunc_N_Given_Daysdeployed_ThreeMonthsToSixMonths = 0.5;
```

Embed C Code

The following Class Label "Device Malfunction" is computed using Naïve Bayesian algorithm to calculate the percentage of probability that a device is malfunction or not (please see Table 14):

Table 14: Naive Bayesian Device Health Prognastics Code

```
void NaiveBayesianDeviceHealthPrognosticsML( intweather,
intsensorreading,
intbatteryChargeMode,
intnumDaysDeployed,
double *prob_DeviceMalFunc_Yes,
double* prob_DeviceMalFunc_No)
{
// Weather Tuples
// 1:Extreme Temperature
// 2:Normal
// 0:Extreme Humidity
 (*prob_DeviceMalFunc_Yes) = 1.0;
 (*prob_DeviceMalFunc_No) = 1.0;
switch (weather)
 {
case 0: // 0:Extreme Humidity
 (*prob_DeviceMalFunc_Yes) = (*prob_DeviceMalFunc_Yes) * Prob_devmalFunc_Y_Given_Weather_ExtremeHumid;
 (*prob_DeviceMalFunc_No) = (*prob_DeviceMalFunc_No) * Prob_devmalFunc_N_Given_Weather_ExtremeHumid;
break;
case 1: // 1:Extreme Temperature
 (*prob_DeviceMalFunc_Yes) = (*prob_DeviceMalFunc_Yes) * Prob_devmalFunc_Y_Given_Weather_ExtremeTemp;
 (*prob_DeviceMalFunc_No) = (*prob_DeviceMalFunc_No) * Prob_devmalFunc_N_Given_Weather_ExtremeTemp;
break;
case 2: // 2:Normal
 (*prob_DeviceMalFunc_Yes) = (*prob_DeviceMalFunc_Yes) * Prob_devmalFunc_Y_Given_Weather_Normal;
 (*prob_DeviceMalFunc_No) = (*prob_DeviceMalFunc_No) * Prob_devmalFunc_N_Given_Weather_Normal;
break;
default:
break;
 }
// SensorReadings Tuples
// 1:Outlier
// 0:Normal
// 0:Normal
switch (sensorreading)
 {
case 0: // 0:Normal
 (*prob_DeviceMalFunc_Yes) = (*prob_DeviceMalFunc_Yes) * Prob_devmalFunc_Y_Given_SensorRead_Normal;
 (*prob_DeviceMalFunc_No) = (*prob_DeviceMalFunc_No) * Prob_devmalFunc_N_Given_SensorRead_Normal;
break;
case 1: // 1:Outlier
 (*prob_DeviceMalFunc_Yes) = (*prob_DeviceMalFunc_Yes) * Prob_devmalFunc_Y_Given_SensorRead_Outlier;
 (*prob_DeviceMalFunc_No) = (*prob_DeviceMalFunc_No) * Prob_devmalFunc_N_Given_SensorRead_Outlier;
```

Table 14 contd. ...

...Table 14 contd.

```
break;
default:
break;
}
// batteryChargeMode Tuples
// 1:Low
// 2:Medium
// 0:Full
switch (batteryChargeMode)
{
case 0: // 0:Full
(*prob_DeviceMalFunc_Yes) = (*prob_DeviceMalFunc_Yes) * Prob_devmalFunc_Y_Given_BatteryCharge_Full;
(*prob_DeviceMalFunc_No) = (*prob_DeviceMalFunc_No) * Prob_devmalFunc_N_Given_BatteryCharge_Full;
break;
case 1: // 1:Low
(*prob_DeviceMalFunc_Yes) = (*prob_DeviceMalFunc_Yes) * Prob_devmalFunc_Y_Given_BatteryCharge_Low;
(*prob_DeviceMalFunc_No) = (*prob_DeviceMalFunc_No) * Prob_devmalFunc_N_Given_BatteryCharge_Low;
break;
case 2: // 1:Medium
(*prob_DeviceMalFunc_Yes) = (*prob_DeviceMalFunc_Yes) * Prob_devmalFunc_Y_Given_BatteryCharge_Medium;
(*prob_DeviceMalFunc_No) = (*prob_DeviceMalFunc_No) * Prob_devmalFunc_N_Given_BatteryCharge_Medium;
break;
default:
break;
}
// NumberOfDaysDeployed Tuples
// 0:3 months - 6 months
// 1:First Three Months
// 2:Greater than 6 months
switch (numDaysDeployed)
{
case 0: // 0:3 months - 6 months
(*prob_DeviceMalFunc_Yes) = (*prob_DeviceMalFunc_Yes) * Prob_devmalFunc_Y_Given_Daysdeployed_
ThreeMonthsToSixMonths;
(*prob_DeviceMalFunc_No) = (*prob_DeviceMalFunc_No) * Prob_devmalFunc_N_Given_Daysdeployed_
ThreeMonthsToSixMonths;
break;
case 1: // 1:First Three Months
(*prob_DeviceMalFunc_Yes) = (*prob_DeviceMalFunc_Yes) * Prob_devmalFunc_Y_Given_Daysdeployede_FirstThreeMonths;
(*prob_DeviceMalFunc_No) = (*prob_DeviceMalFunc_No) * Prob_devmalFunc_N_Given_Daysdeployede_FirstThreeMonths;
break;
case 2: // 1:Greater than 6 months
(*prob_DeviceMalFunc_Yes) = (*prob_DeviceMalFunc_Yes) * Prob_devmalFunc_Y_Given_Daysdeployed_GreatherThanSixMonths;
(*prob_DeviceMalFunc_No) = (*prob_DeviceMalFunc_No) * Prob_devmalFunc_N_Given_Daysdeployed_GreatherThanSixMonths;
break;
default:
break;
}
}
```

The above code computes the probability of device malfunction based on the Sensor reading values, number of days the device was deployed in the field, the battery power level, and any extreme temperatures.

The driver code is as follows (please see Table 15):

Table 15: Naive Bayesian Driver Code

```
double probabilityOfDeviceMalFunctionY = 0.0;
double probabilityOfDeviceMalFunctionN = 0.0;
// Predict Output
// [1, 0, 0, 0]] ) # 1:ExtremeTemperature, 0 : Normal, 0 : Full, 0 : 3 months - 6 months
int weatherlabel = 1; // ExtremeTemperature
int sensor_reading = 0; // Normal
int battery_charge = 0; // Battery Full
int senordeployednumdaysbucket = 0; // 3 months - 6 months
 NaiveBayesianDeviceHealthPrognosticsML(
 weatherlabel,
 sensor_reading,
 battery_charge,
 senordeployednumdaysbucket,
 &probabilityOfDeviceMalFunctionY,
 &probabilityOfDeviceMalFunctionN);
 printf("\nExtreme Temperature --- Normal Sensor Read --- Full Battery Charge --- Device Deployed [3 months to 6 months]\n");
 printf("Probability of Device Malfunction %f Percentage\n", (probabilityOfDeviceMalFunctionY)*100);
 printf("Probability of Device not Malfunction is %f Percentage\n", (probabilityOfDeviceMalFunctionN)*100);
 weatherlabel = 1; // ExtremeTemperature
 sensor_reading = 0; // Normal
 battery_charge = 1; // Battery Low
 senordeployednumdaysbucket = 0; // 3 months - 6 months
 NaiveBayesianDeviceHealthPrognosticsML(
 weatherlabel,
 sensor_reading,
 battery_charge,
 senordeployednumdaysbucket,
 &probabilityOfDeviceMalFunctionY,
 &probabilityOfDeviceMalFunctionN);
 printf("\nExtreme Temperature --- Normal Sensor Read --- Low Battery Charge --- Device Deployed [3 months to 6 months]\n");
 printf("Probability of Device Malfunction %f Percentage\n", (probabilityOfDeviceMalFunctionY) * 100);
 printf("Probability of Device not Malfunction is %f Percentage\n", (probabilityOfDeviceMalFunctionN) * 100);
```

The code, based on the Sensor reading, calls "NaiveBayesianDeviceHealthPrognosticsML" call; based on the probability return, the device performs the next steps.

Time Complexity

Constrained environment considerations for the Naïve Bayesian is very similar to that of evaluation of decision paths. It's very simple and atomic linear call.

Output

Please see Figure 45 for the Device Prognostics Driver code:

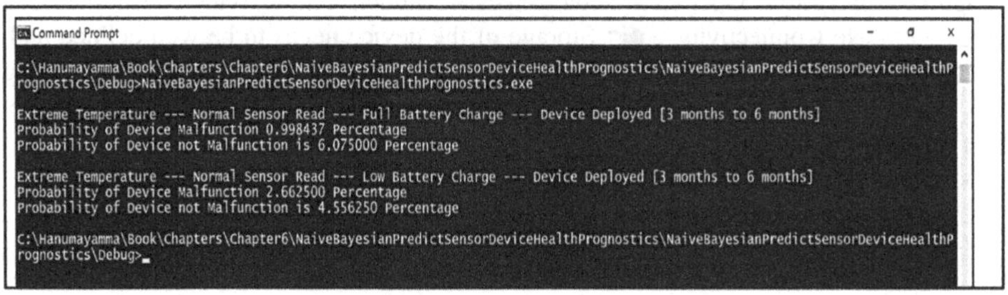

Figure 45: Device Prognostics Driver Output

Hardware Economy—Model Accuracy trade-off

The purpose of the section is to analyze the Naïve Bayesian model deployment in extreme constrained or Tiny Compute hardware.

Modeling no connectivity

Modeling for no connectivity influences model upgradability, storage and power source. Due to no connectivity, the Naïve Bayesian model could only update during new device refresh cycle. Additionally, the lack of connectivity inhibits model upgrade via OTA firmware update.

Space Complexity: constant space
Time Complexity: Linear (O (1)) as the algorithm must predict outcome based on inputs.

Connectivity No Connectivity		Hardware 🔒 Constraint ◯ No Constraint			
		Memory	Power Source	Storage	Computational Power
ML Model Accuracy	Self-Contained & Updated only on Hardware Refresh	Space complexity and time complexity are linear – O (1)	Time complexity is same but increased storage space could exert higher power consumption.	Application data storage could increase due to no connectivity.	Linear time complex.
	OTA Firmware	N/A	N/A	N/A	N/A
	Model Type: Eager Learner	Naïve Bayesian model equation belongs to family of eager learner algorithm and could have less influence memory due to no connectivity – O (1).	Time Complexity: Linear – O (1)	No influence on storage	Linear Time complexity.
	Lazy Learner	Not applicable as decision tree is if-else condition evaluator	Not applicable as decision tree is if-else condition evaluator	Not applicable as decision tree is if-else condition evaluator	Not applicable as decision tree is if-else condition evaluator
	Model Invocation	Inline model invocation O (1)	Inline model invocation O (1) does not demand huge power.	Not applicable	Inline model invocation requires less computational power compared to stack or heap based.

Overall design consideration: Operating Naïve Bayesian Model in an extreme constrained device (xCD) or Tiny device with "No Connectivity", ☞ Storage of the device needs to be well defined and should be sufficient for the SLA time of the device. If the SLA is half-yearly or yearly refresh, the device storage capacity could impact battery power. As far as the model computation and performance, it would not exert pressure on the device footprint.

Modeling Low Power Bluetooth Connectivity

Availability of Bluetooth Low Energy connectivity enables to offload data storage to Edge devices. Only tax is extra power consumption due to BLE power.

Connectivity Bluetooth Low Energy (BLE)		Hardware 🔒 Constraint ⚪ No Constraint			
		Memory	Power Source	Storage	Computational Power
ML Model Accuracy	Self-Contained & Updated only on Hardware Refresh	⚪ Naïve Bayesian is Linear Time Complex O (1)	🔒 Continuous Bluetooth advertisements could increase power needs.	⚪ Data could be offloaded to external devices.	⚪ The model data and storage are offloaded to Edge devices (Smartphone) and thus could reduce hardware footprint.
	OTA Firmware	N/A	N/A	N/A	N/A
	Model Type: Eager Learner	⚪ Linear Model Equations are eager learner with linear time complex O (1)	🔒 Bluetooth advertisements need higher power.	⚪ Periodic offloading of Sensor data to Edge devices lowers the storage needs.	⚪ Lower Data Size and Code size reduce the needs for more computational power.
	Lazy Learner	N/A	N/A	N/A	N/A
	Model Invocation	⚪ Linear time complex O (1) and Inline if-else model invocation.	🔒 Though Linear time complex, continuous Bluetooth advertisements could increase power needs	⚪ Periodic offloading of Sensor data to Edge devices lowers the storage needs.	⚪ Linear time complex & Inline model invocation

Overall design consideration: Operating Naïve Bayesian Machine Learning algorithms in a constrained or Tiny device with "Bluetooth Low Energy (BLE) Connectivity" increases battery power consumption. Model performance & invocation would be linear as time complexity is O (1) with multiplication of probabilities. Only caution to be noted is sustainability of battery for radio advertisements—that is turn on radio & BLE advertisement on a need basis.

Modeling Wi Fi Connectivity

Wi Fi connectivity enables to offload data storage to the Edge devices or call to the central Cloud server. Additionally,it enables to upload Firmware via OTA. The major limitation is many of the constrained devices are not suited for Wi Fi as it drains huge power from the battery.

Connectivity Wi-Fi		Hardware 🔒 Constraint ⚪ No Constraint			
		Memory	Power Source	Storage	Computational Power
ML Model Accuracy	Self-Contained & Updated only on Hardware Refresh	Naïve Bayesian Models are linear Time complex: O(1)	🔒 On board Wi Fi demands more power. For battery power devices, Wi-Fi option is less favorable.	Data could be offloaded to external devices or central Cloud.	The model data and storage are offloaded to Edge devices (Smartphone) and thus could reduce hardware footprint.
	OTA Firmware	OTA allows to deploy the latest and greatest optimized model.	🔒 Onboard Wi Fi demands more power.	Data could be offloaded to external devices or to the central Cloud.	OTA allows to deploy the latest and greatest optimized model.
	Model Type: Eager Learner	Naïve Bayesian Models are eager learner with linear time complex O (1)	🔒 Bluetooth advertisements need higher power.	Periodic offloading of Sensor data to Edge devices lowers the storage needs.	Lower Data Size and Code size reduce the needs for more computational power.
	Lazy Learner	N/A	N/A	N/A	N/A
	Model Invocation	Linear time complex O (1) and Inline multiplicative model.	🔒 Though Linear time complex, continuous Bluetooth advertisements could increase power needs	Periodic offloading of Sensor data to Edge devices lowers the storage needs.	Linear time complex & Inline model invocation

Overall design consideration: Operating Naïve Bayesian Machine Learning algorithms in an extreme constrained or Tiny device with "Wi Fi Connectivity" increases battery power consumption. Model performance & invocation would be linear as time complexity is O (1) with in-line model invocation.

Connectivity—Model Accuracy trade-off

Modeling Memory

Since the model is a linear multiplicative equation and does not require to be memory resident[285] nor heap[286] computation and doesn't require historical data for accuracy inference, the hardware memory is not a constraint for the Model accuracy. That is, the memory (Data movement or access) does not influence power or computational hardware economy.

---

[285] Memory resident - https://www.trendmicro.com/vinfo/us/security/definition/memory-resident
[286] Stack vs. Heap allocation - https://www.geeksforgeeks.org/stack-vs-heap-memory-allocation/

Hardware Memory		Connectivity 🔒 Constraint ⭘ No Constraint		
		No Connectivity	Bluetooth Low Energy (BLE)/LoRa	Wi-Fi
Model Accuracy	Self-Contained & Updated only on Hardware Refresh	○	○ Since model update is cycled on Hardware refresh (SLA half-yearly or yearly), there would be staleness of the model. However, memory constraint has no influence on the Accuracy.	○
	OTA Firmware	○ Since the model gets updated OTA as part of Firmware update, no staleness of model. Memory constraint has no influence on the model accuracy.	○	○
	Model Type: Eager Learner	○	○	○
	Lazy Learner	N/A	N/A	N/A
	Model Invocation	○ In-line model invocation and not memory intense – no memory constraint on the model.	○	○

Design Considerations: Since model update is cycled on hardware refresh (SLA half-yearly or yearly), there would be staleness of the model. However, memory constraint has no influence on the Accuracy. Model is a linear call and will not require data movement between processor & memory space. In-line model invocation and not memory intense—no memory constraint on the model.

Modeling Processing Power

The Model deployed as a linear call and invoked as an in-line function and does not require extra computation power.

Hardware Processing Power		Connectivity 🔒 Constraint ⭘ No Constraint			
		No Connectivity	BLE	Wi-Fi	Design Notes
Model Accuracy	Hardware Refresh	○	○	○	Since model update is cycled on Hardware refresh (SLA half-yearly or yearly), there could be staleness in Model accuracy. This has no influence on the Power.
	OTA Firmware Update	○	○		Processing Power is not a constraint with respect to Model Accuracy.
	Model Type: Eager Learner	○	○	○	Processing Power is not a constraint with respect to Model Accuracy.
	Lazy Learner	N/A	N/A	N/A	
	Model Invocation	○	○	○	Processing Power is not a constraint with respect to Model Accuracy.

Design Considerations: Naïve Bayesian models valuate conditional probabilities and these calculations are very atomic and simple.

Modeling Storage

The Naïve Bayesian Model is deployed as an in-line function and does not require huge storage. Nonetheless, as hardware device collects sensor data, the data needs to be stored on device local storage (offload to Edge Gateway or mobile device) before uploads to Central Storage or Corporate Cloud storage.

Hardware Storage		Connectivity 🔒 Constraint ⭕ No Constraint			
		No Connectivity	BLE	Wi-Fi	Design Notes
Model Accuracy	Hardware Refresh	⭕	🔒	🔒	Since model update is cycled on Hardware refresh (SLA half-yearly or yearly), the data storage needs to be large to store on the device. A flat file-based storage. For BLE and Wi-Fi modes, the storage could be offloaded to gateway or mobile device & Circular storage data pattern could help to concise storage.
	OTA Firmware Update	⭕	🔒	🔒	Since the model gets updated OTA as part of Firmware update, no staleness of model. Storage could be constraint under "No Connectivity" but not a constraint for BLE and Wi-Fi.
	Model Type: Eager Learner	⭕	🔒	🔒	Model is linear equation and will not require data movement between processor & memory space. Storage could be constraint under "No Connectivity" but not a constraint for BLE and Wi-Fi.
	Lazy Learner	N/A	N/A	N/A	
	Model Invocation	⭕	🔒	🔒	In-line model invocation and not memory intense – no memory constraint on the model. Storage could be constraint under "No Connectivity" but not a constraint for BLE and Wi-Fi.

Design Considerations: Having storage that fits the SLA demands for "no-connectivity" is the only important factor that needs to be engineered.

Modeling Environmental Perturbations

Model deployed under geographical conditions with huge temperature or humidity variations, the influence of temperature on the battery could lead to accuracy consideration. Humidity accumulation is also a factor to be considered in the model calculation

Hardware Environmental Perturbations		Connectivity 🔒 Constraint ⭕ No Constraint			
		No Connectivity	BLE	Wi-Fi	Design Notes
Model Accuracy	Hardware Refresh	🔒	🔒	🔒	Model should factor in geographical sensitivity as part of the calculations. The model coefficients need to tame environmental perturbations.
	OTA Firmware Update	🔒	🔒	🔒	The OTA could fine-tune sensitive model coefficients to overcome environmental perturbations.
	Model Type: Eager Learner	N/A	N/A	N/A	No influence of environmental perturbations on data movement.
	Lazy Learner	N/A	N/A	N/A	No influence of environmental perturbations on data movement.
	Model Invocation	N/A	N/A	N/A	No influence of environmental perturbations on data movement.

Connectivity to Hardware trade-off

Modeling: Hardware Refresh

In this mode, the model gets refreshed whenever hardware refresh takes place. Generally, this could be half-yearly or yearly, depending upon the SLA signed with the customer. With this constraint, there is staleness on the model accuracy. That is, adaptive learning is only deferred when a new model gets deployed. The model is a linear function call that is invoked as part of the embedded code. So there is no special requirements for memory, power and computational processing. And hence, no constraints.

Model Accuracy Hardware Refresh	Hardware 🔒 Constraint ◯ No Constraint					
	Power	Memory	Storage	Processing	Environmental	Design Notes
No Connectivity	◯	◯	🔒	◯	◯	Storage needs are higher & inaccuracies due to environmental perturbations.
BLE	N/A	N/A	N/A	N/A	N/A	Due to staleness in model deployment, inaccuracies due to environmental perturbations a possibility.
Wi-Fi	N/A	N/A	N/A	N/A	N/A	

Design Considerations: Since model update is cycled on Hardware refresh (SLA half-yearly or yearly), there could be staleness in model accuracy. This has no influence on the Power. Processing Power is not a constraint with respect to model accuracy.

Modeling Over-The-Air Firmware (OTA) Update

In this mode, the model is only getting refreshed over the air as part of firmware update. Except for no connectivity mode, the model is the latest and learned with the most recent collected data. That is, adaptive learning is current and gets deployed as part of OTA. The model is a linear function that is invoked as part of in-line code. So, there is no special requirements for memory, power, and computational processing. And hence, no constraints.

Model Accuracy OTA	Hardware 🔒 Constraint ◯ No Constraint					
	Power	Memory	Storage	Processing	Environmental	Design Notes
No Connectivity	◯	◯	🔒	◯	🔒	Storage needs are higher & inaccuracies due to environmental perturbations.
BLE	◯	◯	◯	◯	◯	With the model is updated vis OTA, the latest learnings will have higher accuracies.
Wi Fi	◯	◯	◯	◯	◯	

Modeling Active Learner vs. Lazy Learner

The model is an Active learner with no special requirements to be in-memory, memory resident or data movements. The model is a linear function that is invoked as part of in-line code. So, there is no special requirements for memory, power, and computational processing. And hence, no constraints. No restrictions either, except in No Connectivity mode with or without OTA firmware model.

Model Accuracy Learner		Hardware 🔒 Constraint ⬤ No Constraint					
		Power	Memory	Storage	Processing	Environmental	Design Notes
Connectivity	No Connectivity	⬤	⬤	🔒	⬤	🔒	Storage needs are higher & inaccuracies due to environmental perturbations.
	BLE	⬤	⬤	⬤	⬤	⬤	With the model is updated via OTA, the latest learnings will have higher accuracies.
	Wi-Fi	⬤	⬤	⬤	⬤	⬤	

Modeling Model Invocation

The model is invoked as an in-line code, not a heap or memory resident call. Except storage constraint as part of No Connectivity mode, the model has no special restrictions.

Model Accuracy Invocation		Hardware 🔒 Constraint ⬤ No Constraint					
		Power	Memory	Storage	Processing	Environmental	Design Notes
Connectivity	No Connectivity	⬤	⬤	🔒	⬤	🔒	Storage needs are higher & inaccuracies due to environmental perturbations.
	BLE	⬤	⬤	⬤	⬤	⬤	When the model is updated via OTA, the latest learnings will have higher accuracies.
	Wi-Fi	⬤	⬤	⬤	⬤	⬤	

Linear Regression

In statistics, linear regression is an approach for modeling the relationship between a scalar dependent variable *y* and one or more explanatory variables denoted *X*. The case of one explanatory variable is called simple linear regression. When the outcome, or class, is numeric, and all the attributes are numeric, linear regression is a natural technique to consider. This is a staple method in statistics. The idea is to express the class as a linear combination of the attributes, with predetermined weights:

$$x = w_0 + w_1 a_1 + w_2 a_2 + \cdots + w_k a_k$$

where *x* is the class; $a_1, a_2, \ldots, a_k$ are the attribute values; and $w_0, w_1, \ldots, w_k$ are weights.

The weights are calculated from the training data. Here, the notation gets a little heavy, because we need a way of expressing the attribute values for each training instance. The first instance will have a class, say $x(1)$, and attribute values, $a_1^{(1)}, a_2^{(1)}, \ldots, a_k^{(1)}$, where the superscript denotes that it is the first example. Moreover, it is notationally convenient to assume an extra attribute a_0, whose value is always 1.

Crowdedness to Temperature Modeling (Edge State Model)

We have developed two edge analytics models: (1) Crowdedness to temperature and (2) Temperature to Humidity correlation. Both models have used case level applicability to the Smart City edge analytics. Finally, we have developed the Kalman filter and deployed the Machine Learning models in test hardware.

Dataset[i]

We have downloaded Kaggle dataset (see Figure 46) to get crowdedness to temperature machine learning model [11].

---

[i]   https://www.kaggle.com/nsrose7224/crowdedness-at-the-campus-gym

number_people	date	timestamp	day_of_week	is_weekend	is_holiday	temperature	is_start_of_semester	is_during_semester	month	hour
37	2015-08-14 17:00:11-07:00	61211	4	0	0	71.76	0	0	8	17
45	2015-08-14 17:20:14-07:00	62414	4	0	0	71.76	0	0	8	17
40	2015-08-14 17:30:15-07:00	63015	4	0	0	71.76	0	0	8	17
44	2015-08-14 17:40:16-07:00	63616	4	0	0	71.76	0	0	8	17
45	2015-08-14 17:50:17-07:00	64217	4	0	0	71.76	0	0	8	17
46	2015-08-14 18:00:18-07:00	64818	4	0	0	72.15	0	0	8	18
43	2015-08-14 18:20:08-07:00	66008	4	0	0	72.15	0	0	8	18
53	2015-08-14 18:30:09-07:00	66609	4	0	0	72.15	0	0	8	18
54	2015-08-14 18:40:14-07:00	67214	4	0	0	72.15	0	0	8	18
43	2015-08-14 18:50:15-07:00	67815	4	0	0	72.15	0	0	8	18
39	2015-08-14 19:00:16-07:00	68416	4	0	0	69.97	0	0	8	19
38	2015-08-14 19:20:07-07:00	69607	4	0	0	69.97	0	0	8	19
45	2015-08-14 19:30:08-07:00	70208	4	0	0	69.97	0	0	8	19
41	2015-08-14 19:40:14-07:00	70814	4	0	0	69.97	0	0	8	19
36	2015-08-14 19:50:16-07:00	71416	4	0	0	69.97	0	0	8	19
42	2015-08-14 20:00:17-07:00	72017	4	0	0	68.8	0	0	8	20
35	2015-08-14 20:21:14-07:00	73274	4	0	0	68.8	0	0	8	20
36	2015-08-14 20:31:14-07:00	73874	4	0	0	68.8	0	0	8	20
48	2015-08-14 20:41:15-07:00	74475	4	0	0	68.8	0	0	8	20
40	2015-08-14 20:51:17-07:00	75077	4	0	0	68.8	0	0	8	20
49	2015-08-14 21:01:18-07:00	75678	4	0	0	68.04	0	0	8	21
37	2015-08-14 21:20:06-07:00	76806	4	0	0	68.04	0	0	8	21
48	2015-08-14 21:30:07-07:00	77407	4	0	0	68.04	0	0	8	21
45	2015-08-14 21:40:08-07:00	78008	4	0	0	68.04	0	0	8	21
45	2015-08-14 21:50:09-07:00	78609	4	0	0	68.04	0	0	8	21
38	2015-08-14 22:00:10-07:00	79210	4	0	0	67.55	0	0	8	22
41	2015-08-14 22:20:07-07:00	80407	4	0	0	67.55	0	0	8	22
41	2015-08-14 22:30:08-07:00	81008	4	0	0	67.55	0	0	8	22
37	2015-08-14 22:40:09-07:00	81609	4	0	0	67.55	0	0	8	22

Figure 46: Kaggle Excel Data [11]

Dataset: https://www.kaggle.com/nsrose7224/crowdedness-at-the-campus-gym

The dataset consists of 26,000 people counts (about every 10 minutes) over the last year. In addition, extra information was gathered, including weather and semester-specific information that might affect how crowded it is. The label is the number of people, which is predicted, given some subset of the features.

Label:

1. Number of people

Features:

- date (string; datetime of data)
- timestamp (int; number of seconds since beginning of day)
- day_of_week (int; 0 [monday] - 6 [sunday])
- is_weekend (int; 0 or 1) [boolean, if 1, it's either saturday or sunday, otherwise 0]
- is_holiday (int; 0 or 1) [boolean, if 1 it's a federal holiday, 0 otherwise]
- temperature (float; degrees fahrenheit)
- is_start_of_semester (int; 0 or 1) [boolean, if 1 it's the beginning of a school semester, 0 otherwise]
- month (int; 1 [jan] - 12 [dec])
- hour (int; 0 – 23) (see Figure 46)

Model Development

To develop a model equation, we have used the Excel regression data model packet (see Figure 47).

Our goal of the model is to predict the influencing factors that affect the room temperature. The influencing parameters include: Number of people, day of the week, holiday and other dataset parameters (see Figure 48).

Class Variable: Temperature

Model generation: Running data analysis pack with regression parameters creates mode (see Figure 49)

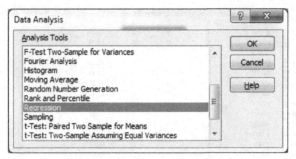

Figure 47: Excel Data Analysis Pack

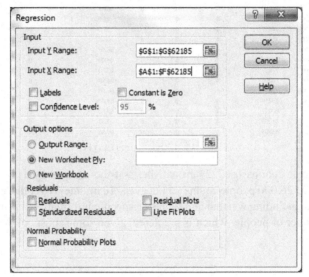

Figure 48: Regression Parameters

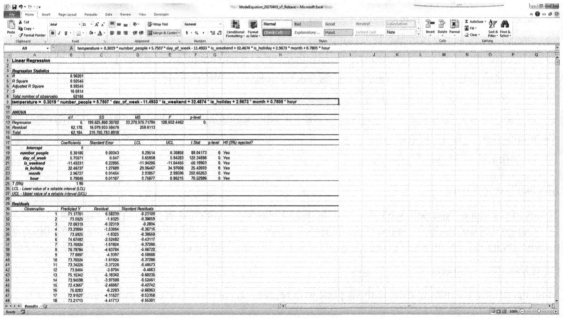

Figure 49: Model Output

Model Validation

F-Test

The F-test for linear regression tests whether any of the independent variables in a multiple linear regression model are significant (see Figures 50 and 51).

ANOVA					
	d.f.	SS	MS	F	p-level
Regression	6.	199,625,860.30702	33,270,976.71784	128,652.4462	0.
Residual	62,178.	16,079,933.58478	258.6113		
Total	62,184.	215,705,793.8918			

Figure 50: ANOVA

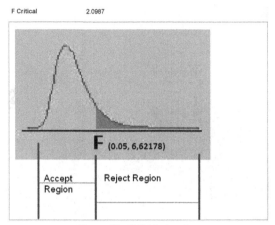

Figure 51: F-Critical region

Since, F value of regression (128652.652) falls in the reject region (> F-Critical), the Null Hypothesis is rejected. That is, the model is valid (see Figure 51).

T-Test

T-Tests are used to conduct hypothesis tests on the regression coefficients obtained in simple linear regression. The goal is to find out the most influencing parameters in the regression model equation (see Figure 52). T Critical = TINV (0.05, 62178) = 1.960002138

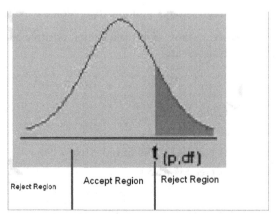

Figure 52: T-Critical region

All the above values fall within the reject region of the t critical. Hence, all the above variables have the explanatory power in explaining the temperature (see Figure 53).

	Coefficients	Standard Error	LCL	UCL	t Stat	p-level	H0 (5%) rejected?
Intercept	0						
number_people	0.30186	0.00343	0.29514	0.30858	88.04173	0.	Yes
day_of_week	5.75071	0.047	5.65858	5.84283	122.34898	0.	Yes
is_weekend	-11.49331	0.22895	-11.94206	-11.04455	-50.19903	0.	Yes
is_holiday	32.46737	1.27689	29.96467	34.97008	25.42693	0.	Yes
month	2.96727	0.01464	2.93857	2.99596	202.65263	0.	Yes
hour	0.78046	0.01107	0.75877	0.80215	70.52986	0.	Yes
T (5%)	1.96						
LCL - Lower value of a reliable interval (LCL)							
UCL - Upper value of a reliable interval (UCL)							

Figure 53: Model Attributes

Data Assumptions

Residually are normally distributed: From the residuals plot, the data points are normally distributed (see Figure 54).

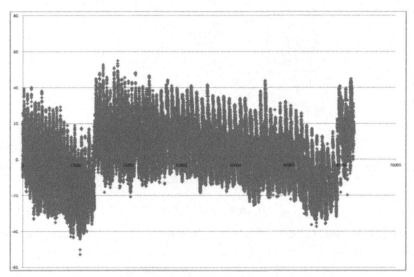

Figure 54: Data Samples

Assumptions 2: Linear - Residuals are independent

To confirm residuals are independent, we need to complete Durbin-Watson test.

A test that the residuals from a linear regression or multiple regression are independent.

Since most regression problems involving time series data exhibit positive autocorrelation, the hypotheses usually considered in the Durbin-Watson test are:

$H_0 : \rho = 0$

$H_1 : \rho > 0$

Durbin-Watson Numerator = 2020.722066

Durbin-Watson Denominator = 62183

Durbin-Watson statics Test (DW-Test) = 2020.722066/62183 = 0.032496

Therefore, the test passes the DW-Test (see Figure 55).

Please note:

As a rough rule of thumb, if Durbin–Watson is less than 1.0, there may be cause for alarm. Small values of d indicate successive error terms are, on average, close in value to one another, or positively correlated. If d > 2, successive error terms are, on average, much more different in value from one another, i.e., negatively correlated. In regressions, this can imply an underestimation of the level of statistical significance.

The test statistic is

$$d = \frac{\sum_{i=2}^{n}(e_i - e_{i-1})^2}{\sum_{i=1}^{n} e_i^2}$$

where $e_i = y_i - \hat{y}_i$ and y_i and $\hat{y}_i$ are, respectively, the observed and predicted values of the response variable for individual i. d becomes smaller as the serial correlations increase. Upper and lower critical values, d_U and d_L have been tabulated for different values of k (the number of explanatory variables) and n:

If $d < d_L$ reject $H_0 : \rho = 0$

If $d > d_U$ do not reject H_0 $\rho = 0$

If $d_L < d < d_U$ test is inconclusive.

Figure 55: Statistical test

Model Equation

From the dataset, we have derived model equation that will calculate the temperature value bases on parameters: Number people, day of the week, is_weekend, is_holiday, month and hour.

Temperature = 0.3019 * number_people + 5.7507 * day_of_week - 11.4933 * is_weekend + 32.4674 * is_holiday + 2.9673 * month + 0.7805 * hour   Eq. (9)

 For Global deployment, the coefficients of independent attributes (NUMBER OF PEOPLE, IS_HOLIDAY, MONTH AND HOUR) are sensitive to the geo location. These coefficients need to be fine-tuned to local environmental factors to maintain explainability[287] and repeatability of the model.

Model Equation & Independent Parameters Coefficients

As provided by the model equation, the independent parameter "number of people" influences on the state variable "temperature". What this means is, if a number of people suddenly entered into a room, keeping all other effect variables constant, the room temperature would increase by a factor of 0.3019.

In other words:

Change of Temperature given number of people in a room is (see Figure 56):

Temperature Change $_{\text{Number of People}}$ = 0.30198

Model Development in Python

Checking Linearity

Before you execute a linear regression model, it is advisable to validate that certain assumptions are met. As noted earlier, you may want to check that a linear relationship exists between the dependent variable and the independent variable/s. In our example, you may want to check that a linear relationship exists between:

- Temperature (dependent variable) and the Number of People (independent variable)

---

[287] Explainable Artificial Intelligence (XAI) - https://www.darpa.mil/program/explainable-artificial-intelligence

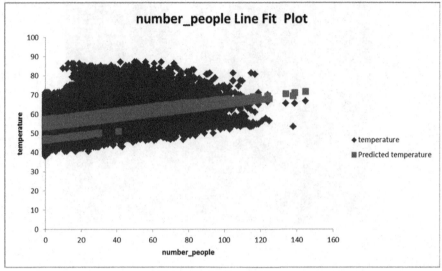

Figure 56: Predicated Values

- Temperature (dependent variable) and the day_of_the_week (independent variable)
- Temperature (dependent variable) and the is_weekend (independent variable)
- Temperature (dependent variable) and the is_holiday (independent variable)
- Temperature (dependent variable) and the month (independent variable)
- Temperature (dependent variable) and the hour (independent variable).

Scatter Diagram

To perform a linearity check, a scatter plot diagram will help (see Figure 57).

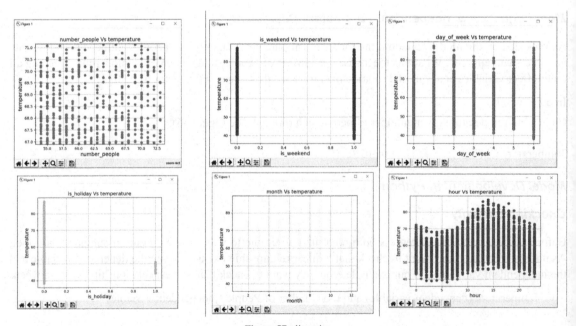

Figure 57: linearity

As you can see from Figure 52, a linear relationship exists in all cases.

The sklearn performs linear regression (see Table 15).

Table 15: sklearn Linear Regression Python Code

```
Stock_Market = pd.read_csv(r'C:\Hanumayamma\CRCBook\Code\MachineLearning\Crowdedness_To_Temperature_20170403.csv')
df = DataFrame(Stock_Market,columns=['number_people','day_of_week','is_weekend','is_holiday','month','hour','temperature'])
X = df[['number_people','day_of_week','is_weekend','is_holiday','month','hour']] # here we have 5 variables for multiple
regression. If you just want to use one variable for simple linear regression, then use X = df['Interest_Rate'] for example. Alternatively,
you may add additional variables within the brackets
Y = df['temperature']
# with sklearn
regr = linear_model.LinearRegression()
regr.fit(X, Y)
# prediction with sklearn
New_number_people = 48
New_day_of_week = 5
New_is_weekend= 0
New_is_holiday = 0
New_month= 9
New_hour=20
print ('Predicted Temperature: \n', regr.predict([[New_number_people ,New_day_of_week, New_is_weekend,New_is_holiday,New_
month,New_hour]]))
# with statsmodels
X = sm.add_constant(X) # adding a constant
model = sm.OLS(Y, X).fit()
predictions = model.predict(X)
print_model = model.summary()
print(print_model)
```

The output contains both model and sklearn output (see Table 16 and Figure 58):

Table 16: Model Output

```
Intercept:
 53.97404514884691
Coefficients:
 [ 0.11713168 0.01120827 1.2402437 -6.95555267 0.17810501 -0.04197969]
Predicted Temperature:
 [60.41575833]
coef std err t P>|t| [0.025 0.975]
--------------------------------------------------------------------------
const 53.9740 0.083 646.981 0.000 53.811 54.138
number_people 0.1171 0.001 92.547 0.000 0.115 0.120
day_of_week 0.0112 0.019 0.587 0.557 -0.026 0.049
is_weekend 1.2402 0.085 14.650 0.000 1.074 1.406
is_holiday -6.9556 0.463 -15.015 0.000 -7.863 -6.048
month 0.1781 0.007 26.171 0.000 0.165 0.191
hour -0.0420 0.004 -10.049 0.000 -0.050 -0.034
====================================================================
Omnibus: 1796.583 Durbin-Watson: 0.025
Prob(Omnibus): 0.000 Jarque-Bera (JB): 3150.086
Skew: 0.248 Prob(JB): 0.00
Kurtosis: 3.984 Cond. No. 786.
====================================================================
```

Figure 58: Python Code

Constrained Environment Considerations

Time Complexity

Constrained Environment considerations for the Linear Regression is very similar to that of evaluation of decision paths for Decision Trees (DTs). It's very simple and linear call.

Pre-Run Compute & Memory Statistics

Please find Device Memory statistics before run:

Memory Details	Used (in Bytes)
Global Used	1644 Bytes
Malloc Used	7740 Bytes
Malloc Available	0 Bytes
System Available Memory	56152 Bytes
Next Heap Pointer	0x10001E3C
Last Stack Back Pointer	0x10001DDC
Last Stack Back Pointer Size	96
Number of Stack Calls	44

Please see Linear Regression Run capture (see Figure 59):

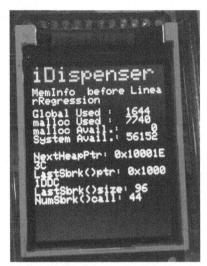

Figure 59: Linear Regression Run Capture

Post Run - Compute & Memory Statistics

Please find Device Memory statistics after run:

Memory Details	Used (in Bytes)
Global Used	1644 Bytes
Malloc Used	7740 Bytes
Malloc Available	0 Bytes
System Available Memory	56152 Bytes
Next Heap Pointer	0x10001E3C
Last Stack Back Pointer	0x10001DDC
Last Stack Back Pointer Size	96
Number of Stack Calls	44

Please see Linear Regression Run capture (see Figure 60):

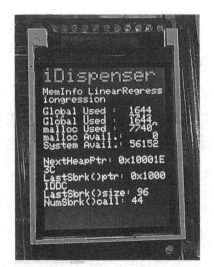

Figure 60: Linear Regression Post Run Screen Capture

As we can see from Figures 59 and 60, there is no computational differences before and after runs (please see Figure 61).

Memory Details	Differentials (in Bytes)
Global Used	0 Bytes
Malloc Used	0 Bytes
Malloc Available	0 Bytes
System Available Memory	0 Bytes
Next Heap Pointer	0 Bytes
Last Stack Back Pointer	0 Bytes
Last Stack Back Pointer Size	0 Bytes
Number of Stack calls	0 Bytes

Linear Regression Run Summary

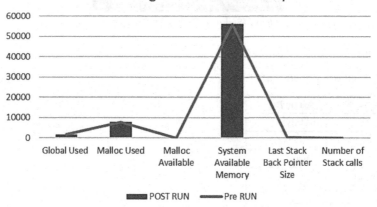

Figure 61: Linear Regression Run Summary

Output

Output of the model (please see Figure 62)

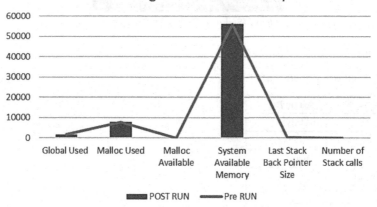

Figure 62: Model Output

Hardware Economy—Model Accuracy trade-off

The purpose of the section is to analyze the Linear regression model equation deployment in constrained or Tiny Compute hardware.

Modeling no connectivity

Modeling for no connectivity influences model upgradability, storage and power source. Since, the no connectivity, the Linear Regression model could only updated during new device refresh cycle. Additionally, the lack of connectivity inhibits model upgrade via OTA firmware update.

Space Complexity: constant space

Time Complexity: Linear (O (1)) as the algorithm must predict outcome based on inputs.

Connectivity No Connectivity		Hardware 🔒 Constraint ⬭ No Constraint			
		Memory	Power Source	Storage	Computational Power
ML Model Accuracy	Self-Contained & Updated only on Hardware Refresh	Space complexity and time complexity are linear – O (1)	Time complexity is same but increased storage space could exert higher power consumption.	Application data storage could increase due to no connectivity.	Linear time complex.
	OTA Firmware	N/A	N/A	N/A	N/A
	Model Type: Eager Learner	Linear Regression model equation belongs to family of eager learner algorithm and could have less influence memory due to no connectivity – O (1).	Time Complexity: Linear – O (1)	No influence on storage	Linear Time complexity.
	Lazy Learner	Not applicable as decision tree is if-else condition evaluator	Not applicable as decision tree is if-else condition evaluator	Not applicable as decision tree is if-else condition evaluator	Not applicable as decision tree is if-else condition evaluator
	Model Invocation	Inline model invocation O (1)	Inline model invocation O (1) does not demand huge power.	Not applicable	Inline model invocation requires less computational power compared to stack or heap based.

Overall design consideration: Operating Linear Regression Model equation in a constrained or Tiny device with "No Connectivity", Storage of the device needs to well defined and should be sufficient for the SLA time of the device. If the SLA is half-yearly or yearly refresh, the device storage capacity could impact battery power. As far as the model computation and performance, it would not exert pressure on the device footprint.

Modeling Low Power Bluetooth Connectivity

Availability of Bluetooth Low Energy connectivity enables to offload Data storage to Edge devices. Only tax is extra power consumption due to BLE power.

Connectivity Bluetooth Low Energy (BLE)		Hardware 🔒 Constraint ⬤ No Constraint			
		Memory	Power Source	Storage	Computational Power
ML Model Accuracy	Self-Contained & Updated only on Hardware Refresh	Linear Regression is Linear Time Complex O (1)	Continuous Bluetooth advertisements could increase power needs.	Data could be offloaded to external devices.	The model data and storage are offloaded to Edge devices (Smartphone) and thus could reduce hardware footprint.
	OTA Firmware	N/A	N/A	N/A	N/A
	Model Type: Eager Learner	Linear Model Equations are eager learner with linear time complex O (1)	Bluetooth advertisements need higher power.	Periodic offloading of Sensor data to Edge devices lowers the storage needs.	Lower Data Size and Code size reduce the needs for more computational power.
	Lazy Learner	N/A	N/A	N/A	N/A
	Model Invocation	Linear time complex O (1) and Inline if-else model invocation.	Though Linear time complex, continuous Bluetooth advertisements could increase power needs	Periodic offloading of Sensor data to Edge devices lowers the storage needs.	Linear time complex & Inline model invocation

Overall design consideration: Operating Linear Regression Machine Learning algorithms in a constrained or Tiny device with "Bluetooth Low Energy (BLE) Connectivity" increases battery power consumption. Model performance & invocation are linear as time complexity is O (1) with if-else invocation. Only caution to be noted is sustainability of Battery for radio advertisements—that is turn on radio & BLE advertisement on a need basis.

Modeling Wi Fi Connectivity

Wi Fi connectivity enables to offload data storage to the Edge devices or call to the central Cloud server. Additionally, it enables to upload Firmware via OTA. The major limitation is many of the constrained devices are not suited for Wi Fi as it drains huge power from the battery.

Connectivity Wi Fi		Hardware 🔒 Constraint ◯ No Constraint			
		Memory	Power Source	Storage	Computational Power
ML Model Accuracy	Self-Contained & Updated only on Hardware Refresh	◯ Linear Regression Models are linear Time complex: O(1)	🔒 On board Wi Fi demands more power. For battery power devices, Wi-Fi option is less favorable.	◯ Data could be offloaded to external devices or central Cloud.	◯ The model data and storage are offloaded to Edge devices (Smartphone) and thus could reduce hardware footprint.
	OTA Firmware	◯ OTA allows to deploy the latest and greatest optimized model.	🔒 Onboard Wi Fi demands more power.	◯ Data could be offloaded to external devices or to the central Cloud.	◯ OTA allows to deploy the latest and greatest optimized model.
	Model Type: Eager Learner	◯ Linear Regression Models are eager learner with linear time complex O (1)	🔒 Bluetooth advertisements need higher power.	◯ Periodic offloading of Sensor data to Edge devices lowers the storage needs.	◯ Lower Data Size and Code size reduce the needs for more computational power.
	Lazy Learner	N/A	N/A	N/A	N/A
	Model Invocation	◯ Linear time complex O (1) and Inline if-else model invocation.	🔒 Though Linear time complex, continuous Bluetooth advertisements could increase power needs	◯ Periodic offloading of Sensor data to Edge devices lowers the storage needs.	◯ Linear time complex & Inline model invocation

Overall design consideration: Operating Linear Regression Machine Learning algorithms in a constrained or Tiny device with "Wi Fi Connectivity" increases battery power consumption. Model performance & invocation would be linear as time complexity is O (1) with in-line model invocation.

Connectivity—Model Accuracy trade-off

Modeling Memory

Since the model is a linear equation and does not require to be memory resident[288] nor heap[289] computation and does not require historical data for accuracy inference, the hardware memory is not a constraint for the model accuracy. That is, the memory (Data movement or access) does not influence power or computational hardware economy.

---

[288] Memory resident - https://www.trendmicro.com/vinfo/us/security/definition/memory-resident
[289] Stack vs. Heap allocation - https://www.geeksforgeeks.org/stack-vs-heap-memory-allocation/

Hardware Memory		Connectivity 🔒 Constraint ⭘ No Constraint		
		No Connectivity	Bluetooth Low Energy (BLE)/LoRa	Wi-Fi
Model Accuracy	Self-Contained & Updated only on Hardware Refresh	⭘	⭘ Since model update is cycled on Hardware refresh (SLA half-yearly or yearly), there would be staleness of the model. However, **memory constraint has no influence on the Accuracy.**	⭘
	OTA Firmware	⭘ Since the model gets updated OTA as part of Firmware update, no staleness of model. Memory constraint has no influence on the model accuracy.	⭘	⭘
	Model Type: Eager Learner	⭘	⭘	⭘
	Lazy Learner	N/A	N/A	N/A
	Model Invocation	⭘ In-line model invocation and not memory intense – no memory constraint on the model.	⭘	⭘

Design Considerations: Since model update is cycled on hardware refresh (SLA half-yearly or yearly), there would be staleness of the model. However, memory constraint has no influence on the Accuracy. Model is a linear call and will not require data movement between processor & memory space. In-line model invocation and not memory intense—no memory constraint on the model.

Modeling Processing Power

The Model deployed as a linear call and invoked as an in-line function and does not require extra computation power.

Hardware Processing Power		Connectivity 🔒 Constraint ⭘ No Constraint			
		No Connectivity	BLE	Wi-Fi	Design Notes
Model Accuracy	Hardware Refresh	⭘	⭘	⭘	Since model update is cycled on Hardware refresh (SLA half-yearly or yearly), there could be staleness in Model accuracy. This has no influence on the Power.
	OTA Firmware Update	⭘	⭘	⭘	Processing Power is not a constraint with respect to Model Accuracy.
	Model Type: Eager Learner	⭘	⭘	⭘	Processing Power is not a constraint with respect to Model Accuracy.
	Lazy Learner	N/A	N/A	N/A	
	Model Invocation	⭘	⭘	⭘	Processing Power is not a constraint with respect to Model Accuracy.

Design Considerations: Since model update is cycled on hardware refresh (SLA half-yearly or yearly), there could be staleness in model accuracy. This has no influence on the Power. Processing Power is not a constraint with respect to Model Accuracy.

Modeling Storage

The model deployed is an in-line function and does not require huge storage. Nonetheless, as hardware device collects sensor data, the data needs to be stored on device local storage (offload to Edge Gateway or Mobile device) before uploads to Central Storage or Corporate Cloud storage.

Hardware Storage		Connectivity ⊕ Constraint ◯ No Constraint			
		No Connectivity	BLE	Wi-Fi	Design Notes
Model Accuracy	Hardware Refresh	◯	◯	◯	Since model update is cycled on Hardware refresh (SLA half-yearly or yearly), the data storage needs to be large to store on the device. A flat file-based storage. For BLE and Wi-Fi modes, the storage could be offloaded to gateway or mobile device & Circular storage data pattern could help to concise storage.
	OTA Firmware Update	◯	◯	◯	Since the model gets updated OTA as part of Firmware update, no staleness of model. Storage could be constraint under "No Connectivity" but not a constraint for BLE and Wi-Fi.
	Model Type: Eager Learner	◯	◯	◯	Model is linear equation and will not require data movement between processor & memory space. Storage could be constraint under "No Connectivity" but not a constraint for BLE and Wi-Fi.
	Lazy Learner	N/A	N/A	N/A	
	Model Invocation	◯	◯	◯	In-line model invocation and not memory intense – no memory constraint on the model. Storage could be constraint under "No Connectivity" but not a constraint for BLE and Wi-Fi.

Design Considerations: Since model update is cycled on hardware refresh (SLA half-yearly or yearly), there could be staleness in Model accuracy. This has no influence on the Power. Processing Power is not a constraint with respect to Model Accuracy.

Modeling Environmental Perturbations

The model deployed under geographical conditions with huge temperature or humidity variations, the influence of temperature on the battery could lead to accuracy consideration. Humidity accumulation is also a factor to be considered in the model calculation

Hardware Environmental Perturbations		Connectivity			
		No Connectivity	BLE	Wi-Fi	Design Notes
Model Accuracy	Hardware Refresh	◯	◯	◯	Model should factor in geographical sensitivity as part of the calculations. The model coefficients need to tame environmental perturbations.
	OTA Firmware Update	◯	◯	◯	The OTA could fine-tune sensitive model coefficients to overcome environmental perturbations.
	Model Type: Eager Learner	N/A	N/A	N/A	No influence of environmental perturbations on data movement.
	Lazy Learner	N/A	N/A	N/A	No influence of environmental perturbations on data movement.
	Model Invocation	N/A	N/A	N/A	No influence of environmental perturbations on data movement.

Connectivity to Hardware Trade-off

Modeling: Hardware Refresh

In this mode, the model is only getting refreshed whenever hardware refresh takes place. Generally, this could be half-yearly or yearly, depending upon the SLA signed with customer. With this constraint, there is staleness on the model accuracy. That is, adaptive learning is only deferred when a new model gets deployed. The model is a linear function call that is invoked as part of the embedded code. So, no special requirements for memory, power, and computational processing. And hence, no constraints.

Model Accuracy Hardware Refresh		Hardware 🔒 Constraint ○ No Constraint					
		Power	Memory	Storage	Processing	Environmental	
Connectivity	No Connectivity	○	○	🔒	○	○	Storage needs are higher & inaccuracies due to environmental perturbations.
	BLE	N/A	N/A	N/A	N/A	N/A	Due to staleness in model deployment, inaccuracies due to environmental perturbations a possibility.
	Wi Fi	N/A	N/A	N/A	N/A	N/A	

Design Considerations: Since model update is cycled on hardware refresh (SLA half-yearly or yearly), there could be staleness in model accuracy. This has no influence on the Power. Processing Power is not a constraint with respect to model accuracy.

Modeling Over-The-Air (OTA) Firmware Update

In this mode, the model is only getting refreshed over-the-air as part of firmware update. Except for no connectivity mode, the model is the latest and learned with the most recent collected data. That is, adaptive learning is current and gets deployed as part of OTA. The model is a linear function that is invoked as part of in-line code. So, no special requirements for memory, power, and computational processing. And hence, no constraints.

Model Accuracy OTA Firmware Update		Hardware 🔒 Constraint ○ No Constraint					
		Power	Memory	Storage	Processing	Environmental	Design Notes
Connectivity	No Connectivity	○	○	🔒	○	🔒	Storage needs are higher & inaccuracies due to environmental perturbations.
	BLE	○	○	○	○	○	With the model is updated vis OTA, the latest learnings will have higher accuracies.
	Wi Fi	○	○	○	○	○	

Modeling Active Learner vs. Lazy Learner

The model is an Active learner with no special requirements to be in-memory, memory resident or data movements. The model is a linear function that is invoked as part of in-line code. So, no special requirements for memory, power, and computational processing. And hence, no constraints. No restrictions except in No connectivity mode with or without OTA firmware model.

Model Accuracy Learner		Hardware 🔒 Constraint ◯ No Constraint					
		Power	Memory	Storage	Processing	Environmental	Design Notes
Connectivity	No Connectivity	◯	◯	🔒	◯	🔒	Storage needs are higher & inaccuracies due to environmental perturbations.
	BLE	◯	◯	◯	◯	◯	With the model is updated via OTA, the latest learnings will have higher accuracies.
	Wi-Fi	◯	◯	◯	◯	◯	

Modeling Model Invocation

The model is invoked as an in-line code and not a heap or memory resident call. Except storage constraint as part of No Connectivity mode, the model has no special restrictions.

Model Accuracy Invocation		Hardware 🔒 Constraint ◯ No Constraint					
		Power	Memory	Storage	Processing	Environmental	Design Notes
Connectivity	No Connectivity	◯	◯	🔒	◯	🔒	Storage needs are higher & inaccuracies due to environmental perturbations.
	BLE	◯	◯	◯	◯	◯	With the model is updated via OTA, the latest learnings will have higher accuracies.
	Wi Fi	◯	◯	◯	◯	◯	

Kalman Filter

The Kalman filter theory has been proved an efficient tool for correcting systematic forecast errors, combining observations with model forecasts. Applications of the Kalman filter theory for the correction of surface temperature forecasts are based either only on temperature information where other parameters, such as relative humidity and surface winds, are used. A complete description of the Kalman filter can be found in Kalman. For the convenience of the reader, some basic notions of the general Kalman filter theory are presented in the following paragraphs [12][13].

In the Kalman filter, the fundamental concept is the notion of the state. By this is meant, intuitively, some quantitative information (a set of numbers, a function, etc.) that is the least amount of data one has to know about the past behavior of the system in order to predict its future behavior. The dynamics are then described in terms of state transitions, i.e., one must specify how one state is transformed into another as time passes [12][13][14].

In order to use the Kalman filter to estimate the internal state of a process given only a sequence of noisy observations, one must model the process in accordance with the framework of the Kalman filter. This means specifying the following matrices:

- F_k, the state-transition model;
- H_k, the observation model;
- Q_k, the covariance of the process noise;
- R_k, the covariance of the observation noise; and
- Sometimes B_k, the control-input model, for each time-step, k, as described below.

The Kalman filter model assumes the true state at time k is evolved from the state at $(k - 1)$ according to [12][13][14]

$$x_k = F_k x_{k-1} + B_k u_k + w_k \qquad \text{Eq. (11)}$$

where

- F_k is the state transition model which is applied to the previous state x_{k-1}
- B_k is the control-input model which is applied to the control vector u_k
- w_k is the process noise which is assumed to be drawn from a zero mean.

At time k an observation (or measurement) z_k of the true state x_k is made according to

$$z_k = H_k x_k + v_k \qquad \text{Eq. (12)}$$

where

- H_k is the observation model which maps the true state space into the observed space and
- v_k is the observation noise which is assumed to be zero mean Gaussian.

The goal of the Kalman filter is to generate next predicted values and compare the predicted value to Sensor computed value. The deviated values are notified (see Figure 63).

Time = k – 1 → Historical value

Time = k → Current Value

Time = k+1 → Future value

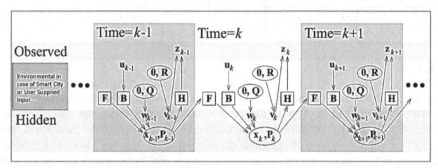

Figure 63: Kalman Over three intervals [12]

Kalman Filter Block Diagram Representation

The Kalman filter forms an optimal state estimate for a state-space plant model in the presence of process and measurement noise. The optimality criterion for the steady-state Kalman filter is the minimization of the steady-state error covariance matrix P shown

$$P = \lim_{t \to \infty} E(\{\hat{x} - x\}\{\hat{x} - x\}^T) \qquad \text{Eq. (13)}$$

Where E represents the expectation (or mean) of the parenthesized expression, $\hat{x}$ is the state estimate, and x is the true state vector. The term $\{-\hat{x}\,x\}$ is the state estimation error. The diagonal elements of P contain the variance (the square of the standard deviation) of the state estimation error. The off-diagonal terms represent the correlation between the errors of the state vector elements.

The integrator stands for n integrators such that the output of each is a state variable; $F(t)$ indicates how the outputs of the integrators are fed back to the inputs of the integrators. Thus $f_{ij}(t)$ is the coefficient with which the output of the jth integrator is fed back to the input of the ith integrator. It is not hard to relate this formalism to more conventional methods of linear system analysis (see Figure 64) [12][13][14].

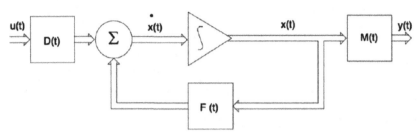

Figure 64: Kalman Model

where

- $u(t)$ is an m-vector ($m \le n$) representing the inputs to the system

Kalman filter is a recursive data processing algorithm that generates the best measurement based on all previous measurements. It basically consists of two methods, prediction and correction. Thus, it is good fit to predict anomaly behavior in the system at the edge level. Another use of the Kalman filter is automatic re-calibration of the failed sensor from the neighboring sensor. The reasons behind choosing the Kalman filter is as follows:

- It is light weight
- It works efficiently for real-time processing
- Removes Gaussian noise and gives good results.

Kalman Filter for Smart City

Kalman filtering is useful for analyzing real-time inputs about a system that can exist in certain states. Typically, there is an underlying model of how the various components of the system interact with and affect each other. A Kalman filter processes the various inputs, attempts to identify the errors in the input, and predicts the current state. For example, a Kalman filter in our Smart Cities Ambiance Maintenance System can process various inputs, such as number of people, day of the week, month, hour and location, and update the estimate of the room temperature.

In our system, the model is:

> temperature = 0.3019 * number_people + 5.7507 * day_of_week - 11.4933 * is_weekend + 32.4674 * is_holiday + 2.9673 * month + 0.7805 * hour (See Equation 10)

> **Please note:** day_of_week, is_weekend, is_holiday and month parameters are computed by tapping into onboard embedded real-time clock. Since, these values are critical to the embedded system, we have not modeled filtering for these parameters as part of the Kalman Filter design.

The use of Kalman filters is motivated by: (a) their support for streaming analysis using only current input measurements (therefore making the solution more memory efficient), (b) they do not require matrix calculations (therefore, the solution is more computationally efficient), (c) the ease of the algorithm tuning process, and (d) their implementation simplicity.

The number of people can be retrieved either interfacing with Venue system or by adding data from strip sensor.

During the initialization process the parameters which need tuning are the process noise covariance q, the sensor measurement noise covariance r, the initial estimated error covariance p and an initial measurement x.[290]

---

[290] Filters - http://commons.apache.org/proper/commons-math/userguide/filter.html

State Transition Matrix :: x = [0.3019 * NUMBER OF PEOPLE]

Control Input Matrix :: B = NULL;

$$
\text{Measurement Matrix:: H} =
\begin{bmatrix}
0.3019 * \text{NUMBER OF PEOPLE}_1 \\
0.3019 * \text{NUMBER OF PEOPLE}_2 \\
0.3019 * \text{NUMBER OF PEOPLE}_3 \\
0.3019 * \text{NUMBER OF PEOPLE}_4
\end{bmatrix}
$$

Process Noise Covariance Matrix:: q = [0.25%] or [0.0025]

As per Si 7020 – A10 Sensor Data Sheet - the sensor shows an accuracy of 0.25%RH[291] [16][17].

Measurement Noise Covariance Matrix :: r = [0.005]

Kalman Filter implementation for Smart City (Number of people in avenue vs. Temperature) – see Table 17:

Table 17: Kalman Filter Code—Smart City

```
# -*- coding: utf-8 -*-
"""
Created on Wed Nov 28 18:37:06 2018
@author: cvuppalapati
"""
# Kalman filter example demo in Python
# A Python implementation of the example given in pages 11-15 of "An
# Introduction to the Kalman Filter" by Greg Welch and Gary Bishop,
# University of North Carolina at Chapel Hill, Department of Computer
# Science, TR 95-041,
# https://www.cs.unc.edu/~welch/media/pdf/kalman_intro.pdf
# by Andrew D. Straw
# https://scipy-cookbook.readthedocs.io/items/KalmanFiltering.html
import numpy as np
import matplotlib.pyplot as plt
plt.rcParams['figure.figsize'] = (10, 8)
# intial parameters
n_iter = 50
sz = (n_iter,) # size of array
x = 0.3019 # Temperature state Transition matrix 0.3019 * NUMBER OF PEOPLE z = np.random.normal(x,0.1,size=sz) # observa-
tions (normal about x, sigma=0.1)
Q = 0.0025 # process variance
# allocate space for arrays
xhat=np.zeros(sz) # a posteri estimate of x
P=np.zeros(sz) # a posteri error estimate
xhatminus=np.zeros(sz) # a priori estimate of x
Pminus=np.zeros(sz) # a priori error estimate
K=np.zeros(sz) # gain or blending factor
R = 0.1**2 # estimate of measurement variance, change to see effect
# intial guesses
xhat[0] = 0.0
P[0] = 1.0
for k in range(1,n_iter):
 # time update
 xhatminus[k] = xhat[k-1]
 Pminus[k] = P[k-1]+Q
 # measurement update
 K[k] = Pminus[k]/( Pminus[k]+R )
 xhat[k] = xhatminus[k]+K[k]*(z[k]-xhatminus[k])
 P[k] = (1-K[k])*Pminus[k]
```

Table 17 contd. ...

---

[291] Si7020 – A 10 http://www.mouser.com/ds/2/368/Si7020-272416.pdf

...Table 17 contd.

```
plt.figure()
plt.plot(z,'k+',label='noisy measurements')
plt.plot(xhat,'b-',label='a posteri estimate')
plt.axhline(x,color='g',label='truth value')
plt.legend()
plt.title('Estimate vs. iteration step', fontweight='bold')
plt.xlabel('Iteration')
plt.ylabel('Temperature')
plt.figure()
valid_iter = range(1,n_iter) # Pminus not valid at step 0
plt.plot(valid_iter,Pminus[valid_iter],label='a priori error estimate')
plt.title('Estimated $\it{\mathbf{a \ priori}}$ error vs. iteration step', fontweight='bold')
plt.xlabel('Iteration')
plt.ylabel('$(Temperature)^2$')
plt.setp(plt.gca(),'ylim',[0,.01])
plt.show()
```

The goal of the Kalman filter is to predict the deviations in the actual temperature vs. predicted values (see Figure 65).

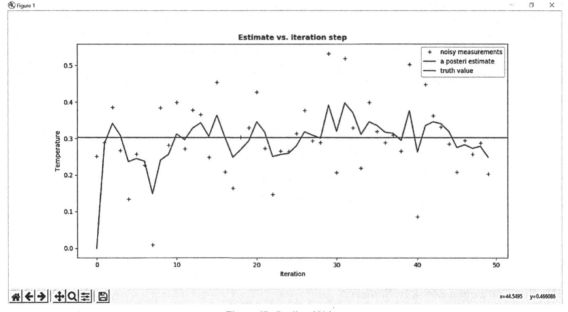

Figure 65: Predicted Values

Constrained Environment Considerations

Let's consider the computational complexity of Kalman Filter. The computational requirements of an algorithm can be estimated by the number of multiplications per cycle.
Kalman filtering is a 2-step process (please see Figure 66):

- The filter predicts the next position and its estimated error based on the previous position and its estimated error with a movement model

- The filter calculates the new position and its estimated error by updating the predicted position using frequency measurements.

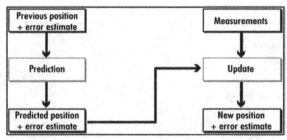

Figure 66: Kalman Filter can be understood [18] as a loop[292]

Computational Complexity

Kalman Filter works on prediction-correction model used for linear and time-variant or time-invariant systems [19]. Prediction model involves the actual system and the process noise. The update model involves updating the predicated or the estimated value with the observation noise. Kalman gain is calculated based on Recursive Least Squares (RLS) algorithm in order to reach the optimal value within less amount of time [19].

Recursive Least Squares (RLS) is based on weighted least squares in which previous values taken in account for determining the future value. Each weight is exponentially assigned to each previous value of the actual system. The weights are updated recursively based on memory.

The computational complexity of Kalman Filter contains two cycles: one cycle of a Kalman filter[293] can be divided into predict part (subscript p) and update part (subscript u). Using the standard formula, the number of multiplications M is computed, assuming that the components of the measurement are independent and taking advantage of the symmetric matrices, i.e., only the upper triangular matrix has to be computed [19].

The computational complexity of RLS is $O(N^2)$ and Computational Time Complexity of Kalman Filter is $O(N^3)$ [19][20].

Output

We ran Kalman Filter on an embedded device with the following technical specifications:

Board Specification	SJ One Board - LPC1758 ARM Cortex M3, 512K ROM, 64K RAM.
Operating System	Real-time operating system - FreeRTOS. The RTOS kernel requires 5 to 10K ROM. The scheduler consumes 236 bytes and the task/thread consumes 64 bytes.
Bluetooth BLE module	Adafruit Bluefruit LE. Connected to SJOne board via UART at baud rate of 9600 bps.Has Raytac's MDBT40 which is BT 4.0 & BT 4.1 stack BLE module. Requires power supply of 3.3V.
Power	Supply of 5V.

Pre-Run Compute & Memory Statistics

Please find memory captured—running Kalman Filter

Memory Details	Used (in Bytes)
Global Used	1644 Bytes
Malloc Used	7740 Bytes
Malloc Available	0 Bytes
System Available Memory	56152 Bytes
Next Heap Pointer	0x10001E3C
Last Stack Back Pointer	0x10001DDC
Last Stack Back Pointer Size	96
Number of Stack calls	44

---

[292] Kalman Filter Image - https://www.argos-system.org/manual/3-location/32_principle.htm
[293] Kalman Filter Configurations for Multiple Radar Systems - https://apps.dtic.mil/dtic/tr/fulltext/u2/a026367.pdf

Figure 67: Kalman Filter Output

Please see the output of Kalman (Figure 67):

Post-Run Compute & Memory Statistics

Memory Details	Used (in Bytes)
Global Used	1644 Bytes
Malloc Used	8528 Bytes
Malloc Available	0 Bytes
System Available Memory	55364 Bytes
Next Heap Pointer	0x10002150
Last Stack Back Pointer	0x1000212C
Last Stack Back Pointer Size	36
Number of Stack calls	87

The output of Kalman after the run is complete (please see Figure 68)

Figure 68: Kalman Output

As you can see from the Pre-Run and Post-Run tables, the memory used for 10 iteration Kalman costed hardware 788 bytes. For a 64KB module, an approximate 1 kb hit every time Kalman Filter runs can be expensive both in terms of hardware and power computation (please see Figure 69).

Memory Details	Differentials (in Bytes)
Global Used	0 Bytes
Malloc Used	788 Bytes
Malloc Available	0 Bytes
System Available Memory	−788 Bytes
Next Heap Pointer	0X314 Pointer advancement
Last Stack Back Pointer	0x350 Pointer advancement
Last Stack Back Pointer Size	−60
Number of Stack calls	43

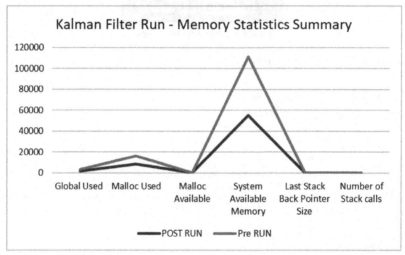

Figure 69: Kalman Filter Run Memory Statistics Summary

Hardware Economy—Model Accuracy tradeoff

The purpose of this section is to analyze the decision tree deployment in constrained or Tiny Compute hardware.

Modeling for no Connectivity

Modeling for no connectivity influences model upgradability, storage and power source. Since there is no connectivity environment, the Kalman Filter coefficients could only update during new device refresh cycle. Additionally, the lack of connectivity inhibits model upgrade via OTA firmware update.

Connectivity No Connectivity		Memory	Power Source	Storage	Computational Power
ML Model Accuracy	Self-Contained & Updated only on Hardware Refresh	◯ The Time Complexity – $O(N^3)$ where N is number of iterations	🔒 Time complexity is $O(N^3)$ and increased storage space due to no connectivity could exert higher power consumption.	🔒 Application data storage could increase due to no connectivity.	🔒 High computational power for Kalman Filter - $O(N^3)$
	OTA Firmware	N/A	N/A	N/A	N/A
	Model Type: Eager Learner	Not applicable as Kalman Filter is a Recursive Least Squares algorithm.	Not applicable as Kalman Filter is a Recursive Least Squares algorithm.	Not applicable as Kalman Filter is a Recursive Least Squares algorithm.	Not applicable as Kalman Filter is a Recursive Least Squares algorithm.
	Lazy Learner	Not applicable as Kalman Filter is a Recursive Least Squares algorithm.	Not applicable as Kalman Filter is a Recursive Least Squares algorithm.	Not applicable as Kalman Filter is a Recursive Least Squares algorithm.	Not applicable as Kalman Filter is a Recursive Least Squares algorithm.
	Model Invocation	◯ Prediction-correction model used for linear and time-variant or time-invariant systems – $O(N^3)$	🔒 Prediction-correction model used for linear and time-variant or time-invariant systems – $O(N^3)$	Not applicable	🔒 Inline model invocation requires more computational power for $O(N^3)$

Hardware 🔒 Constraint ◯ No Constraint

Overall design consideration: Operating Kalman Filter is an extremely constrained device (xCD) or Tiny device with "No Connectivity", Storage of the device needs to be well defined and should be sufficient for the SLA time of the device. If the SLA is half-yearly or yearly refresh, the device storage capacity could impact battery power. As far as the model computational performance, Kalman Filter with higher iterations (N) could impact better performance.

 For Outlier detection algorithms such as Kalman Filter, the impact of battery power consumption is based on the number of iterations that RLS minimizes the predicted to actual output error.

Modeling Low Power Bluetooth Connectivity

Bluetooth Low Energy connectivity availability enables to offload data storage to Edge devices. Only tax is extra power consumption due to BLE power. Additionally, computational power levy penalties on battery power source by Kalman Filter that have higher iterations (N) or complex state model (K) defined.

Connectivity Bluetooth Low Energy (BLE)		Hardware 🔒 Constraint ⭕ No Constraint			
		Memory	Power Source	Storage	Computational Power
ML Model Accuracy	Self-Contained & Updated only on Hardware Refresh	⭕ The Time Complexity – $O(N^3)$ where N is number of iterations	🔒 Continuous Bluetooth advertisements and time complexity could increase power needs.	⭕ Data could be offloaded to external devices.	🔒 Higher computational power for Kalman Filter – depending upon Number of iterations and Complex state model – $O(N^3)$
	OTA Firmware	N/A	N/A	N/A	N/A
	Model Type: Eager Learner	N/A	N/A	N/A	N/A
	Lazy Learner	N/A	N/A	N/A	N/A
	Model Invocation	⭕ Prediction-correction model used for linear and time-variant or time-invariant systems – $O(N^3)$.	🔒 Prediction-correction model used for linear and time-variant or time-invariant systems – $O(N^3)$ continuous Bluetooth advertisements could increase power needs	⭕ Periodic offloading of Sensor data to Edge devices lowers the storage needs.	🔒 Prediction-correction model used for linear and time-variant or time-invariant systems – $O(N^3)$ – Complexity depends upon State Model & Iterations

Overall design consideration: Operating Kalman Filter in an extremely constrained device (xCD) or Tiny device with "Bluetooth Low Energy (BLE) Connectivity" increases battery power consumption. Model performance & invocation are N cube time complexity with **$O(N^3)$** with Prediction-correction model.

 For Kalman Filter deployments in extremely constrained devices (xCD), the impact of BLE connectivity & model invocation, i.e., the number of iterations count and state model complexity will have considerable battery power implications.

Modeling Wi-Fi Connectivity

Wi-Fi connectivity enables to offload data storage to the Edge devices or call to the central Cloud Server. Additionally, enables to download Firmware via OTA. The major limitation is many of the constrained devices are not suited for Wi-Fi as it drains huge power from the battery.

Connectivity Wi-Fi		Hardware 🔒 Constraint ⭘ No Constraint			
		Memory	Power Source	Storage	Computational Power
ML Model Accuracy	Self-Contained & Updated only on Hardware Refresh	⭘ The Time Complexity – **O (N³)** where N is number of iterations	🔒 On board Wi-Fi demands more power. For battery power devices, Wi-Fi option is less favorable.	⭘ Data could be offloaded to external devices or central Cloud.	🔒 The Time Complexity – **O (N³)** where N is number of iterations to reduce error deviation between actual and predicted state model values.
	OTA Firmware	⭘ OTA allows to deploy the latest and greatest optimized Kalman Filter with updated Kalman coefficients.	🔒 Onboard Wi-Fi demands more power & Time Complexity – **O (N³)** where N is number of iterations to reduce error deviation between actual and predicted state model values.	⭘ Data could be offloaded to external devices or to the central Cloud.	🔒 OTA allows to deploy the latest and greatest optimized model. However, Time complexity demands high computational power.
	Model Type: Eager Learner	N/A	N/A	N/A	N/A
	Lazy Learner	N/A	N/A	N/A	N/A
	Model Invocation	⭘ The Time Complexity – *O (N³)* where *N* is number of iterations to reduce error deviation between actual and predicted state model values.	🔒 The Time Complexity – *O (N³)* where *N* is number of iterations to reduce error deviation between actual and predicted state model values.	⭘ Periodic offloading of Sensor data to Edge devices lowers the storage needs.	🔒 The Time Complexity – *O (N³)* where *N* is number of iterations to reduce error deviation between actual and predicted state model values.

Overall design consideration: Operating Kalman Filter algorithm in an extremely constrained devices (xCD) or Tiny device with "Wi Fi Connectivity" increases battery power consumption. Model performance and invocation is $O(N^3)$ with complexity increase based on N is number of iterations that reduce error deviation between actual and predicted state model values.

Connectivity—Model Accuracy trade-off

Modeling Memory

Since the model is a prediction-correction state model and does require memory resident[294] or heap[295] computation, the hardware memory is a constraint for the Model accuracy & performance. That is, the memory (Data movement or access) does influence power or computational hardware economy.

---

[294] Memory resident - https://www.trendmicro.com/vinfo/us/security/definition/memory-resident
[295] Stack vs. Heap allocation - https://www.geeksforgeeks.org/stack-vs-heap-memory-allocation/

Hardware Memory		No Connectivity	Bluetooth Low Energy (BLE)/LoRa	Wi Fi
Model Accuracy	Self-Contained & Updated only on Hardware Refresh	🔒	🔒 Since model update is cycled on Hardware refresh (SLA half-yearly or yearly), there would be staleness of the model. Memory constraint has influence on the Accuracy.	🔒
	OTA Firmware	🔒 Since the model gets updated OTA as part of Firmware update, no staleness of Kalman coefficients. Memory constraint has influence on the model accuracy.	🔒	🔒
	Model Type: Eager Learner	N/A	N/A	N/A
	Lazy Learner	N/A	N/A	N/A
	Model Invocation	🔒 Higher computational power for Kalman Filter – depending upon Number of iterations and Complex state model – $O(N^3)$	🔒 Higher computational power for Kalman Filter – depending upon Number of iterations and Complex state model – $O(N^3)$	🔒 Higher computational power for Kalman Filter – depending upon Number of iterations and Complex state model – $O(N^3)$

Connectivity 🔒 Constraint ⬭ No Constraint

Design Considerations: Since model update is cycled on Hardware refresh (SLA half-yearly or yearly), there would be staleness of the model. And memory constraint has influence on the Accuracy. Since the model is a Prediction-correction model and will require data movement between processor and memory space for higher iterative complex state models.

 Kalman Filter Loop needs to store predicted values in a non-local variable and hence exerts more computational cycles. In general, Use of Local variables over global variables where possible preferred. Local variables are contained in the CPU while global variables are stored in the RAM, the CPU accesses local variables faster.[296]

Modeling Processing Power

The Kalman filter is a prediction-correction model and invoked as an in-line function and does require extra computation power.

---

[296] Minimizing Power Consumption in Microcontrollers - https://medium.com/@emmaodunlade/minimizing-power-consumption-in-microcontrollers-b3eb9593c893

Hardware Processing Power	Connectivity 🔒 Constraint ⭘ No Constraint			
	No Connectivity	BLE	Wi-Fi	Design Notes
Hardware Refresh	🔒	🔒	🔒	Since model update is cycled on Hardware refresh (SLA half-yearly or yearly), there could be staleness in Model accuracy. Additionally, due to $O(N^3)$ time complexity, the power consumption is higher.
OTA Firmware Update	🔒	🔒	🔒	$O(N^3)$ time complexity – dependent upon Number iterations & complex model statement – increases processing power needs.
Model Type: Eager Learner	N/A	N/A	N/A	
Lazy Learner	N/A	N/A	N/A	
Model Invocation	🔒	🔒	🔒	$O(N^3)$ time complexity – dependent upon Number iterations & complex model statement – increases processing power needs.

(Left vertical label: Model Accuracy)

Design Considerations: Processing power, given $O(N^3)$ time complexity, is higher and the model complexity increases $O(N^3)$ with N is number of iterations to reduce error deviation between actual and predicted state model values.

Modeling Storage

The model deployed as an in-line function and does not require huge storage. Nonetheless, as hardware device collects sensor data, the data needs to be stored on device local storage (offload to Edge Gateway or Mobile device) before uploads to Central Storage or Corporate Cloud storage.

Hardware Storage	Connectivity 🔒 Constraint ⭘ No Constraint			
	No Connectivity	BLE	Wi-Fi	Design Notes
Hardware Refresh	🔒	⭘	⭘	Since model update is cycled on Hardware refresh (SLA half-yearly or yearly), the data storage needs to be large to store on the device. For BLE and Wi Fi connectivity modes, the storage could be offloaded to gateway or mobile device and could help to concise storage.
OTA Firmware Update	🔒	⭘	⭘	Since the model gets updated OTA as part of Firmware update, no staleness of model. Storage could be constraint under "No Connectivity" but not a constraint for BLE and Wi Fi.
Model Type: Eager Learner	N/A	N/A	N/A	
Lazy Learner	N/A	N/A	N/A	
Model Invocation	🔒	🔒	🔒	Higher computational power for Kalman Filter – depending upon Number of iterations and Complex state model – $O(N^3)$

(Left vertical label: Model Accuracy)

Design Considerations: Since model update is cycled on Hardware refresh (SLA half-yearly or yearly), there could be staleness in model accuracy.

Modeling Environmental Perturbations

Model deployed under geographical conditions with huge temperature or humidity variations, the influence of temperature on the battery could lead to accuracy consideration. Humidity accumulation is also a factor to be considered in the model calculation

Hardware Environmental Perturbations	Connectivity 🔒 Constraint ○ No Constraint			
	No Connectivity	BLE	Wi-Fi	Design Notes
Model Accuracy — Hardware Refresh	🔒 Staleness of the model is a constraint in taming (adaptive) environmental perturbations.	○	○	Model should factor in geographical sensitivity as part of the calculations. The model coefficients need to tame environmental perturbations. Given Model complexity is $O(N^3)$, the power consumption is higher and computational power requirement are higher.
OTA Firmware Update	○	○	○	The OTA could fin-tune sensitive model coefficients to overcome environmental perturbations.
Model Type: Eager Learner	N/A	N/A	N/A	No influence of environmental perturbations on data movement.
Lazy Learner	N/A	N/A	N/A	No influence of environmental perturbations on data movement.
Model Invocation	🔒	🔒	🔒	Model exhibits $O(N^3)$ Time complexity – the model time complexity depends upon state model & number of iterations to reduce predicted to actual values.

Design Considerations: Environmental perturbations increase the complexity.

 Kalman Filter exhibits higher computational and power processing under environmental perturbations.

Connectivity to Hardware trade-off

Modeling Hardware Refresh

In this mode, the model gets refreshed only when hardware refresh takes place. Generally, this could be half-yearly or yearly, depending upon the SLA signed with customer. With this constraint, there is staleness on the model accuracy. That is, adaptive learning of Kalman gain is only deferred when a new model gets deployed. The model is a Prediction-correction function with $O(N^3)$ time complex that is invoked as part of in-line code. So the model requires higher memory, power and computational processing.

Model Accuracy Hardware Refresh	Hardware 🔒 Constraint ○ No Constraint					
	Power	Memory	Storage	Processing	Environmental	Design Notes
Connectivity — No Connectivity	🔒	🔒	🔒	🔒	🔒	Model is a prediction-correction function with $O(N^3)$ time complex.
BLE	🔒	🔒	○	🔒	🔒	Model is a prediction-correction function with $O(N^3)$ time complex.
Wi-Fi	🔒	🔒	○	🔒	🔒	

Design Considerations: Since model update is cycled on hardware refresh (SLA half-yearly or yearly), there could be staleness in model accuracy. This has influence on Power, Processing Power and Environmental perturbations.

Modeling Over-The-Air (OTA) Firmware Update

In this mode, the model is only getting refreshed over the air as part of firmware update. Except for no connectivity mode, the model is the latest and learned with the most recent collected data. That is, adaptive learning is current and gets deployed as part of OTA. The model is a $O(N^3)$ function that is invoked as part of in-line code. So, extra care is required to handle power and computational processing.

Model Accuracy OTA Firmware Update		Hardware 🔒 Constraint ⚪ No Constraint					
		Power	Memory	Storage	Processing	Environmental	Design Notes
Connectivity	No Connectivity	🔒	🔒	🔒	🔒	🔒	Storage needs are higher & inaccuracies due to environmental perturbations.
	BLE	🔒	🔒	⚪	🔒	🔒	With the model is updated vis OTA, the latest learnings will have higher accuracies.
	Wi-Fi	🔒	🔒	⚪	🔒	🔒	

Design Considerations: Firmware updates over- the- air enables deployment of the latest and greatest model – Kalman Filter gain & coefficients. This would reduce the need to have higher hardware capacities.

 Over- the - air Firmware updates enable the model to be equipped with most recent learnings.

Modeling Active Learner vs. Lazy Learner

Design Considerations: Kalman filter is an outlier detection algorithm and model gets computed as part of the inference. So, Active or Lazy learner is not applicable.

Modeling Model Invocation

The model is a Prediction-correction function with $O(N^3)$ time complex that is invoked as part of in-line code.

Model Accuracy Invocation		Hardware 🔒 Constraint ⚪ No Constraint					
		Power	Memory	Storage	Processing	Environmental	Design Notes
Connectivity	No Connectivity	🔒	🔒	🔒	🔒	🔒	Storage needs are higher & inaccuracies due to environmental perturbations.
	BLE	🔒	🔒	⚪	🔒	🔒	With the model is updated via OTA, the latest learnings will have higher accuracies.
	Wi-Fi	🔒	🔒	⚪	🔒	🔒	

Design Considerations: Model invocation is not subject to change for different connectivity modes and hardware modes.

Here is the summary of algorithms discussed (please see Table 18):

Table 18: Summary of Supervised algorithms

Algorithm	Time Complexity	Constrained Design Considerations		Description
Decision Tree	O (1)	Memory ☐	ML Model Accuracy ☐	
		Battery Power ☒	Environmental Perturbations ☐	
		Computational Constraint ☐	Connectivity ☐	
		Storage ☐		
XGB	Dependent upon Number of trees (K), depth (d), Number of features (q) – **O (K d ‖x‖0 log q)**	Memory ☒	ML Model Accuracy ☐	XGBoost algorithm demands more computational and memory capacities as compared to decision trees - impact the extremely constrained devices (xCDs) performance.
		Battery Power ☒	Environmental Perturbations ☒	
		Computational Constraint ☒	Connectivity ☐	
		Storage ☒		
Naïve Bayesian	O (1)	Memory ☐	ML Model Accuracy ☐	Very similar to that of Decision Tree. The best algorithm for constrained applications.
		Battery Power ☐	Environmental Perturbations ☐	
		Computational Constraint ☐	Connectivity ☐	
		Storage ☐		
Linear Regression	O (1)	Memory ☐	ML Model Accuracy ☐	Regression classifier is in the form of equation (y=mx+c). Performance impact on constrained devices is minimal.
		Battery Power ☐	Environmental Perturbations ☐	
		Computational Constraint ☐	Connectivity ☐	
		Storage ☐		
Kalman Filter	$O(N^3)$	Memory ☒	ML Model Accuracy ☒	Kalman Filter demands more computational and memory capacities $O(N^3)$, impacting the extremely constrained devices (xCDs) performance.
		Battery Power ☒	Environmental Perturbations ☒	
		Computational Constraint ☒	Connectivity ☒	
		Storage ☒		

After reading this chapter, you should be able to design, develop, implement, and deploy the following Machine Learning Algorithms into extremely constrained devices (xCDs) and mobile platforms. In other words, the chapter enables you to ML deployment in xCDs.

- Decision Trees
- Random Forest
- Extreme Gradient Boosting (XGB)
- Linear Regression
- Kalman Filter
- Naïve Bayesian

References

1.   Jiawei Han, Micheline Kamber and Jian Pei. Data Mining: Concepts and Techniques. Morgan Kaufmann; 3 edn (July 6, 2011), ISBN-10: 1558609016.
2.   Barry de Ville and Padraic Neville. Decision Trees for Analytics: Using SAS Enterprise Miner. SAS Institute (July 10, 2013), ISBN-10: 1612903150.

3. David Wolpert and Paul Ampadu. Managing Temperature Effects in Nanoscale Adaptive Systems, Springer; 2012 edition (August 31, 2011), ISBN-10: 1461407478.

4. Jason Brownlee. A Gentle Introduction to the Gradient Boosting Algorithm for Machine Learning. September 9, 2016, URL: https://machinelearningmastery.com/gentle-introduction-gradient-boosting-algorithm-machine-learning/, Access Date: September 18, 2019.

5. Tianqi Chen and Carlos Guestrin. XGBoost: Reliable Large-scale Tree Boosting System. 2015. URL: http://learningsys.org/papers/LearningSys_2015_paper_32.pdf, Access Date: December 24, 2019.

6. Tianqi Chen and Carlos Guestrin. 2016. XGBoost: A Scalable Tree Boosting System. In Proceedings of the 22nd ACM SIGKDD International Conference on Knowledge Discovery and Data Mining (KDD'16). Association for Computing Machinery, New York, NY, USA, 785–794. DOI: https://doi.org/10.1145/2939672.2939785.

7. Jason Brownlee. A Gentle Introduction to Bayes Theorem for Machine Learning. October 4, 2019/Last Updated on December 4, 2019, URL: https://machinelearningmastery.com/bayes-theorem-for-machine-learning/, December 25, 2019.

8. C.D. Manning, P. Raghavan and H. Schuetze. 2008. Introduction to Information Retrieval. Cambridge University Press; Illustrated edition (July 7, 2008), pp. 234–265.

9. Kedari, S., Vuppalapati, J.S., Ialapakurti, A., Kedari, S., Vuppalapati, R. and Vuppalapati, C. 2018. Adaptive edge analytics—a framework to improve performance and prognostics capabilities for dairy iot sensor. *In*: Karwowski, W. and T. Ahram (Eds.). Intelligent Human Systems Integration. IHSI 2018. Advances in Intelligent Systems and Computing, vol. 722. Springer, Cham.

10. Vuppalapati, C., Ilapakurti, A., Kedari, S., Vuppalapati, J., Kedari, S. and Vuppalapati, R. Democratization of Artificial Intelligence (AI) to Small Scale Farmers: A Framework to Deploy AI Models to Tiny IoT Edges That Operate in Constrained Environments. DOI: 10.5220/0009358706520657, In Proceedings of the 9th International Conference on Pattern Recognition Applications and Methods (ICPRAM 2020), pages 652–657, ISBN: 978-989-758-397-1.

11. Nick Rose. Crowdedness at the Campus Gym. 2016. https://www.kaggle.com/nsrose7224/crowdedness-at-the-campus-gym, Access Date, 7 December 2018.

12. Kalman, R.E. 1960. A new approach to linear filtering and prediction problems. Transactions of the American Society of Mechanical Engineers, D 82, pp. 35– 45.

13. Priestley, M.B. 1981. Spectral Analysis and Time Series. Academic Press, London, UK, pp. 807–815.

14. Mariken Homleid. Diurnal Corrections of Short-Term Surface Temperature Forecasts Using the Kalman Filter. http://www.leg.ufpr.br/~eder/Artigos/Homleid_1995.pdf , Access date: May 05, 2017.

15. Gelb, A. 1974. Applied Optimal Estimation. MIT Press, Cambridge, Massachusetts, USA. 374 pp.

16. Atmel Corporation. 2016. Atmel AVR 8-bit and 32-bit Microcontrollers. http://www.atmel.com/products/microcontrollers/avr/, Access date: April 30, 2017 [39].

17. Atmel Corporation. 2016. 8-bit "AVR Microcontrollers – Datasheet Complete," http://www.atmel.com/Images/Atmel-42735-8-bit-AVR-Microcontroller-ATmega328-328P_Datasheet.pdf, Access Date: May 18, 2017.

18. Argos User Manual. Location Calculations. URL: https://www.argos-system.org/manual/index.html#3-location/32_principle.htm, Access Date: March 10, 2020.

19. D. Willner, C.B Chang and K.P Dunn. Kalman Filter Configurations for Multiple Radar Systems. 14 April 1976, URL: https://apps.dtic.mil/dtic/tr/fulltext/u2/a026367.pdf, Access Date: November 22, 2019.

20. Sharath Srini. The Kalman Filter: An Algorithm for Making Sused sensor insight. April 18, 2018, URL: https://towardsdatascience.com/kalman-filter-an-algorithm-for-making-sense-from-the-insights-of-various-sensors-fused-together-ddf67597f35e, Access Date: January 08, 2020.

## CHAPTER 7

Unsupervised Models

"Every company has big data in its future and every company will eventually be in the data business..."

Thomas H. Davenport

The chapter starts with Unsupervised models design and apply Model Connectivity Hardware Design as part of the Machine Learning models. Unsupervised models discussed as part of the chapter includes K-Means Cluster and Hierarchical Clusters. Each of the Unsupervised models discussed as part of the chapter analyzed from extremely Constrained Devices (xCDs) and environment perspective and the goes in depth to optimize the Machine Learning model to be effectively functional in the constrained environment. The constrained environment modeling includes Hardware Connectivity Trade-offs, Connectivity Model Trade-offs, and Hardware Connectivity Trade-offs.

By following the procedures and the frameworks described as part of the chapter, the Machine Learning and Embedded Engineers can develop machine learning models and analyze holistically to successfully deploy in resource constrained environments.

Hierarchical Clustering

Two of the most common data clustering algorithms are hierarchical clustering and non-hierarchical clustering (please see Figure 1). In hierarchical clustering, successive clusters are discovered using previously established clusters, while clusters are established all at once using non-hierarchical algorithms. Hierarchical algorithms can be further divided into divisive algorithms and agglomerative algorithms. Divisive clustering begins with all data elements in one cluster, which is later successively divided into smaller ones. On the other hand, agglomerative clustering begins with single element clusters and works by successively merging them into larger ones [1].

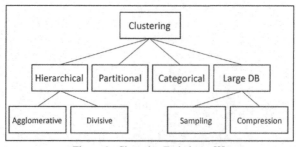

Figure 1: Clustering Techniques [2]

As for non-hierarchical clustering, some of the common algorithms are K-means and fuzzy C means. Other less conventional clustering algorithms include ant-based clustering, which is inspired by the collective behavior in decentralized, self-organized systems. In addition to the clustering algorithm, a function or measure

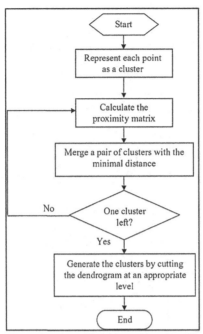

Figure 2: Hierarchical Cluster Flowchart [3]

for determining the similarity, distance or relatedness between elements is essential. In conventional clustering, feature-based measures such as cosine similarity, Jaccard distance or Dice similarity [3] are employed to support the use of hierarchical and non-hierarchical clustering.

Flowchart of the agglomerative hierarchical clustering algorithm is shown in Figure 2. Agglomerative clustering considers each data point as a cluster in the beginning. Two clusters are then merged in each step until all objects are forced into the same group [3].

Here is the summary of hierarchical clustering:

- Clusters are created in levels creating sets of clusters at each level
- Agglomerative
  - o Initially each item in its own cluster
  - o Iteratively clusters are merged together
  - o Bottom Up
- Divisive
  - o Initially all items in one cluster
  - o Large clusters are successively divided
  - o Top Down

Merging Cluster Techniques

Single Link: Distance between two clusters is the distance between the closest points [2]. Also called "neighbor joining" (Please see Figure 3).

Average Link: Distance between clusters is distance between the cluster centroids [2]. Please see Figure 4.

Complete Link: Distance between clusters is the distance between the farthest pair of points [2]. (Please see Figure 5.)

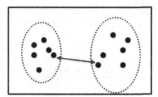

Figure 3: Single Link

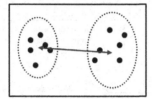

Figure 4: Average Link

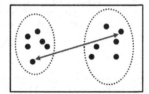

Figure 5: Complete Link

Agglomerative Cluster (Python) Code

The following code creates the Agglomerative Cluster Code:

```
# -*- coding: utf-8 -*-
"""
Created on Mon Mar 23 15:58:05 2020
@author: CHVUPPAL
"""
import numpy as np
import sklearn
import matplotlib.pyplot as plt
from sklearn.feature_extraction.text import TfidfVectorizer
from sklearn.metrics.pairwise import cosine_similarity
from sklearn.cluster import AgglomerativeClustering
from scipy.cluster.hierarchy import dendrogram
def get_distances(X,model,mode='l2'):
    distances = []
    weights = []
    children=model.children_
    dims = (X.shape[1],1)
    distCache = {}
    weightCache = {}
    for childs in children:
```

```
c1 = X[childs[0]].reshape(dims)
c2 = X[childs[1]].reshape(dims)
c1Dist = 0
c1W = 1
c2Dist = 0
c2W = 1
if childs[0] in distCache.keys():
    c1Dist = distCache[childs[0]]
    c1W = weightCache[childs[0]]
if childs[1] in distCache.keys():
    c2Dist = distCache[childs[1]]
    c2W = weightCache[childs[1]]
d = np.linalg.norm(c1-c2)
cc = ((c1W*c1)+(c2W*c2))/(c1W+c2W)
X = np.vstack((X,cc.T))
newChild_id = X.shape[0]-1
# How to deal with a higher level cluster merge with lower distance:
if mode=='l2': # Increase the higher level cluster size suing an l2 norm
    added_dist = (c1Dist**2+c2Dist**2)**0.5
    dNew = (d**2 + added_dist**2)**0.5
elif mode == 'max': # If the previrous clusters had higher distance, use that one
    dNew = max(d,c1Dist,c2Dist)
elif mode == 'actual': # Plot the actual distance.
    dNew = d
wNew = (c1W + c2W)
distCache[newChild_id] = dNew
weightCache[newChild_id] = wNew
distances.append(dNew)
weights.append( wNew)
return distances, weights
# Check version
print(sklearn.__version__)
```

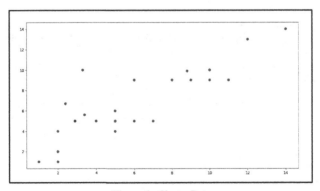

Figure 6: Cluster Data

The following code creates Cluster data (please see Figure 7) and creates model (please see Figure 8):

```
X = np.array([[1.0,1.0],
 [2.0,1.0],
 [2.0,2.0],
 [4.0,5.0],
 [5.0,4.0],
 [5.0,5.0],
 [5.0,6.0],
 [6.0,5.0],
 [9.0,9.0],
 [10.0,9.0],
 [10.0,10.0],
 [11.0,9.0],
 [2.0,4.0],
 [6.0,9.0],
 [12.0,13.0],
 [14.0,14.0],
 [10.0,10.0],
 [2.0,2.0],
 [7.0,5.0],
 [3.4,5.6],
 [2.4,6.7],
 [8.0,9.0],
 [8.8,9.9],
 [3.3,10.00],
 [5.0,5.0],
 [2.9,5.0],])

# Plot the clusters
colors = ['r']*25 + ['b']*25 + ['g']*25 + ['y']*25
plt.scatter(X[:,0],X[:,1],c=colors)

model = AgglomerativeClustering(linkage="single")
model.fit(X)

distance, weight = get_distances(X,model)
linkage_matrix = np.column_stack([model.children_, distance, weight]).astype(float)
plt.figure(figsize=(20,10))

dendrogram(linkage_matrix)
plt.show()
```

Agglomerative Cluster Jupyter Notebook Code (Climate Change Data)

Agglomerative Hierarchical Code in C

The distance metrics is based on the Euclidean distance. The distance between two points: (x_1, y_1) and (x_2, y_2) is. The distance between two points:

$$\sqrt{(x2 - x1)^2 + (y2 - y1)^2}$$
                                                                          Eq. (1)

The C Code (please see Table 1 – Eq. 1)

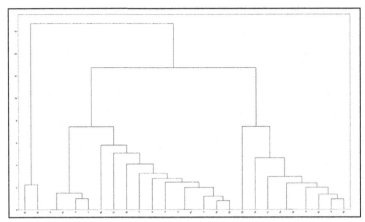

Figure 7: Cluster Dendrograms

Table 1: Euclidean Distance

```
float euclidean_distance(constcoord_t* a, constcoord_t* b)
{
return sqrt(pow(a->x - b->x, 2) + pow(a->y - b->y, 2));
}
```

Single Linkage Distance Formula

The Single Linkage distance calculates the shortest distance between two cluster centroids (please see Table 2):

Table 2: Single Link

```
float single_linkage(float** distances, constinta[],
constintb[], intm, intn)
{
float min = FLT_MAX, d;
for (int i = 0; i <m; ++i)
for (int j = 0; j <n; ++j) {
 d = distances[a[i]][b[j]];
if (d < min)
 min = d;
 }
return min;
}
```

Distances is calculated by (please see Table 3):

Table 3: Distance calculations

```
float get_distance(cluster_t* cluster, intindex, inttarget)
{
/* if both are leaves, just use the distances matrix */
if (index<cluster->num_items &&target<cluster->num_items)
returncluster->distances[index][target];
else {
cluster_node_t* a = &(cluster->nodes[index]);
cluster_node_t* b = &(cluster->nodes[target]);
if (distance_fptr == centroid_linkage)
return euclidean_distance(&(a->centroid),
&(b->centroid));
elsereturn distance_fptr(cluster->distances,
 a->items, b->items,
 a->num_items, b->num_items);
 }
}
```

The agglomerate Cluster code (please see Table 4):

Table 4: Agglomerate Cluster

```
cluster_t* agglomerate(intnum_items, item_t* items)
{
cluster_t* cluster = alloc_mem(1, cluster_t);
if (cluster) {
 cluster->nodes = alloc_mem(2 * num_items - 1, cluster_node_t);
if (cluster->nodes)
init_cluster(cluster, num_items, items);
else {
alloc_fail("cluster nodes");
goto cleanup;
 }
 }
else
alloc_fail("cluster");
goto done;
cleanup:
 free_cluster(cluster);
 cluster = NULL;
done:
return cluster;
 }
```

Visual Studio Code (please see Figure 8):

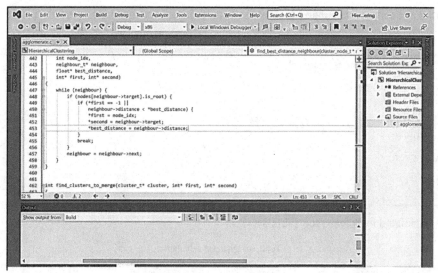

Figure 8: Microsoft Visual Studio

Time Complexity

Hierarchical Cluster with N Node and K Clusters, the time complexity[297] of hierarchical algorithms is O (KN^2) [5]. Agglomerative clustering is more extensively researched than divisive clustering. Yet, the quantum variant is more apt for the divisive type. The classical divisive clustering algorithm begins by placing all data instances in a single cluster C0. Then, it chooses the data instance whose average dissimilarity from all the other instances is the largest. This is the computationally most expensive step, having $\Omega(N^2)$ complexity in general[6].

---

[297] Hierarchical Clustering - https://www.sciencedirect.com/topics/computer-science/hierarchical-clustering

Pre-Run Compute & Memory Statistics:

Please find memory captured—running Hierarchical Cluster

Memory Details	Used (in Bytes)
Global Used	1652 Bytes
Malloc Used	7740 Bytes
Malloc Available	0 Bytes
System Available Memory	56144 Bytes
Next Heap Pointer	0x10001E3C
Last Stack Back Pointer	0x10001DDC
Last Stack Back Pointer Size	96
Number of Stack calls	44

Please see the output of Hierarchical (Figure 9):

Figure 9: Hierarchical Cluster Filter Output

Post-Run Compute & Memory Statistics

Memory Details	Used (in Bytes)
Global Used	1652 Bytes
Malloc Used	7888 Bytes
Malloc Available	5332 Bytes
System Available Memory	50664 Bytes
Next Heap Pointer	0x100033A4
Last Stack Back Pointer	0x1000336C
Last Stack Back Pointer Size	56
Number of Stack calls	238

The output of hierarchical cluster after the run complete (please see Figure 10).

As you can see from Pre-Run and Post-Run tables, the memory used for 76 nodes (N) three cluster (K) Hierarchical Cluster cost hardware 5480 bytes. For a 64KB module, an approximate 5 kb hit every time hierarchical cluster runs can be expensive, both in terms of hardware and power computation (please see Figure 11).

Figure 10: Hierarchical Cluster

Memory Details	Differentials (in Bytes)
Global Used	0 Bytes
Malloc Used	148 Bytes
Malloc Available	5332 Bytes
System Available Memory	−5480 Bytes
Next Heap Pointer	0X1568 Pointer advancement
Last Stack Back Pointer	0x1590 Pointer advancement
Last Stack Back Pointer Size	−60
Number of Stack calls	43

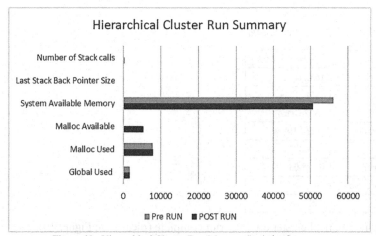

Figure 11: Hierarchical Cluster Run Memory Statistics Summary

Hardware Economy—Model Accuracy tradeoff

The purpose of the section is to analyze the hierarchical cluster deployment in constrained or Tiny Compute hardware.

Modeling no Connectivity

Modeling for no connectivity influences model upgradability, storage and power source. Since, no connectivity environment, the *Hierarchical Cluster* could only update during new device refresh cycle. Additionally, the lack of connectivity inhibits model upgrade via OTA firmware update.

Connectivity **No Connectivity**		Hardware 🔒 Constraint ◯ No Constraint			
		Memory	Power Source	Storage	Computational Power
ML Model Accuracy	Self-Contained & Updated only on Hardware Refresh	🔒 The Time Complexity – *O (KN²)*, where *K* is number of Clusters and *N* is number of Nodes. The memory consumption is higher for the large clusters with higher nodes.	🔒 The Time Complexity – *O (KN²)*, where *K* is number of Clusters and *N* is number of Nodes - no connectivity could exert higher power consumption.	🔒 Given **O (KN²) *time complexity*,** the storage needs are higher.	🔒 Higher computational power due to *O (KN²)*
	OTA Firmware	N/A	N/A	N/A	N/A
	Model Type: Eager Learner	N/A	N/A	N/A	N/A
	Lazy Learner	🔒	🔒	🔒	🔒
	Model Invocation	🔒	🔒	🔒	🔒

Overall design consideration: Operating Hierarchical Cluster model in an extremely constrained device (xCD) or Tiny device with "No Connectivity" requires the following design considerations:

- Well-defined Storage—Storage of the device needs to be well defined and should be enough for the SLA time of the device. If the SLA calls for half-yearly or yearly refresh, the device storage capacity needs to be higher and could impact battery power.
- Clusters with large number of Data Points (*N*) and higher clusters *K* will have more performance impact than clusters with lower number of nodes.

 For Hierarchical Cluster algorithms such as Agglomerative, the impact of battery power consumption is based on the number Nodes (*N*) in the cluster plus cluster size (*K*).

Modeling Low Power Bluetooth Connectivity

Bluetooth Low Energy connectivity availability enables to offload data storage to Edge devices. Only tax is extra power consumption due to BLE power. Additionally, computational power levy penalties on battery power source by Hierarchical Cluster that have higher Nodes (*N*) or number of Clusters (*K*) defined.

Connectivity Bluetooth Low Energy (BLE)	Hardware 🔒 Constraint ⬤ No Constraint			
	Memory	Power Source	Storage	Computational Power
Self-Contained & Updated only on Hardware Refresh	🔒 The Time Complexity – $O\,(KN^2)$, where K is number of Clusters and N is number of Nodes.	🔒 Continuous Bluetooth advertisements and time complexity could increase power needs.	🔓 Data could be offloaded to external devices.	🔒 Higher computational power for Hierarchical Clustering – depending upon Number of Clusters K and Number of Cluster Nodes N - $O\,(KN^2)$
OTA Firmware	N/A	N/A	N/A	N/A
Model Type: Eager Learner	N/A	N/A	N/A	N/A
Lazy Learner	🔒 The Time Complexity – $\boldsymbol{O\,(KN^2)}$, where K is number of Clusters and N is number of Nodes.	🔒 Continuous Bluetooth advertisements and time complexity could increase power needs.	🔓 Data could be offloaded to external devices.	🔒 Higher computational power for Hierarchical Clustering – depending upon Number of Clusters K and Number of Cluster Nodes N - $\boldsymbol{O\,(KN^2)}$
Model Invocation	🔒 The Time Complexity – $O\,(KN^2)$,where K is number of Clusters and N is number of Nodes.	🔒 Continuous Bluetooth advertisements and time complexity could increase power needs.	🔓 Data could be offloaded to external devices.	🔒 Higher computational power for Hierarchical Clustering - depending upon Number of Clusters K and Number of Cluster Nodes N - $\boldsymbol{O\,(KN^2)}$

(Row group label, left vertical: ML Model Accuracy)

Overall design consideration: Operating Hierarchical Clustering in an extremely constrained device (xCD) or Tiny device with "Bluetooth Low Energy (BLE) Connectivity" increases battery power consumption. ☛ Model performance and invocation are dependent upon number of Clusters (K) and number of Nodes (z^2).

 For Hierarchical Cluster deployments in extremely constrained devices (xCD), the impact of BLE connectivity and model invocation, i.e., the number of nodes and cluster size will have considerable battery power implications. Higher the number of nodes the greater battery power drain.

Modeling Wi-Fi Connectivity

Wi Fi connectivity enables to offload data storage to the Edge devices or call to the Central Cloud server. Additionally, it enables to download Firmware via OTA. The major limitation is many of the constrained devices are not suited for Wi Fi as it drains huge power from the battery.

Connectivity Wi Fi		Hardware Constraint / No Constraint			
		Memory	Power Source	Storage	Computational Power
ML Model Accuracy	Self-Contained & Updated only on Hardware Refresh	The Time Complexity – $O\ (KN^2)$ where N is number of Nodes and K is the number of Clusters.	On board Wi Fi demands more power. For battery power devices, Wi Fi option is less favorable.	Data could be offloaded to external devices or Central Cloud.	The Time Complexity – $O\ (KN^2)$ where N is number of Nodes and K is the number of Clusters.
	OTA Firmware	The Time Complexity – $O\ (KN^2)$ where N is number of Nodes and K is the number of Clusters.	On board Wi Fi demands more power. For battery power devices, Wi-Fi option is less favorable.	Data could be offloaded to external devices or central Cloud.	The Time Complexity – $O\ (KN^2)$ where N is number of Nodes and K is the number of Clusters.
	Model Type: Eager Learner	N/A	N/A	N/A	N/A
	Lazy Learner	The Time Complexity – $O\ (KN^2)$ where N is number of Nodes and K is the number of Clusters.	On board Wi Fi demands more power. For battery power devices, WiFi option is less favorable.	Data could be offloaded to external devices or Central Cloud.	The Time Complexity – $O\ (KN^2)$ where N is number of Nodes and K is the number of Clusters.
	Model Invocation	Given higher time complexity, the number of stack calls are higher in Hierarchical Cluster based ML Models.	Given higher time complexity, the number of stack calls are higher in Hierarchical Cluster based ML Models.	Periodic offloading of Sensor data to Edge devices lowers the storage needs.	Given higher time complexity, the number of stack calls are higher in Hierarchical Cluster based ML Models.

☞ **Overall design consideration:** Operating Hierarchical Cluster algorithm in an extremely constrained devices (xCD) or Tiny device with "Wi Fi Connectivity" increases battery power consumption. Model performance and invocation is $O(KN^2)$ with complexity increase based on N is number of Nodes and Cluster size (K).

Connectivity—Model Accuracy Tradeoff

Modeling Memory

Since the model is a lazy learner and does require memory resident[298] or heap[299] computation, the hardware memory is a constraint for the model accuracy and performance. That is, as the number of Nodes (N) and Cluster (K) increases, the memory requirements go higher—thus, influence power or computational hardware economy.

---

[298] Memory resident - https://www.trendmicro.com/vinfo/us/security/definition/memory-resident
[299] Stack vs. Heap allocation - https://www.geeksforgeeks.org/stack-vs-heap-memory-allocation/

Hardware Memory	Connectivity 🔒 Constraint No Constraint		
	No Connectivity	Bluetooth Low Energy (BLE)/LoRa	Wi Fi
Self-Contained & Updated only on Hardware Refresh	🔒 The model requires higher memory – $O(KN^2)$.	🔒 The model requires higher memory – $O(KN^2)$.	🔒 The model requires higher memory – $O(KN^2)$.
OTA Firmware	🔒 The model requires higher memory – $O(KN^2)$.	🔒 The model requires higher memory – $O(KN^2)$.	🔒 The model requires higher memory – $O(KN^2)$.
Model Type: Eager Learner	N/A	N/As	N/A
Lazy Learner	🔒 The model requires higher memory – $O(KN^2)$.	🔒 The model requires higher memory – $O(KN^2)$.	🔒 The model requires higher memory – $O(KN^2)$.
Model Invocation	🔒 Higher computational power for Hierarchical Cluster - depending upon Number of Nodes and Clusters - **O (K N²)**	🔒 Higher computational power for Hierarchical Cluster - depending upon Number of Nodes and Clusters - **O (K N²)**	🔒 Higher computational power for Hierarchical Cluster - depending upon Number of Nodes and Clusters - **O (K N²)**

(left axis label: Model Accuracy)

Design Considerations: Since model update is cycled on Hardware refresh (SLA half-yearly or yearly), there would be staleness of the model. And memory constraint has influence on the Accuracy. Since the model is a lazy learner and will require data movement between processor & memory space for higher iterative complex state models.

 The new data points added as part of Hierarchical Cluster are stored as part of memory (in local or global variables). In general, Use of Local variables over global variables where possible preferred. Local variables are contained in the CPU while global variables are stored in the RAM, the CPU accesses local variables faster.[300]

Modeling Processing Power

The Kalman filter is a Lazy Learner model and invoked as an in-line function and does require extra computation power.

Hardware Processing Power	Connectivity 🔒 Constraint No Constraint			
	No Connectivity	BLE	Wi-Fi	Design Notes
Hardware Refresh	🔒	🔒	🔒	Since model update is cycled on Hardware refresh (SLA half-yearly or yearly), there could be staleness in Model accuracy. Additionally, due to $O(KN^2)$ time complexity, the power consumption is higher.
OTA Firmware Update	🔒	🔒	🔒	$O(KN^2)$ time complexity – dependent upon number Nodes & Cluster size K – increases processing power needs.
Model Type: Eager Learner	N/A	N/A	N/A	
Lazy Learner	🔒	🔒	🔒	
Model Invocation	🔒	🔒	🔒	$O(KN^2)$ time complexity – dependent upon number Nodes & Cluster size K – increases processing power needs.

(left axis label: Model Accuracy)

---

[300] Minimizing Power Consumption in Microcontrollers - https://medium.com/@emmaodunlade/minimizing-power-consumption-in-microcontrollers-b3eb9593c893

Design Considerations: Processing power, given $O(KN^2)$ time complexity, is higher and the model complexity increases $O(KN^2)$ with number of nodes (N) and Cluster Size (K).

Modeling Storage

The model deployed as a lazy learner with $O(KN^2)$ time complexity and does require considerable storage for no connectivity mode. Nonetheless, as hardware device collects sensor data, the data needs to be stored on device local storage (offload to Edge Gateway or Mobile device) before uploads to Central Storage or Corporate Cloud storage.

Hardware Storage		Connectivity 🔒 Constraint ⊘ No Constraint			
		No Connectivity	BLE	Wi Fi	Design Notes
Model Accuracy	Hardware Refresh	🔒	🔒	🔒	Since model update is cycled on Hardware refresh (SLA half-yearly or yearly), the data storage needs to be large to store on the device. For BLE and Wi Fi connectivity modes, the storage could be offloaded to gateway or mobile device and could help to concise storage.
	OTA Firmware Update	🔒	🔒	🔒	Since the model gets updated OTA as part of Firmware update, no staleness of model. Storage could be constraint under "No Connectivity" but not a constraint for BLE and Wi Fi.
	Model Type: Eager Learner	N/A	N/A	N/A	
	Lazy Learner	🔒	🔒	🔒	
	Model Invocation	🔒	🔒	🔒	Higher computational power for Hierarchical Cluster - $O(KN^2)$ time complexity - dependent upon number Nodes & Cluster size K - increases processing power needs.

Design Considerations: Since model update is cycled on Hardware refresh (SLA half-yearly or yearly), there could be staleness in model accuracy.

Modeling Environmental Perturbations

Model deployed under geographical conditions with huge temperature or humidity variations, the influence of temperature on the battery could lead to accuracy consideration. Humidity accumulation is also a factor to be considered in the model calculation

Hardware Perturbations	Connectivity 🔒 Constraint ⬤ No Constraint			
	No Connectivity	BLE	Wi-Fi	Design Notes
Hardware Refresh	🔒 Staleness of the model is a constraint in taming (adaptive) environmental perturbations.	🔒	🔒	Model should factor in geographical sensitivity as part of the calculations. The model coefficients need to tame environmental perturbations. Given Model complexity is $O(N^3)$, the power consumption is higher and computational power requirement are higher.
OTA Firmware Update	🔒	🔒	🔒	The OTA could fin-tune sensitive model coefficients to overcome environmental perturbations.
Model Type: Eager Learner	N/A	N/A	N/A	No influence of environmental perturbations on data movement.
Lazy Learner	🔒	🔒	🔒	No influence of environmental perturbations on data movement.
Model Invocation	🔒	🔒	🔒	Model exhibits $O(N^3)$ Time complexity- the model time complexity depends upon state model & number of iterations to reduce predicted to actual values.

(Left side vertical label: Model Accuracy)

Design Considerations: Environmental perturbations increase the complexity.

 Hierarchical Clustering exhibits higher computational and power processing under environmental perturbations.

Connectivity to Hardware Trade-off

Modeling Hardware Refresh

In this mode, the model is only getting refreshed whenever hardware is refreshed. Generally, this could be half-yearly or yearly, depending on the SLA signed with customer. With this constraint, there is staleness on the model accuracy. That is, adaptive learning of Hierarchical Cluster is only installed when a new model gets deployed. The Hierarchical Cluster model with $O(KN^2)$ time complexity with in-line code invocation requires higher memory, power, and computational processing.

Hardware Refresh	Hardware 🔒 Constraint ⬤ No Constraint					
	Power	Memory	Storage	Processing	Environmental	Design Notes
No Connectivity	🔒	🔒	🔒	🔒	🔒	Model is a lazy learner-correction function with $O(KN^2)$ time complex.
BLE	🔒	🔒	🔒	🔒	🔒	Model is a lazy learner-correction function with $O(KN^2)$ time complex.
Wi-Fi	🔒	🔒	🔒	🔒	🔒	

(Left side vertical label: Connectivity)

Design Considerations: Since model update is cycled on hardware refresh (SLA half-yearly or yearly), there could be staleness in model accuracy. This has influence on Power, Processing Power and environmental perturbations.

Modeling Over-The-Air (OTA) Firmware Update

In this mode, the model only gets refreshed OTA as part of firmware update. Except for no connectivity mode, the model is the latest and learned with the most recent collected data. That is, adaptive learning is current and gets deployed as part of OTA. The model is an $O(KN^2)$ function that is invoked as part of in-line code. So, extra care is required to handle power and computational processing.

Hardware OTA		Hardware 🔒 Constraint ⭘ No Constraint					
		Power	Memory	Storage	Processing	Environmental	Design Notes
Connectivity	No Connectivity	🔒	🔒	🔒	🔒	🔒	Storage needs are higher & inaccuracies due to environmental perturbations.
	BLE	🔒	🔒	🔓	🔒	🔒	With the model is updated vis OTA, the latest learnings will have higher accuracies.
	Wi-Fi	🔒	🔒	🔓	🔒	🔒	

Design Considerations: Firmware updates over the air enables deployment of the latest and greatest model— Hierarchical Cluster. This would reduce the need to have higher hardware capacities.

	Over the air Firmware updates enable the model to be equipped with the most recent learnings.

Modeling Active Learner vs. Lazy Learner

The model is a Lazy learner with higher requirements of memory, compute and power.

ML Active vs. Lazy		Hardware 🔒 Constraint ⭘ No Constraint					
		Power	Memory	Storage	Processing	Environmental	Design Notes
Connectivity	No Connectivity	🔒	🔒	🔒	🔒	🔒	Storage needs are higher for no connectivity mode & power requirements are higher due to environmental perturbations.
	BLE	🔒	🔒	🔓	🔒	🔒	When the model is updated via OTA, the latest learnings will have higher accuracies.
	Wi Fi	🔒	🔒	🔓	🔒	🔒	

Modeling Model Invocation

The model is invoked as an in-line code with Lazy learner. The number of call stack and memory allocations/ free are higher. The environment with connectivity would provide much favorable runtime execution over no connectivity.

ML Model invocation		Hardware 🔒 Constraint 🔓 No Constraint					
		Power	Memory	Storage	Processing	Environmental	Design Notes
Connectivity	No Connectivity	🔒	🔒	🔒	🔒	🔒	Model invocation is Lazy Learner.
	BLE	🔒	🔒	🔓	🔒	🔒	
	Wi-Fi	🔒	🔒	🔓	🔒	🔒	

Deployment of Climate Models in Extremely Constrained Devices (xCDs)

The Dataset, the Earth Surface Temperature Data, is from the Berkeley Earth, which is affiliated with Lawrence Berkeley National Laboratory. The dataset is downloaded from Kaggle Website.[301]

Annual Average Global Temperatures,[302] as shown in the Figure 12, both panels are calculated using a 1951–80 period and represent the average Global Temperature index in degrees C. At the poles where ice cover varies over the record, we provide two cases. For the baseline case the air temperature over ice is used for the average and in the alternative case in the bottom panel Sea Temperature under the ice cover is used. –1.8 C is used for this value in all cases.

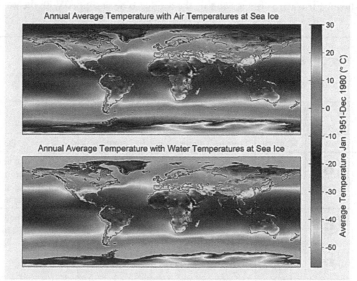

Figure 12: Annual Average Global Temperatures

Data Attributes

In this dataset, "GlobalLandTemperaturesByCity.csv", the following fields (see Figure 13):

- Date: starts in 1750 for average land temperature and 1850 for max and min land temperatures and global ocean and land temperatures
- AverageTemperature: global average land temperature in Celsius

---

[301] Kaggle Climate Change: Earth Surface Data - https://www.kaggle.com/berkeleyearth/climate-change-earth-surface-temperature-data
[302] Annual Average Global Temperatures - http://berkeleyearth.org/wp-content/uploads/2015/03/land-and-ocean-baseline-comparison-map-large.png

dt	AverageTemperature	AverageTemperatureUncertainty	City	Country	Latitude	Longitude
1796-01-01	20.412	2.280	Visakhapatnam	India	18.48N	82.82E
1796-02-01	22.944	1.474	Visakhapatnam	India	18.48N	82.82E
1796-03-01	25.528	2.371	Visakhapatnam	India	18.48N	82.82E
1796-04-01	28.911	1.754	Visakhapatnam	India	18.48N	82.82E
1796-05-01	30.428	1.431	Visakhapatnam	India	18.48N	82.82E
...	...	...	...	...	...	...
2013-05-01	31.887	0.590	Visakhapatnam	India	18.48N	82.82E
2013-06-01	28.059	0.951	Visakhapatnam	India	18.48N	82.82E
2013-07-01	26.139	0.482	Visakhapatnam	India	18.48N	82.82E
2013-08-01	26.605	0.602	Visakhapatnam	India	18.48N	82.82E
2013-09-01	NaN	NaN	Visakhapatnam	India	18.48N	82.82E

2613 rows × 6 columns

Figure 13: Global City Temperatures

- AverageTemperatureUncertainty: the 95% confidence interval around the global average land and ocean temperature
- City: Major City
- Country: Country name
- Latitude: Location Latitude
- Longitude: Location Longitude.

Local Temperatures

Using the above dataset, we have extracted local temperatures to model the Hierarchical Cluster to store as part of the device. The extremely constrained devices (xCDs) can calibrate based on the local temperature using the dataset we have exported to the device.

```
spring_temperatures_India_Hyderabad = []
summer_temperatures_India_Hyderabad = []
autumn_temperatures_India_Hyderabad = []
winter_temperatures_India_Hyderabad = []
rainy_temperatures_India_Hyderabad = []
pre_winter_temperatures_India_Hyderabad = []

min_year = df_India['year'].min()
max_year = df_India['year'].max()
years_India = range(min_year, max_year + 1)
```

The above code groups local temperature based on season and outputs to CSV file.

Hyderabad India, Climate Data (From 1850 to 2012)

Please find Hyderabad Summer Data from 1850 to 2012 (please see Figure 14):

Hyderabad (India) Winter Data (from 1850–2012)

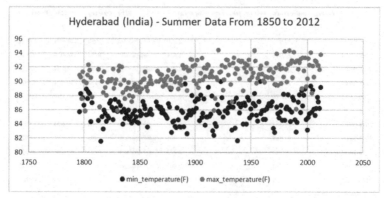

Figure 14: Hyderabad, India - Temperature from 1750 to 2011 (Summer)

Hyderabad, India Winter temperature (see Figure 15)

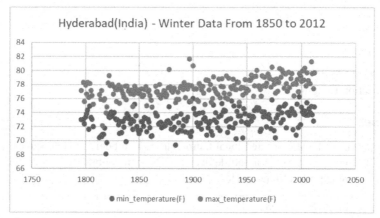

Figure 15: Hyderabad (India) - From 1850 to 2012

San Francisco Climate Data (From 1850 to 2012)

Here is temperature of San Francisco from 1850 to 2011:

Winter Data (1850 to 2012)

Please find San Francisco Winter Data from 1850 to 2012 (see Figure 16):

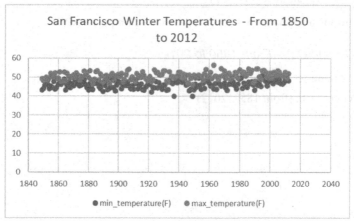

Figure 16: San Francisco Winter Data - 1850 to 2012

Summer Data (1850 to 2012)

Please find San Francisco Summer Data from 1850 to 2012 (see Figure 17):

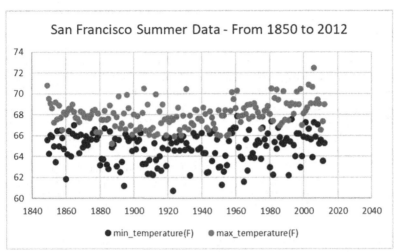

Figure 17: San Francisco Summer Data

Climate Change—Hierarchical Cluster

The next step is to create Hierarchal Cluster with local temperature dataset (see Figure 18):

```
data = dfIndia_final_spring_Hyderabad.values
from sklearn.cluster import AgglomerativeClustering
import scipy.cluster.hierarchy as shc
plt.figure(figsize=(10, 7))
plt.title("Customer Dendograms")
dend = shc.dendrogram(shc.linkage(data, method='single'))
```

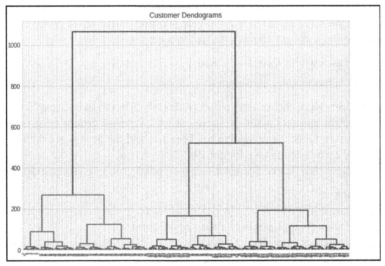

Figure 18: Global City Temperature Clusters

```
from sklearn.cluster import AgglomerativeClustering
cluster = AgglomerativeClustering(n_clusters=4, affinity='euclidean', linkage='ward')
cluster.fit_predict(data)
```

Clusters to C Array

Convert Hierarchical cluster data into location index with minimum and maximum temperatures. This way, when a new temperature is read by the sensor, the real-time embedded device could locate temperature coefficient by identifying the cluster in which the new readings fall:

```
array([3, 3, 3, 3, 3, 3, 3, 3, 3, 3, 3, 3, 3, 3, 3, 3, 3, 3, 3, 3, 3, 3, 3,3, 3, 3, 3, 3, 3, 3, 3, 3, 3, 3, 3, 1, 1, 1, 1, 1, 1, 1,
1, 1, 1, 1,1, 1, 1, 1, 1, 1, 1, 1, 1, 1, 1, 1, 1, 1, 1, 1, 1, 1, 1, 1, 1, 1, 1,1, 1, 1, 1, 1, 1, 1, 1, 1, 1, 1, 1, 2, 2, 2,
2, 2, 2, 2, 2,2, 2, 2, 2, 2, 2, 2, 2, 2, 2, 2, 2, 2, 2, 2, 2, 2, 2, 2, 2, 2, 2, 2, 2, 2, 2, 2, 2, 2, 2, 2, 2, 2, 2, 2, 2,
2, 2, 2, 2, 2, 2, 2, 2, 2, 2, 2, 2, 2, 0, 0, 0, 0, 0, 0, 0, 0, 0, 0, 0, 0, 0, 0, 0, 0, 0, 0, 0, 0, 0, 0, 0, 0, 0, 0, 0,
0, 0, 0, 0, 0, 0, 0, 0, 0, 0, 0, 0, 0, 0, 0, 0, 0, 0, 0, 0, 0, 0, 0, 0, 0, 0, 0, 0, 0, 0, 0, 0, 0, 0, 0, 0, 0, 0, 0, 0])
```

Location array is de-normalized form of Clusters mined. The values 0, 1 and 2 represent three clusters.

K-Means Clustering

K-means clustering[303] is a type of unsupervised learning which is used when you have unlabeled data (i.e., data without defined categories or groups). The goal of this algorithm is to find groups in the data, with the number of groups represented by the variable *K*. The algorithm works iteratively to assign each data point to one of K groups based on the features that are provided. Data points are clustered based on feature similarity. The results of the *K*-means clustering algorithm are [7]:

- The centroids of the *K* clusters, which can be used to label new data
- Labels for the training data (each data point is assigned to a single cluster).

Rather than defining groups before looking at the data, clustering allows you to find and analyze the groups that have formed organically. The "Choosing *K*" section below describes how the number of groups can be determined.

Each centroid of a cluster is a collection of feature values which define the resulting groups. Examining the centroid feature weights can be used to qualitatively interpret what kind of group each cluster represents.

This introduction to the *K*-means clustering algorithm covers:

- Common business cases where *K*-means is used
- The steps involved in running the algorithm.

The *K*-means clustering algorithm is used to find groups which have not been explicitly labeled in the data. This can be used to confirm business assumptions about what types of groups exist or to identify unknown groups in complex data sets. Once the algorithm has been run and the groups are defined, any new data can be easily assigned to the correct group (see Figure 19).

This is a versatile algorithm that can be used for any type of grouping. Some examples of use cases are:

- Behavioral segmentation:
  - o Segment by purchase history
  - o Segment by activities on application, website, or platform
  - o Define personas based on interests
  - o Create profiles based on activity monitoring

---

[303] K Means Clustering - https://www.datascience.com/blog/k-means-clustering

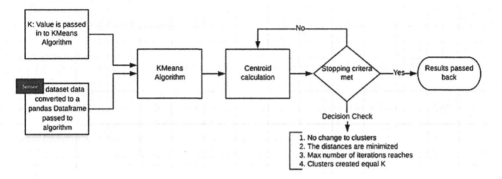

Figure 19: *K*-Means Sequence Flow

- Sorting sensor measurements:
  - o Detect activity types in motion sensors
  - o Group images
  - o Separate audio
  - o Identify groups in health monitoring

The algorithm works with the following steps:

1. Data assignment step
2. Centroid calculation step.

Time Complexity of K-Means

Analysis of time complexity[304] is very important as it provides a measure algorithm computation and processing power. Nearest Neighbor is a learning-by-memorization type of rule.[305] It requires the entire training data set to be stored, and at test time, we need to scan the entire data set in order to find the neighbors. The time of applying the NN rule is therefore Θ $(d\ m)$ [7][8]. This leads to expensive computation at test time. When d is small, several results from the field of computational geometry have proposed data structures that enable to apply the NN rule in time $O(d\,O(1)\,\log(m))$. However, the space required by these data structures is roughly $m\ O(d)$, which makes these methods impractical for larger values of d. To overcome this problem, it was suggested to improve the search method by allowing an approximate search. Formally, an r-approximate search procedure is guaranteed to retrieve a point within distance of at most r times the distance to the nearest neighbor. Three popular approximate algorithms for NN are the kd-tree, balltrees, and locality-sensitive hashing (LSH).

The k-means algorithm is known to have a time complexity of $O(n^2)$, where n is the input data size. This quadratic complexity debars the algorithm from being effectively used in large applications [9] and extremely constrained environments (xCDs). Looking at these notes time complexity of Lloyds algorithm for k-means clustering is given as: $O(n * K * I * d)$ where

n : number of points

K : number of clusters

I : number of iterations

d : number of attributes

[304] Clustering Algorithms: K Means - https://www.cs.princeton.edu/courses/archive/spring08/cos435/Class_notes/clustering2_toPost.pdf

[305] Understanding ML Theory and Algorithm - https://www.cs.huji.ac.il/~shais/UnderstandingMachineLearning/understanding-machine-learning-theory-algorithms.pdf

Generally, number of attributes (in the case of points, x & y) is less than number of Nodes. Same goes true for number of Clusters. This makes number of points and number of iterations as the driver of the k-means algorithm. If the number of Clusters are high, then K-means algorithm could be $O(n^3)$.

Use Case: Sensor Signal and Data Interference & Machine Learning

In Cluster Sensor Deployments, for instance, the distance between Sensors and Signal Strengths, the edge device has to make the decision by taking the most appropriate cluster of sensors with high accuracy (see Figure 20). For example, in the figure below , Dairy Cow Necklaces are tied and the edge device, mobile, could get the data from all the sensors. To connect to appropriate Cow Necklace, the K-means clustering is used:

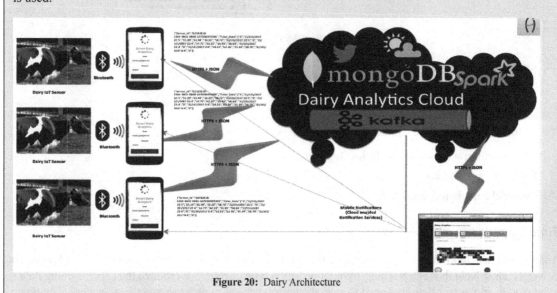

Figure 20: Dairy Architecture

K-Means Example

As a simple illustration of a k-means algorithm, consider the following data set consisting of the signal strength (Received Signal Strength Indicator) and distance from seven sensors:

Sensors	RSSI	Distance
1	1.0	1.0
2	1.5	2.0
3	3.0	4.0
4	5.0	7.0
5	3.5	5.0
6	4.5	5.0
7	3.5	4.5

The goal at the Edge level is to pick the sensor value that is more accurate and weed out any signal or sensory interferes. Here, we can apply K means to separate the sensor-based on RSSI and distance parameters: Let the RSSI and distance parameters values of the two sensors furthest apart (using the Euclidean distance measure), define the initial cluster means, giving:

	Sensors	Mean Vector (Centroid)
Group 1	1	(1.0,1.0)
Group 2	4	(5.0, 7.0)

The remaining sensors are now examined in sequence and allocated to the cluster to which they are closest, in terms of Euclidean distance to the cluster mean. The mean vector is recalculated each time a new member is added. This leads to the following series of steps:

Step	Cluster 1 (Closer Sensors)		Cluster 2 (Further Sensors)	
	Sensors	Mean Vector (Centroid)	Sensors	Mean Vector (Centroid)
1	1	(1.0, 1.0)	4	(5.0, 7.0)
2	1,2	(1.2, 1.5)	4	(5.0, 7.0)
3	1,2,3	(1.8,2.3)	4	(5.0, 7.0)
4	1,2,3	(1.8,2.3)	4,5	(4.2,6.0)
5	1,2,3	(1.8,2.3)	4,5,6	(4.3,5.7)
6	1,2,3	(1.8,2.3)	4,5,6,7	(4.1,5.4)

Now the initial partition has changed, and the two clusters at this stage have the following characteristics:

	Sensors	Mean Vector (Centroid)
Cluster 1	1,2,3	(1.8, 2.3)
Cluster 2	4,5,6,7	(4.1, 5.4)

But we cannot yet be sure that each sensor has been assigned to the right cluster. So we compare each sensor distance to its own cluster mean and to that of the opposite cluster. And we find:

Distance between Sensor 1 & Mean Vector 1
$= SQRT ((1.8 - 1.0)^2 + (2.3 - 1)^2)$
Distance between Sensor 1 & Mean Vector 2
$= SQRT ((4.1 - 1.0)^2 + (5.4 - 1.0)^2)$

Sensor	Distance mean (Centroid) of Cluster 1	Distance mean (Centroid) of Cluster 2
1	1.5	5.4
2	0.4	4.3
3	2.1	1.8
4	5.7	1.8
5	3.2	0.7
6	3.8	0.6
7	2.8	1.1

Only Sensor 3 is nearer to the mean of the opposite cluster (Cluster 2) than its own (Cluster 1). In other words, each sensor's distance to its own cluster mean should be smaller than the distance to the other cluster's mean (which is not the case with Sensor 3). Thus, Sensor 3 is relocated to Cluster 2, resulting in the new partition:

	Sensor	Mean Vector (Centroid)
Cluster 1	1,2	(1.3, 1.5)
Cluster 2	3,4,5,6,7	(3.9,5.1)

Through *K*-means, now, we can separate sensors (RSSI and Distance) in two clusters and take appropriate actions.

K-Means Clustering—Python Code

The following code performs the *K*-means clustering that we have performed in the section above:

Step 1: Plot existing Sensor Deployment (see Figure 21)

```
x1 = np.array([1.0, 1.5, 3.0, 5.0, 3.5, 4.5, 3.5])
x2 = np.array([1.0, 2.0, 4.0, 7.0, 5.0, 5.0, 4.5])
plt.plot()
plt.xlim([0, 10])
plt.ylim([0, 10])
plt.title('Dataset')
plt.scatter(x1, x2)
plt.show()
# create new plot and data
plt.plot()
X = np.array(list(zip(x1, x2))).reshape(len(x1), 2)
colors = ['b', 'g', 'r','y']
markers = ['o', 'v', 's','+']
```

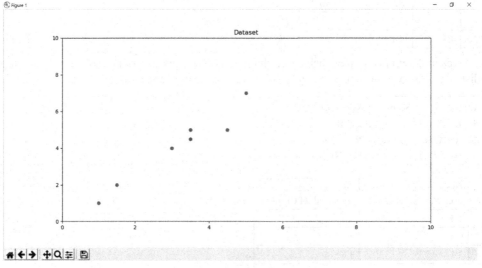

Figure 21: *K*-means

Step 2: perform *K*-means clustering

```
# KMeans algorithm
K = 4
kmeans_model = KMeans(n_clusters=K).fit(X)
plt.plot()
for i, l in enumerate(kmeans_model.labels_):
 plt.plot(x1[i], x2[i], color=colors[l], marker=markers[l],ls='None')
 plt.xlim([0, 10])
 plt.ylim([0, 10])
plt.show()
```

You can perform *K* with different values ($k = 2,3,$ or 4).

Choose appropriate cluster:

$K = 2$

Cluster 1 (Triangles) and Cluster 2 (Circles) (see figure K = 2 Cluster)

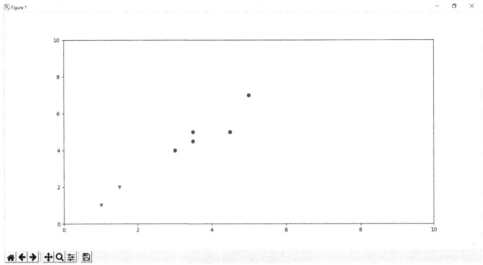

Figure 22: K = 2 Clustering

$K = 3$

Cluster 1 (Triangles), Cluster 2 (Circles), Cluster 3 (Squares) (see figure K = 3 cluster)

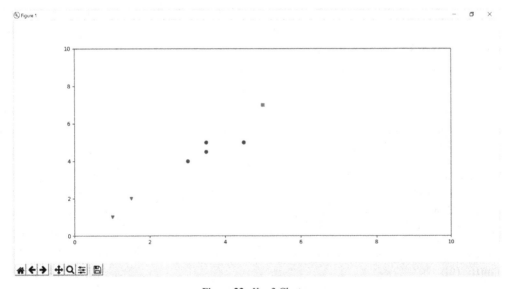

Figure 23: K = 3 Cluster

K = 4

Cluster 1 (Triangles), Cluster 2 (Circles), Cluster 3 (Squares) and Cluster 4 (see figure K = 4 cluster) (Plus Sign)

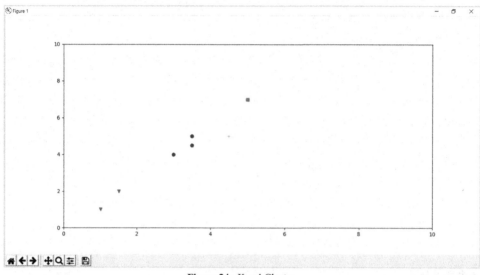

Figure 24: K = 4 Cluster

The complete source code:

```
# -*- coding: utf-8 -*-
"""
Created on Wed Oct 10 17:33:07 2018
@author: cvuppalapati
"""
# -*- coding: utf-8 -*-
"""
Created on Wed Oct 10 17:31:34 2018
@author: cvuppalapati
"""
# clustering dataset
from sklearn.cluster import KMeans
from sklearn import metrics
import numpy as np
import matplotlib.pyplot as plt
x1 = np.array([1.0, 1.5, 3.0, 5.0, 3.5, 4.5, 3.5])
x2 = np.array([1.0, 2.0, 4.0, 7.0, 5.0, 5.0, 4.5])
plt.plot()
plt.xlim([0, 10])
plt.ylim([0, 10])
plt.title('Dataset')
plt.scatter(x1, x2)
plt.show()
# create new plot and data
```

```
plt.plot()
X = np.array(list(zip(x1, x2))).reshape(len(x1), 2)
colors = ['b', 'g', 'r','y']
markers = ['o', 'v', 's','+']
# KMeans algorithm
K = 4
kmeans_model = KMeans(n_clusters=K).fit(X)
plt.plot()
for i, l in enumerate(kmeans_model.labels_):
 plt.plot(x1[i], x2[i], color=colors[l], marker=markers[l],ls='None')
 plt.xlim([0, 10])
 plt.ylim([0, 10])
 plt.show()
```

In the next section, we will evaluate the modeling of *K*-means Clustering in Constrained environment.

 K-means clusters fall into Lazy learners,[306] in which the learner waits until the last minute before doing any model construction to classify a given test tuple. That is, when given a training tuple, RSSI-Sensor, a Lazy learner simply stores it (or does only a little minor processing) and waits until it is given a test tuple. Only when it sees the test tuple does it perform generalization to classify the tuple based on its similarity to the stored training tuples. When making a classification or prediction, lazy learners can be computationally expensive.[307] They require efficient storage techniques and are well-suited to implementation on parallel hardware.

Computational Complexity

Looking at these notes time complexity of Lloyds algorithm for *k*-means clustering is given as: $O(n * K * I * d)$ where

n = number of points

K = number of clusters

I = number of iterations

d = number of attributes

Generally, number of attributes (in the case of points, x & y) is less than number of nodes. Same goes true for number of clusters. This makes number of points and number of iterations are the driver of the k-means algorithm. If the number of clusters are high, then K means algorithm could be $O(n^3)$.

Output

We ran *K*-means on embedded device with the following technical specifications:

Board Specification	SJ One Board - LPC1758 ARM Cortex M3, 512K ROM, 64K RAM.
Operating System	Real-time operating system - FreeRTOS. The RTOS kernel requires 5 to 10K ROM. Thescheduler consumes 236 bytes and the task/thread consumes 64 bytes.
Bluetooth BLE module	Adafruit Bluefruit LE. Connected to SJOne board via UART at baud rate of 9600 bps.Has Raytac's MDBT40 which is BT 4.0 & BT 4.1 stack BLE module. Requires power supply of 3.3V.
Power	Supply of 5V.

[306] Eager   vs.   Lazy   Learners   -   http://eacharya.inflibnet.ac.in/data-server/eacharya-documents/53e0c6cbe413016f23443704_INFIEP_33/89/LM/33-89-LM-V1-S1__db_knn.pdf

[307] Data Mining: Concepts and Techniques, Third Edition - Morgan Kaufmann; 3 edition (June 15, 2011), ISBN-10: 9780123814791

Hardware Economy—Model Accuracy trade-off

The purpose of this section is to analyze the decision tree deployment in constrained or Tiny compute hardware.

Modeling no Connectivity

Modeling for no connectivity influences model upgradability, storage and power source. Since, no connectivity in environment, the K-means could only update during new device refresh cycle. Additionally, the lack of connectivity inhibits model upgrade via OTA firmware update.

Connectivity No Connectivity		Memory	Power Source	Storage	Computational Power
ML Model Accuracy	Self-Contained & Updated only on Hardware Refresh	The Time Complexity – $O(n * K * I * d)$ where n is number of points, K is number of clusters, I is number of iterations, and d is attribute dimensions. Generally, number of iterations equal to nodes in the cluster.	The Time Complexity – $O(n * K * I * d)$ where n is number of points, K is number of clusters, I is number of iterations, and d is attribute dimensions.	Application data storage could increase due to no connectivity.	The Time Complexity – $O(n * K * I * d)$ where n is number of points, K is number of clusters, I is number of iterations, and d is attribute dimensions.
	OTA Firmware	N/A	N/A	N/A	N/A
	Model Type: Eager Learner	N/A	N/A	N/A	N/A
	Lazy Learner	As the number of nodes in the cluster increase, the performance impact on the hardware become steep. It will have impact on Memory.	As the number of nodes in the cluster increase, the performance impact on the hardware become steep. It will have impact on Memory.		As the number of nodes in the cluster increase, the performance impact on the hardware become steep. It will have impact on Memory.
	Model Invocation	Model invocation is iterative.		Not applicable	

Header: Hardware 🔒 Constraint / 🔓 No Constraint

Overall design consideration: Operating K-means cluster in an extremely constrained device (xCD) or Tiny device is very expensive. The Time Complexity is $O(n * K * I * d)$ where n is number of points, K is number of clusters, I is number of iterations, and d is attribute dimensions. 🔒 As the number of nodes (n) increase or as the number of clusters grow (k), the deployment of K-means cluster is prohibitively expensive!

The K-means algorithm is known to have a time complexity of $O(n^2)$, where n is the input data size. This quadratic complexity debars the algorithm from being effectively used in large applications [9] and extremely constrained environments (xCDs).

Modeling Low Power Bluetooth Connectivity

Bluetooth Low Energy connectivity availability enables to offload data storage to Edge devices. Additionally, computational power levy penalties on battery power source by K-means cluster especially that have higher number of nodes (n) or higher iterations (I) or large number of clusters (K).

Connectivity Bluetooth Low Energy (BLE)	Hardware 🔒 Constraint ⭕ No Constraint			
	Memory	Power Source	Storage	Computational Power
ML Model Accuracy — Self-Contained & Updated only on Hardware Refresh	🔒 The Time Complexity – $O(n * K * I * d)$ where n is number of points, K is number of clusters, I is number of iterations, and d is attribute dimensions. Generally, number of iterations equal to nodes in the cluster.	🔒 Continuous Bluetooth advertisements and time complexity could increase power needs.	⭕ Data could be offloaded to external devices.	🔒 The Time Complexity – $O(n * K * I * d)$ where n is number of points, K is number of clusters, I is number of iterations, and d is attribute dimensions. Generally, number of iterations equal to nodes in the cluster.)
OTA Firmware	N/A	N/A	N/A	N/A
Model Type: Eager Learner	N/A	N/A	N/A	N/A
Lazy Learner	🔒	🔒	⭕	🔒
Model Invocation	🔒 The Time Complexity – $O(n * K * I * d)$ where n is number of points, K is number of clusters, I is number of iterations, and d is attribute dimensions. Generally, number of iterations equal to nodes in the cluster.	🔒 Continuous Bluetooth advertisements and time complexity could increase power needs.	⭕ Data could be offloaded to external devices.	🔒 The Time Complexity – $O(n * K * I * d)$ where n is number of points, K is number of clusters, I is number of iterations, and d is attribute dimensions. Generally, number of iterations equal to nodes in the cluster.)

Overall design consideration: Operating K-means cluster in an extremely constrained device (xCD) or Tiny device with BLE connectivity increases battery power consumption. The model performance & invocation are N cube time complexity with $O(n * K * I * d)$.

Modeling Wi-Fi Connectivity

Wi Fi connectivity enables to offload data storage to the Edge devices or call to the central Cloud server. Additionally, enables to download Firmware via OTA. The major limitation is many of the constrained devices are not suited for Wi Fi as it drains huge power from the battery.

Connectivity Wi-Fi	Hardware 🔒 Constraint ⭕ No Constraint			
	Memory	Power Source	Storage	Computational Power
ML Model Accuracy — Self-Contained & Updated only on Hardware Refresh	🔒 The Time Complexity – $O(n * K * I * d)$ where n is number of points, K is number of clusters, I is number of iterations, and d is attribute dimensions.	🔒 On board Wi-Fi demands more power. For battery power devices, Wi-Fi option is less favorable.	⭕ Data could be offloaded to external devices or central Cloud.	🔒 The computational power is higher due to Time Complexity – $O(n * K * I * d)$.
OTA Firmware	🔒	🔒	⭕	🔒
Model Type: Eager Learner	N/A	N/A	N/A	N/A
Lazy Learner	🔒	🔒	⭕	🔒
Model Invocation	🔒	🔒	⭕	🔒

Overall design consideration: Operating K-means algorithm in an extremely constrained devices (xCD) or Tiny device with "Wi Fi Connectivity" increases battery power consumption. Model time complexity $O(n * K * I * d)$ increases with increased (n) number of points, increased (K) number of clusters, (I) iteration loop and distance matrix.

Connectivity—Model Accuracy tradeoff

Modeling Memory

Since the model is a Lazy learner and does require memory resident[308] or heap[309] computation, the hardware memory is a constraint for the model accuracy and performance. That is, the memory (Data movement or access) does influence power or computational hardware economy.

Hardware Memory	Connectivity 🔒 Constraint ⚪ No Constraint		
	No Connectivity	Bluetooth Low Energy (BLE)/LoRa	Wi-Fi
Self-Contained & Updated only on Hardware Refresh	🔒	🔒 Since model update is cycled on Hardware refresh (SLA half-yearly or yearly), there would be staleness of the model. Memory constraint has influence on the Accuracy.	🔒
OTA Firmware	🔒 Since the model gets updated OTA as part of Firmware update, no staleness of Kalman coefficients. Memory constraint has influence on the model accuracy.	🔒	🔒
Model Type: Eager Learner	N/A	N/A	N/A
Lazy Learner	N/A	N/A	N/A
Model Invocation	🔒 Higher computational power for K means cluster - depending upon $O(n * K * I * d)$ where n is number of points, K is number of clusters, I is number of iterations, and d is the attribute dimensions.	🔒 Higher computational power for K means cluster - depending upon $O(n * K * I * d)$ where n is number of points, K is number of clusters, I is number of iterations, and d is the attribute dimensions.	🔒

Design Considerations: Since model update is cycled on hardware refresh (SLA half-yearly or yearly), there would be staleness of the model. And memory constraint has influence on the Accuracy. Since the model is a Lazy learner type and will require data movement between processor and memory space for higher iterative complex state models.

Modeling Processing Power

The K-means cluster is a Lazy learner type and invoked as an in-line function and does require extra computation power.

---

[308] Memory resident - https://www.trendmicro.com/vinfo/us/security/definition/memory-resident
[309] Stack vs. Heap allocation - https://www.geeksforgeeks.org/stack-vs-heap-memory-allocation/

Hardware Processing Power		Connectivity			
		No Connectivity	BLE	Wi-Fi	Design Notes
Model Accuracy	Hardware Refresh	🔒	🔒	🔒	Since model update is cycled on Hardware refresh (SLA half-yearly or yearly), there could be staleness in Model accuracy. Additionally, due to *O*, $O(n * K * I * d)$ time complexity, the power consumption is higher.
	OTA Firmware Update	🔒	🔒	🔒	$O(n * K * I * d)$ where *n* is number of points, *K* is number of clusters, *I* is number of iterations, and *d* is attribute dimensions - increases processing power needs.
	Model Type: Eager Learner	N/A	N/A	N/A	
	Lazy Learner	🔒	🔒	🔒	
	Model Invocation	🔒	🔒	🔒	$O(n * K * I * d)$ where *n* is number of points, *K* is number of clusters, *I* is number of iterations, and ***d*** is attribute dimensions

Design Considerations: Higher processing power, given $O(n * K * I * d)$ time complexity.

Modeling Storage

The model is deployed as an in-line function and does not require huge storage. Nonetheless, as hardware device collects sensor data, the data needs to be stored on device local storage (offload to Edge Gateway or a mobile device) before uploads to Central Storage or Corporate Cloud storage.

Hardware Storage		Connectivity 🔒 Constraint 🔓 No Constraint			
		No Connectivity	BLE	Wi-Fi	Design Notes
Model Accuracy	Hardware Refresh	🔒	🔓	🔓	Since model update is cycled on Hardware refresh (SLA half-yearly or yearly), the data storage needs to be large to store on the device. For BLE and Wi Fi connectivity modes, the storage could be offloaded to gateway or mobile device and could help to concise storage.
	OTA Firmware Update	🔒	🔓	🔓	Since the model gets updated OTA as part of Firmware update, no staleness of model. Storage could be constraint under "No Connectivity" but not a constraint for BLE and WinFi.
	Model Type: Eager Learner	N/A	N/A	N/A	
	Lazy Learner	N/A	N/A	N/A	
	Model Invocation	🔒	🔒	🔒	Higher computational power for K- means cluster – depending upon Number of iterations and n is number of points, *K* is number of clusters, I is number of iterations, and d is attribute dimensions – $O(n * K * I * d)$

Design Considerations: Since model update is cycled on Hardware refresh (SLA half-yearly or yearly), there could be staleness in Model accuracy.

Modeling Environmental Perturbations

Model deployed under geographical conditions with huge temperature or humidity variations, the influence of temperature on the battery could lead into accuracy consideration. Humidity accumulation is also a factor to be considered in model calculation.

Hardware Environmental Perturbations		Connectivity			
		No Connectivity	BLE	Wi-Fi	Design Notes
Model Accuracy	Hardware Refresh	🔒 Staleness of the model is a constraint in taming (adaptive) environmental perturbations.	🔒	🔒	Model should factor in geographical sensitivity as part of the calculations.
	OTA Firmware Update	🔓	🔓	🔓	The OTA could fin-tune sensitive model coefficients to overcome environmental perturbations.
	Model Type: Eager Learner	N/A	N/A	N/A	No influence of environmental perturbations on data movement.
	Lazy Learner	🔒	🔒	🔒	No influence of environmental perturbations on data movement.
	Model Invocation	🔒	🔒	🔒	Model exhibits $O(n * K * I * d)$ where n is number of points, K is number of clusters, I is number of iterations, and d is attribute dimensions - increases processing power needs.

Design Considerations: Environmental perturbations increase the complexity.

 K-means cluster exhibits higher computational and power processing under higher environmental perturbations.

Connectivity to Hardware trade-off

Modeling Hardware Refresh

In this mode, the model is only getting refreshed whenever hardware refresh takes place. Generally, this could be half-yearly or yearly, depending upon the SLA signed with the customer. With this constraint, there is staleness on the model accuracy or number of points or number of clusters could be different. So, the model requires higher memory, power and computational processing.

Model Accuracy Hardware Refresh		Hardware 🔒 Constraint 🔓 No Constraint					
		Power	Memory	Storage	Processing	Environmental	Design Notes
Connectivity	No Connectivity	🔒	🔒	🔒	🔒	🔒	Model is a Lazy Learner with $O(n * K * I * d)$ time complex.
	BLE	🔒	🔒	🔓	🔒	🔒	Model is a Lazy Learner with $O(n * K * I * d)$ time complex.
	Wi-Fi	🔒	🔒	🔓	🔒	🔒	

Design Considerations: Since model update is cycled on hardware refresh (SLA half-yearly or yearly), there could be staleness in model accuracy. This has influence on Power, Processing Power and environmental perturbations.

Modeling Over-The-Air (OTA) Firmware Update

In this mode, the model is only getting refreshed over the air as part of firmware update. Except for no connectivity mode, the model is the latest and learned with the most recent collected data. That is, adaptive learning is current and gets deployed as part of OTA. The model is an $O(n * K * I * d)$ function that is invoked as part of in-line code. So, extra care required to handle power and computational processing.

Model Accuracy Invocation		Hardware 🔒 Constraint ⭕ No Constraint					
		Power	Memory	Storage	Processing	Environmental	Design Notes
Connectivity	No Connectivity	🔒	🔒	🔒	🔒	🔒	Storage needs are higher & inaccuracies are there due to environmental perturbations.
	BLE	🔒	🔒	⭕	🔒	🔒	With the model updated via OTA, the latest learnings will have higher accuracies.
	Wi-Fi	🔒	🔒	⭕	🔒	🔒	

Design Considerations: Firmware updates over the air (OTA) enables deployment of the latest and greatest model—with optimized n and K. This would reduce the need to have higher hardware capacities.

 Over the air Firmware updates enable the model to be equipped with most recent learnings.

Modeling Active Learner vs. Lazy Learner

Design Considerations: K-means is a Lazy learner cluster algorithm and the model gets computed as part of the inference.

Modeling Model Invocation

The model is a Lazy learner function with $O(n * K * I * d)$ time complex that is invoked as part of in-line code.

Accuracy Invocation		Hardware 🔒 Constraint ⭕ No Constraint					
		Power	Memory	Storage	Processing	Environmental	Design Notes
Connectivity	No Connectivity	🔒	🔒	🔒	🔒	🔒	Storage needs are higher & inaccuracies due to environmental perturbations.
	BLE	🔒	🔒	⭕	🔒	🔒	With the model updated via OTA, the latest learnings will have higher accuracies.
	Wi-Fi	🔒	🔒	⭕	🔒	🔒	

Design Considerations: Model invocation is not subject to change for different connectivity modes and hardware modes.

Here is the summary of the algorithms discussed (please see Table 5)

Table 5: Summary of Unsupervised algorithms

Algorithm	Time Complexity	Constrained Design Considerations		Description
Hierarchical Clustering	$O(KN^2)$	Memory ☒	ML Model Accuracy ☒	The Time Complexity – $O (KN^2)$ where K is number of Clusters and N is number of Nodes.
		Battery Power ☒	Environmental Perturbations ☒	
		Computational Constraint ☒	Connectivity ☒	
		Storage ☒		
K-Means Cluster	$O(n * K * I * d)$ where n is number of points, K is number of clusters, I is number of iterations, and d is attribute dimensions	Memory ☒	ML Model Accuracy ☒	K means cluster algorithm demands more computational and memory capacities - impact the extremely constrained devices (xCDs) performance.
		Battery Power ☒	Environmental Perturbations ☒	
		Computational Constraint ☒	Connectivity ☐	
		Storage ☒		

After reading this chapter, you should be able to design, implement, and deploy the following Machine Learning Algorithms into extremely constrained devices (xCDs) and mobile platforms.

• *K*-means Cluster

• Hierarchical Clusters

References

1.   Min Song (Author, Editor). Yi-Fang Brook Wu (Editor), Handbook of Research on Text and Web Mining Technologies. Information Science Reference (September 30, 2008), ISBN-10: 1599049902.
2.   Jiawei Han, Micheline Kamber and Jian Pei. Data Mining: Concepts and Techniques. Morgan Kaufmann; 3 edn (July 6, 2011), ISBN-10: 1558609016.
3.   Rui Xu and Don Wunsch. Clustering. Wiley-IEEE Press (October 24, 2008), ISBN-10: 0470276800.
4.   L. Minh Dang, Md. Jalil Piran, Dongil Han, Kyungbok Min and Hyeonjoon Moon. A Survey on Internet of Things and Cloud Computing for Healthcare. https://www.mdpi.com/2079-9292/8/7/768/htm, 9 July 2019, Access Date: 01/08/2020.
5.   S.G. Roy, A. Chakrabarti. A novel graph clustering algorithm based on discrete-time quantum random walk. 2017. Quantum Inspired Computational Intelligence, 2017, https://www.sciencedirect.com/science/article/pii/B9780128044094000115.
6.   PeterWittek. Unsupervised Learning. Publisher: Elsevier, Date: 2014, https://www.sciencedirect.com/science/article/pii/B9780128009536000050.
7.   Shai Shalev-Shwartz (Author), Shai Ben-David (Author). Understanding Machine Learning: From Theory to Algorithms, Cambridge University Press; 1st edn (May 19, 2014), ISBN-10: 1107057132.
8.   Sanjeev Arora and Boaz Barak. Computational Complexity: A Modern Approach, Cambridge University Press; 1st edn. (April 20, 2009), ISBN-10: 0521424267.
9.   M. K. Pakhira. A Linear Time-Complexity K-Means Algorithm Using Cluster Shifting. 2014 International Conference on Computational Intelligence and Communication Networks, Bhopal, 2014, pp. 1047–1051.

SECTION-IV

Democratization & Future of AI

CHAPTER 8

National Strategies

"We are what we repeatedly do. Excellence, then, is not an act, but a habit..."

Aristotle

 Artificial Intelligence is here to stay. It has become a strategic asset of many companies. Private and public companies have updated their existing business models, incorporated new blue-ocean models, and developed data monetization services to better serve their customers. The main purpose is to serve the customer and humanity at large.

The Non-Governmental Organizations (NGOs), small and large, have included AI or on the trajectory to include Machine Learning based algorithms as part of their day-to-day business processes to gain operational efficiency, to exclusively serve people, and better forecast the future needs of the people.

Governments across the world have made AI as a strategic digital initiative and have included AI services as part of governmental core-service to better govern people, better utilize scarce resources to fulfill national plans, better provision of health services to meet current and future healthcare services to citizens, prudent allocation of important and scarce resources to serve well with few services, adaptive law and order services to control and mitigate any risks posed to public in large, and better inform citizens on the future of government vision. It is evidently clear that the AI is not only a powerful data-enabled futuristic business engine but also a strategic governing tool that could influence international politics[310] and can make a kingmaker [1].

In this chapter, I would like to introduce AI strategy of countries that are at the forefront of AI in terms of the impact based on their economic footprint, Gross Domestic Product (GDP), and the contributions of those countries' engineers and researchers in terms of knowledge enablement of AI through research. The chapter concludes with AI strategy of the United Nations [2].

> "...the one who becomes the leader in this sphere [AI] will be the ruler of the world."
> Vladimir Putin[311] (The International Forecaster) [2]

---

[310] How will AI change international politics? - https://www.japantimes.co.jp/opinion/2019/01/15/commentary/japan-commentary/will-ai-change-international-politics/#.XeFy27xKg2w

[311] The International Forecaster - https://theinternationalforecaster.com/topic/international_forecaster_weekly/when_robots_rule_the_markets

National Technology Strategies for Serving People

Artificial Intelligence (AI) holds tremendous promise to benefit nearly all aspects of society, including the economy, healthcare, security, law, transportation, even technology itself. Nations across the world have developed AI strategic directives to infuse AI and its design strategies into the DNA of governance to overcome AI challenges in massive scale adoption.

Let us look at the top seven world's biggest (2019 and 2020) economies' AI strategies as they would be pivotal for the future of AI and enablement of AI services for the humanity and future generations[312] (Figure 1) [3].

Another selection measure,[313] in addition to breadth of the economies [4], is citable documents published from these countries in the field of AI from 2010–2016 (source: Scimago Journal and Country Rank (SJR)[314]). There is a correlation between size of economies and research work that these countries has produced (Figure 2).

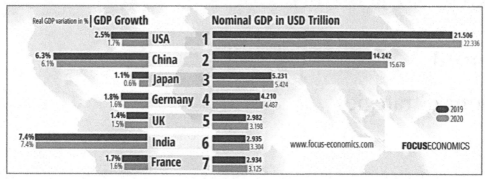

Figure 1: Top 7 World's Biggest Economies of 2019 and 2020

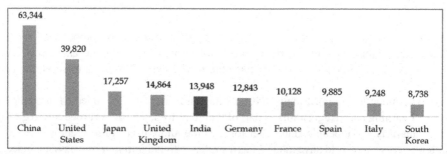

Figure 2: Citable Documents in AI (2010–2016) [4]

Here is the list of countries: [3][4]

- United States of America
- China
- Japan
- Germany
- UK
- India
- France

---

[312] The World's Biggest Economies - https://www.focus-economics.com/blog/the-largest-economies-in-the-world

[313] National Strategy for AI – India – Discussion Paper - https://niti.gov.in/writereaddata/files/document_publication/NationalStrategy-for-AI-Discussion-Paper.pdf

[314] Scimago Journal and Country Rank - https://www.scimagojr.com/

A shortage of engineers trained in artificial intelligence

A McKinsey Global Institute report from 2011 had already warned of a deficit of 190,000 data scientists in 2018, as well as 1.5 million managers and analysts capable, quite simply, of understanding the ins and outs and of making decisions in the AI context.[315]

The study published in early 2017 by Burning Glass Technologies, BHEF and IBM, meanwhile, predicts a 28% rise in the number of Data Scientist and Data Analyst jobs worldwide over the next five years, to a total 2,720,000, and that 39% of these jobs require a Masters or PhD [5].

Lastly, in December 2017, according to a study compiled by Tencent Research Institute, there are just 300,000 "AI researchers and practitioners" worldwide at present—when the market demand is for millions of roles (even if there is not much detail on how such figures were reached). Tencent suggests that the bottleneck in this case is education. Incidentally, the study identifies the US, China, Japan, and the UK as the top contending countries in the AI race, with special mention made of Canada and Israel particularly in terms of education.

Artificial Intelligence of the American People[316]

The age of artificial intelligence (AI) has arrived and is transforming everything from healthcare to transportation to manufacturing.[317]

"Continued American leadership in Artificial Intelligence is of paramount importance to maintaining the economic and national security of the United States."

– President Donald J. Trump

America has long been the global leader in this new era of AI and is poised to maintain this leadership going forward because of our strong innovation ecosystem. Realizing the full potential of AI for the nation requires the combined efforts of industry, academia, and government. The Administration has been active in developing policies and implementing strategies that accelerate AI innovation in the U.S. for the benefit of the American people. These activities align with several areas of emphasis: AI for American Innovation, AI for American Industry, AI for the American Worker, and AI with American Values. This AI.gov website provides a portal for exploring these activities in more depth, and serves as a resource for those who want to learn more about how to take full advantage of the opportunities of AI.

Full report is available at [6]:

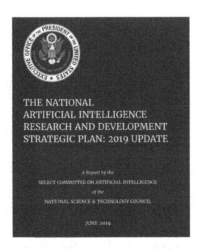

THE NATIONAL
ARTIFICIAL INTELLIGENCE
RESEARCH AND DEVELOPMENT
STRATEGIC PLAN: 2019 UPDATE

A Report by the
SELECT COMMITTEE ON ARTIFICIAL INTELLIGENCE
of the
NATIONAL SCIENCE & TECHNOLOGY COUNCIL

JUNE 2019

[315] For a meaningful Artificial Intelligence - https://www.aiforhumanity.fr/pdfs/MissionVillani_Report_ENG-VF.pdf
[316] AI for American People - https://www.whitehouse.gov/ai/
[317] Source: https://www.whitehouse.gov/ai/, date of access November 27, 2019 7:00 pm Pacific

The National AI R&D Strategic Plan: 2019 Update establishes a set of objectives for Federally funded AI research, identifying the following eight strategic priorities:[318],[319]

Strategy 1: Make long-term investments in AI research [6]. Prioritize investments in the next generation of AI that will drive discovery and insight and enable the United States to remain a world leader in AI.
Strategy 2: Develop effective methods for human-AI collaboration [6]. Increase understanding of how to create AI systems that effectively complement and augment human capabilities
Strategy 3: Understand and address the ethical, legal, and societal implications of AI [6]. Research AI systems that incorporate ethical, legal, and societal concerns through technical mechanisms.
Strategy 4: Ensure the safety and security of AI systems. Advance knowledge of how to design AI systems that are reliable, dependable, safe, and trustworthy.
Strategy 5: Develop shared public datasets and environments for AI training and testing. Develop and enable access to high-quality datasets and environments, as well as to testing and training resources.
Strategy 6: Measure and evaluate AI technologies through standards and benchmarks. Develop a broad spectrum of evaluative techniques for AI, including technical standards and benchmarks.
Strategy 7: Better understand the national AI R&D workforce needs. Improve opportunities for R&D workforce development to strategically foster an AI-ready workforce
Strategy 8: Expand public-private partnerships to accelerate advances in AI. Promote opportunities for sustained investment in AI R&D and for transitioning advances into practical capabilities, in collaboration with academia, industry, international partners, and other non-Federal entities

Artificial intelligence (AI) holds tremendous promise to benefit nearly all aspects of society, including the economy, healthcare, security, the law, transportation, even technology itself.[320]

Finally, please find Word Cloud that was extracted from Executive Summary Section of US National Artificial Intelligence Research and Development Strategic Plan – 2019 Update **[6]**:

As is clear from the Word Cloud (please see Figure 3), AI Executive Summary clearly calls out strategic investment in AI, workforce development, research, and national AI strategies.

Figure 3: The USA AI Strategic Plan - Executive Summary Word Cloud [6]

---

[318] AI for the American People - https://www.whitehouse.gov/ai/ai-american-innovation/

[319] Source:https://www.whitehouse.gov/wp-content/uploads/2019/06/National-AI-Research-and-Development-Strategic-Plan-2019-Update-June-2019.pdf

[320] https://www.whitehouse.gov/wp-content/uploads/2019/06/National-AI-Research-and-Development-Strategic-Plan-2019-Update-June-2019.pdf

China's New Generation of Artificial Intelligence Development Plan[321]

China has been the breeding ground for AI technology and AI businesses. With a digital-savvy population of 1.4 billion, consumers are demanding ever more powerful digital services than before. The local e-commerce market alone is expected to rise to $1.7 trillion by 2020. That is a big opportunity for the new age companies in China taking to AI to generate higher traffic and higher revenues [7]. The rapid development of AI will profoundly change human social life and the world. To seize major strategic opportunities for the development of AI, China wants to be the number one p layer in the AI space.[322]

Chinese government is promoting usage of AI on a larger scale to realize the larger objective of gaining the top spot and becoming a superpower in AI [7]. It has divided the timeline into three segments for achieving certain specific targets. To promote AI businesses, China has some control over its national champions. Alibaba is making a great push in smart cities, while Tencent is focusing on computer vision for medical diagnostics. Baidu is specializing in autonomous driving and iFlyTek in voice intelligence. Chinese government has announced that China's open AI platforms will rely on Alibaba for smart cities, Tencent for AI in healthcare, imaging and diagnostics and Baidu for autonomous vehicles[8].

Strategic Goals

Strategic goals are divided in three steps:

First Step: by 2020
  To keep up with advanced level of AI in the world.

- A new generation of AI theory and technology made important progress in the areas of large-scale analytics, cross-media intelligence, hybrid intelligence and other advanced techniques
- AI industry competitiveness develops into the first international phalanx
- Further optimize AI development environment.

Second Step: by 2025
  To develop AI at world leading level.

- We shall make AI industry into global high-end value chain
- We shall make initial establishment of artificial intelligence laws and regulations, ethical norms and policy systems, and form artificial intelligence safety assessment and control capabilities.

Third Step by: 2030
  We shall make AI theory, technology and application achieve the world's leading level to be the major artificial intelligence innovation center of the world, intelligent economy, intelligent society achieve remarkable results, and lay an important foundation for China's entry into the forefront of the innovative countries and economic powers.[323]

National policy: from IoT to big data to AI Since 2009, China's AI policy has undergone five stages[324] with changing keywords which reflect the different priorities in each stage, with the focus shifting from basic research in such categories as IoT, information security and database in the early period, to big data and infrastructure in the middle period, to AI itself and also intellectual property protection after 2017. Overall, China's AI policy mainly focuses on six categories: "made in China", innovation-driven development, IoT,

---

[321] China's New Generation of AI Development Plan - https://flia.org/notice-state-council-issuing-new-generation-artificial-intelligence-development-plan/

[322] A New Generation of Artificial Intelligence Development Plan. It is translated by Flora Sapio (FLIA Scholar), Weiming Chen (FLIA Research Assistant), and Adrian Lo (FLIA Research Intern) - https://flia.org/wp-content/uploads/2017/07/A-New-Generation-of-Artificial-Intelligence-Development-Plan-1.pdf

[323] https://flia.org/wp-content/uploads/2017/07/A-New-Generation-of-Artificial-Intelligence-Development-Plan-1.pdf

[324] China AI Development Report 2018 - http://www.sppm.tsinghua.edu.cn/eWebEditor/UploadFile/China_AI_development_report_2018.pdf

Internet+, big data, and scientific and technological R&D. Please find executive strategy Word Cloud from China AI Development report, 2018 [9].

As it is clear from strategic directives from China AI development 2018 and Word Cloud (please see Figure 4), China is aiming AI-China, Chinese research, universities policies, World, market, and AI China Development [9].

Figure 4: China Executive Strategy Word Cloud [9]

Japan Strategic Council for AI & Strategy

The AI Technology Strategic Council[325] has formulated an Industrialization Roadmap in March 2017. The council has outlined the following major four areas to infuse AI in business & industry [10]:

1. Present Conditions surrounding Data, AI and Computing
2. Governmental Promotional Structures related to AI
3. Fusion of AI & Industrial Roadmap
4. R&D and Promotion of AI.

The roadmap has three phase implementations of AI across Industrial Productivity, Health, Medical Care & Welfare and Mobility (please see Figure 5):

*   Phase 1: Utilization and application of Data Driven AI developed in various domains
*   Phase 2: Public use of AI and data across various domains
*   Phase 3: Integration of multiple domains.

The goal of the Council is to use AI to a promise of a new industrial revolution—one that marries advanced manufacturing techniques with the Internet of Things to create manufacturing systems that are not only interconnected, but communicate, analyze, and use information to drive further intelligent action back in the physical world.[326] In order to realize the roadmap, the Council has identified the following approaches [11]:

*   Research and Development
*   Fostering of Human Resources

---

[325] AI Strategies and Related activities in Japan - http://events.science-japon.org/dlai17/doc/MIC%20-%20France-Japan%20 Symposium%2020171025.pdf

[326] Industry 4.0 - https://www2.deloitte.com/us/en/insights/focus/industry-4-0.html

- Improvement for Environment for Data and Tools
- Support for Startups.

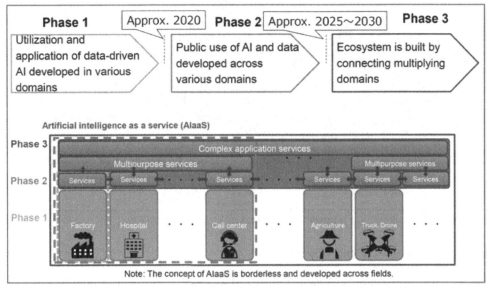

Figure 5: AI Development Phases [10]

The roadmap objectives in the three identified areas include [10][11]:

Productivity	Healthcare, Medical Care, Welfare	Mobility
Enhancement of People's Creativity	High speed telecommunication & Diagnostic equipment	AI based supply and demand matching
Infrastructure for Data Use	Image & Anomaly detection	Real-time collection of location & road conditions
AI Based prediction & Supply Chain demand	Vital Sign Sensor	Infrastructure network
AI, Robots	Data Collection & Data Preparation	Autonomous driving (ground)
Real Time assessment of Operation & Status	Drug Discovery	Transportation Devices
	Regenerative Medicine	Autonomous control (air)
	Image recognition & tactile sensor	Air personal data
	AI, medical elderly care robots	VR, communication environment
	Voice recognition	

Future of Hardware and Communication technologies need to be devised with new compute architectures to improve power consumption and Edge processing of AI.

> "Data is indispensable to use and application of AI technology and data itself may become competitive power."
>
> Japan Strategic Council for AI [10]

The extract from Japan's Artificial Intelligence Technology Strategy Report,[327] conditions surrounding AI, section Word Cloud clearly emphasize the lack of existing human personnel to develop AI. Additionally, the report calls for collaboration of universities and industry, and active contributions to research by publishing high quality papers (please see Figure 6).

---

[327] Japan AI Technology Strategy Report - https://www.nedo.go.jp/content/100865202.pdf

Figure 6: Japan Technology Roadmap – AI [10]

Germany Government—AI Strategy

In November 2018, the German government published its strategy for Artificial Intelligence[328] [12]. The Federal Government's Artificial Intelligence (AI) strategy[329] was jointly developed by the Federal Ministry of Education and Research, the Federal Ministry for Economic Affairs and Energy, and the Federal Ministry of Labour and Social Affairs based on suggestions taken from a nationwide online consultation. The Federal Government AI strategy is based on the "weak" AI [12].

In highly abstract terms, AI researchers can be assigned into two groups: "strong" and "weak" AI. "Strong" AI means that AI systems have the same intellectual capabilities as humans, or even exceed them. "Weak" AI is focused on the solution of specific problems using methods from mathematics and computer science, whereby the systems developed are capable of self-optimization. To this end, aspects of human intelligence are mapped and formally described, and systems are designed to simulate and support human thinking. The Federal Government is oriented its strategy to the use of AI to solve specific problems, i.e., to the "weak" approach [12]:

1. Deduction systems, machine-based proofs

2. Knowledge-based systems

3. Pattern analysis and pattern recognition

4. Robotics

5. Smart multimodal human-machine interaction

The Federal Republic of Germany pursues 14 goals with its national AI strategy: the purpose of the goals is to develop AI for the serve the good of society [12]. In addition,

1. We want to make Germany and Europe a leading center for AI and thus help safeguard Germany's competitiveness in the future.

2. We want a responsible development and use of AI which serves the good of society.

3. We will integrate AI in society in ethical, legal, cultural and institutional terms in the context of a broad societal dialogue and active political measures.

---

[328] AI Strategy - https://www.ki-strategie-deutschland.de/home.html?file=files/downloads/Nationale_KI-Strategie_engl.pdf

[329] Germany AI Strategy - https://ec.europa.eu/knowledge4policy/publication/germany-artificial-intelligence-strategy_en

The fourteen goals include[330] [12]:

1. Artificial Intelligence (AI) made in Germany should become the globally recognized seal of approval.

2. Germany should expand its strong position in Industry 4.0 and become a leader in AI applications in this area. Our strong medium-sized companies should also benefit from AI applications.

3. Germany should become an attractive location for the smartest AI minds in the world.

4. The focus should always be on the benefits of AI for citizens.

5. In Germany, data should only be used for the benefit of society, the environment, the economy. The basis for new AI applications is to be created with a new infrastructure for real-time data transmission.

6. AI in Germany should be flanked by a high level of IT security.

7. We want to find a European answer to data-based business models and new ways of data-based value creation that corresponds to our economic, value and social structure.

8. Even in the world of work, AI should always be for the benefit of all employed people.

9. The potential of AI should make life and work areas safer, more efficient and more sustainable.

10. AI is intended to promote social participation, freedom of action and self-determination of citizens.

11. The potential of AI should be used for sustainable development and thus contribute to the sustainability goals of the 2030 Agenda.

12. Framework conditions are set for AI applications that create and maintain diversity and guarantee the space available for the development of cultural and media freedoms.

The most notable aspects of AI strategy are applicability and benefits of AI for society and people! The Summary "AI Made in Germany", highlights AI contribution to the society. The Federal Government will take on the task of providing a policy response to the rapid advances in the field of AI and will make comprehensive use of the innovations triggered by the technology for the benefit of society at large. We want to safeguard Germany's outstanding position as a research center, to build up the competitiveness of German industry, and to promote the many ways to use AI in all parts of society in order to achieve tangible progress in society in the interest of its citizens. We will focus on the benefits for people and the environment and continue the intensive dialogue with all sections of society (please see Figure 7). Germany is already extremely well positioned in many areas of AI. This strategy builds on existing strengths and transfers them to areas where no or little use has been made of the potential.

Figure 7: AI Made in Germany Word Cloud [12]

---

[330] AI Strategies - https://www.ki-strategie-deutschland.de/home.html

UK AI Strategy

UK AI Strategy [13] reinforces the 5 foundations of the Industrial Strategy:[331]

- Ideas
- People
- Infrastructure
- Business environment
- Places

The aim of Industrial strategy is creating an economy that boosts productivity and earning power throughout the UK. The Industrial Strategy is built on 5 foundations [13]:

- Ideas – the world's most innovative economy
- People – good jobs and greater earning power for all
- Infrastructure – a major upgrade to the UK's infrastructure
- Business environment – the best place to start and grow a business
- Places – prosperous communities across the UK

UK is eager on advancing and setting leadership for industries of the future, the grand challenge:[332] [14]

- AI and Data Economy – We will put the UK at the forefront of the artificial intelligence and data revolution
- Future of Mobility – We will become a world leader in the way people, goods and services move
- Clean Growth – We will maximize the advantages for UK industry from the global shift to clean growth
- Ageing Society – We will harness the power of innovation to help meet the needs of an ageing society

Key Policies: [13] [14]

Ideas

- Raise total research and development (R&D) investment to 2.4% of GDP by 2027
- Increase the rate of R&D tax credit to 12%
- Invest £725 million in new Industrial Strategy Challenge Fund programs to capture the value of innovation.

People

- Establish a technical education system that rivals the best in the world to stand alongside our world-class higher education system
- Invest an additional £406 million in mathematics, digital and technical education, helping to address the shortage of science, technology, engineering and mathematics (STEM) skills
- Create a new National Retraining Scheme that supports people to re-skill, beginning with a £64 million investment for digital and construction training.

Infrastructure

- Increase the National Productivity Investment Fund to £31 billion, supporting investments in transport, housing and digital infrastructure
- Support electric vehicles through £400 million charging infrastructure investment and an extra £100 million to extend the plug-in car grant

---

[331] UK AI Sector – Policy Paper - https://www.gov.uk/government/publications/artificial-intelligence-sector-deal/ai-sector-deal#:~:text=A%20key%20ambition%20of%20the,of%20AI%20across%20the%20country.

[332] The Grand Challenge - https://www.gov.uk/government/publications/industrial-strategy-the-grand-challenges/industrial-strategy-the-grand-challenges

- Boost our digital infrastructure with over £1 billion of public investment, including £176 million for 5G and £200 million for local areas to encourage roll out of full-fiber networks.

Business environment

- Launch and roll out Sector Deals – partnerships between government and industry aiming to increase sector productivity. The first sector deals are in life sciences, construction, artificial intelligence and the automotive sector
- Drive over £20 billion of investment in innovative and high potential businesses, including through establishing a new £2.5 billion Investment Fund, incubated in the British Business Bank
- Launch a review of the actions that could be most effective in improving the productivity and growth of small and medium-sized businesses, including how to address what has been called the '**long tail**' of lower productivity firms [13][14].

Places

- Agree Local Industrial Strategies that build on local strengths and deliver on economic opportunities
- Create a new Transforming Cities fund that will provide £1.7 billion for intra-city transport. This will fund projects that drive productivity by improving connections within city regions
- Provide £42 million to pilot a Teacher Development Premium. This will test the impact of a £1,000 budget for high-quality professional development for teachers working in areas that have fallen behind.

A key ambition of the industrial strategy is for the UK to be the world's most innovative economy—this Sector Deal aims to attract and retain both domestic and global AI talent; deliver major upgrades to our digital and data infrastructure; ensure that the UK is the best place to start and grow an AI business; and contribute to communities' prosperity by spreading the benefits of AI across the country [14].

UK AI Strategy is very clear: they would like to be forefront of contributions to AI and would like to pursue to continue the legacy of Turing.

Finally, it has been estimated that AI could add an additional USD $814 billion (£630bn) to the UK economy by 2035, increasing the annual growth rate of GVA from 2.5 % to 3.9%. The Word Cloud view UK AI Strategy also emphasizes the value add of AI to the UK society[333] (please see Figure 8) [15].

Figure 8: UK AI Strategy - Executive Summary Word Cloud [14] [15]

---

[333] Growing the AI Industry in UK - https://assets.publishing.service.gov.uk/government/uploads/system/uploads/attachment_data/file/652097/Growing_the_artificial_intelligence_industry_in_the_UK.pdf

National Institution for Transforming India Aayog National strategy for AI[334]

Technology disruptions like AI are once in a generation phenomenon and large-scale deployments of AI should strike balance between commercial aspirations to social good.[335] AI, as broadly described by National Institution for Transforming India (NITI) Aayog,[336] refers to the ability of machines to perform cognitive tasks like thinking, perceiving, learning, problem solving and decision making. Initially conceived as a technology that could mimic human intelligence, AI has evolved in ways that far exceeds its original conception. With incredible advances made in data collection, processing and computation power, intelligent systems can now be deployed to take over a variety of tasks, enable connectivity and enhance productivity. As AI's capabilities have dramatically expanded, so have its utility in a growing number of fields. NITI Aayog outlined the following five primary sectors that have a huge positive impact on Indian citizens and world at large [4]:

- Healthcare: increased access and affordability of quality healthcare
- Agriculture: enhanced farmers' income, increased farm productivity and reduction of wastage,
- Education: improved access and quality of education,
- Smart Cities and Infrastructure: efficient and connectivity for the burgeoning urban population, and
- Smart Mobility and Transportation: smarter and safer modes of transportation and better traffic and congestion problems.

The report identifies the following barriers to large-scale adoption of AI: [4]

(a) Lack of broad-based expertise in research and application of AI
(b) Absence of enabling data ecosystems – access to intelligent data
(c) High resource cost and low awareness for adoption of AI
(d) Privacy and security, including a lack of formal regulations around anonymization of data, and
(e) Absence of collaborative approach to adoption and application of AI.

Finally, as per the National Strategy For AI – Discussion Paper,[337] the strategy should strive to leverage AI for economic growth, social development and inclusive growth, and finally as a "Garage" for emerging

Figure 9: Strategic AI Paper India [4]

[334] National Strategy for AI - https://niti.gov.in/writereaddata/files/document_publication/NationalStrategy-for-AI-Discussion-Paper.pdf
[335] National Strategy for AI – India – Discussion Paper - https://niti.gov.in/writereaddata/files/document_publication/NationalStrategy-for-AI-Discussion-Paper.pdf
[336] NITI Aayog - https://niti.gov.in/
[337] AI Discussion Paper - https://niti.gov.in/writereaddata/files/document_publication/NationalStrategy-for-AI-Discussion-Paper.pdf

and developing economies. While AI has the potential to provide large incremental value to a wide range of sectors, adoption till date has been driven primarily from a commercial perspective. Technology disruptions like AI are once-in-a generation phenomenon, and hence large-scale adoption strategies, especially national strategies, need to strike a balance between narrow definitions of financial impact and the greater good. For the Word Cloud view of executive summary please see Figure 9: [4].

French Strategy for Artificial Intelligence—AI for Humanity[338]

The President of the French Republic presented his vision and strategy to make France a leader in artificial intelligence at the Collège de France on 29 March 2018.

Artificial Intelligence often sounds like a promise for the future, but we must not fool ourselves: this revolution is happening here and now. This radical transformation is both an unprecedented opportunity and an immense responsibility.

We must fully seize the opportunities offered by artificial intelligence now, while designing the framework to regulate it. That is the ambition of the President and he is committed to (please see Figure 10): [5]

1. Bet on French talent

2. Pooling our assets

3. Establishing an ethical framework

Since 2013, the U.S., Germany, the UK, Japan and China have rolled out their AI strategies and policies, each with their own priorities, with the U.S. focusing on the impact of AI on economic growth, technology development and national security, the EU on the ethical risks brought by AI in such aspects as security, privacy and human dignity, Japan on building "Society 5.0", and China on industrialization of AI applications in the service of its "Manufacturing Power" strategy. This leads to remarkable differences among the countries in their AI research priorities and application areas.[339]

Figure 10: France AI Executive Summary Word Cloud [5]

The United Nations AI Technology Strategy

The United Nations AI Technology Strategy focuses on the development of data science and analytics for creating sustainable development [16]. Please see Figure 11.

---

[338] French Strategy for Humanity - https://www.aiforhumanity.fr/en/

[339] China AI Development Report - http://www.sppm.tsinghua.edu.cn/eWebEditor/UploadFile/China_AI_development_report_2018.pdf

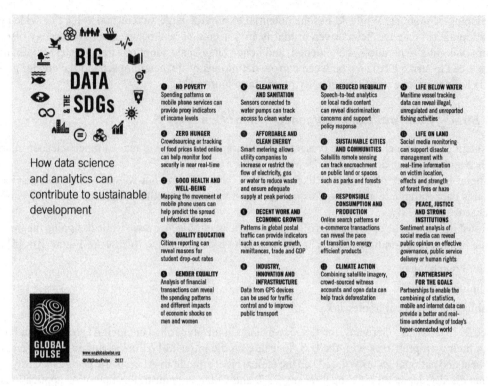

Figure 11: The Big Data & SDGs [16]

Area Focus:[340]

1. No Poverty
2. Zero Hunger
3. Good Health and Well Being
4. Quality Education
5. Gender Equality
6. Clean Water and Sanitation
7. Affordable and Clean Energy
8. Decent Work and Economic Growth
9. Industry Innovation and Research
10. Reduced Inequality
11. Sustainable Cities and Communities
12. Responsible Consumption and Production
13. Climate Action
14. Life Below Water
15. Life on Land
16. Peace, Justice and Strong Institutions
17. Partnerships for the Growth

---

[340] The Big Data & SDGs - https://www.un.org/en/pdfs/Bigdata_SDGs_single_spread_2017.pdf

The United Nations (UN) General Recommendations calls for the following areas [16]:

- Big Data
- The Data Revolution
- Opportunities
- Risks
- The Big Data Development and Humanitarian Actions.

The Role of the UN

One of the key roles of the UN and other international or regional organizations is setting principles and standards to guide collective action around the safe use of big data for development and humanitarian action within a global community and according to common norms. These standards seek to increase the usefulness of data through a much greater degree of openness and transparency, avoid invasion of privacy and abuse of human rights from misuse of data on individuals and groups, and minimize inequality in production, access to and use of data. Achievement of the SDGs in our digital world will require recognition of the need not only to prevent misuse of data, but also to ensure that when data can be used responsibly for the public good, it is.[341]

Recommendations:

1. Fostering and promoting innovation to fill data gaps
2. Mobilizing resources to overcome inequalities between developed and developing countries and between data-poor and data-rich people
3. Leadership and coordination to enable the data revolution to play its full role in the realization of sustainable development.

AI in the Hands of People

Artificial Intelligence holds tremendous promise to benefit nearly all aspects of society, including the economy, healthcare, security, law, transportation, even technology itself.

Increased use of Artificial Intelligence can bring major social and economic benefits to the society. With AI, computers can analyze and learn from information at higher accuracy and speed than humans can. AI offers massive gains in efficiency and performance to most or all industrial sectors, from drug discovery to logistics. AI is a software that can be integrated into existing processes, improving them, scaling them, and reducing their costs, by making or suggesting more accurate decisions through better use of information.

Initially conceived as a technology that could mimic human intelligence, AI has evolved in ways that far exceed its original conception. With incredible advances made in data collection, processing and computation power, intelligent systems can now be deployed to take over a variety of tasks, enable connectivity and enhance productivity. As AI's capabilities have dramatically expanded, so have its utility in a growing number of fields.

While AI has the potential to provide large incremental value to a wide range of sectors, adoption till date has been driven primarily from a commercial perspective. Technology disruptions like AI are once-in-a-generation phenomenon, and hence large-scale adoption strategies, especially national strategies, need to strike a balance between narrow definitions of financial impact and the greater good. The strategy should strive to leverage AI for economic growth, social development and inclusive growth, and finally as a "Garage" for emerging and developing economies.

The AI technology that is currently progressing, is a specialized AI technology for carrying out specialized tasks and is used only to supplement human capabilities. Based on the progression of AI technology, various inferences have become possible from past data, image recognition, language recognition, etc. By using and applying AI technology as a service based on data, the capabilities of human beings are drawn out to the fullest

---

extent, human society has become abundant, including sustainability of society and approaches to social issues such as environmental problems, and economic and industrial benefits are yielded.

Finally, Technologies and Innovations for serving humanity and beyond should be the main moto of Artificial Intelligence and AI should be used in augmentation of human needs and should serve humanity to its fullest capacity.

After reading this chapter, you should be able to design, implement, and deploy the following Machine Learning algorithms into small embedded devices and mobile platforms.

• Decision Trees

• Kalman Filter

• *K*-means clustering

• Sliding Window

• Clustering

• Neural Network for Sound Classification

• Fuzzy Logic and Re-enforcement Learning

References

1.   Kuni Miyake, M. How will AI Change International Politics? January 15, 2019, URL: https://www.japantimes.co.jp/opinion/2019/01/15/commentary/japan-commentary/will-ai-change-international-politics/#.XnjYBYhKj4d, Access Date: September 18, 2019.

2.   James Corbett. When Robots Rule the Markets. November 18, 2017, URL: https://theinternationalforecaster.com/topic/international_forecaster_weekly/when_robots_rule_the_markets , Access Date: November 22, 2019.

3.   Focus Economics. The World's Top 5 Largest Economies in 2023. January 27, 2020, URL: https://www.focus-economics.com/blog/the-largest-economies-in-the-world, Access Date: March 1, 2020.

4.   Anna Roy. National Strategy for Artificial Intelligence. June 2018, URL: https://niti.gov.in/writereaddata/files/document_publication/NationalStrategy-for-AI-Discussion-Paper.pdf, Access Date: December 23, 2019.

5.   Cédric Villani. For a Meaningful Artificial Intelligence. 8 March 2018, URL: https://www.aiforhumanity.fr/pdfs/MissionVillani_Report_ENG-VF.pdf, Access Date: March 16, 2020.

6.   Michael Kratsios. The National Artificial Intelligence Research and Development Strategic Plan: 2019 Update. June 21, 2019, URL: https://www.whitehouse.gov/wp-content/uploads/2019/06/National-AI-Research-and-Development-Strategic-Plan-2019-Update-June-2019.pdf, Access Date: November 22, 2019.

7.   Dr. Prabhat Kumar. Artificial Intelligence: Reshaping Life and Business. BPB Publications; 1st edn (August 17, 2019), ISBN-10: 9388511077.

8.   State Council. Notice of the State Council Issuing the - New Generation of Artificial Intelligence Development Plan. July 8, 2017. URL: https://flia.org/wp-content/uploads/2017/07/A-New-Generation-of-Artificial-Intelligence-Development-Plan-1.pdf, Access Date: September 18, 2019.

9.   China Institute for Science and Technology Policy at Tsinghua University. China AI Development Report 2018. July 2018, URL: http://www.sppm.tsinghua.edu.cn/eWebEditor/UploadFile/China_AI_development_report_2018.pdf, Access Date: December 25, 2019.

10. Artificial Intelligence Technology Council. Artificial Intelligence Technology Strategy. March 31, 2017. URL: http://events.science-japon.org/dlai17/doc/MIC%20-%20France-Japan%2020171025.pdf, Access Date: January 08, 2020.

11. Deloitte Touche Tohmatsu Limited. Industry 4.0. January 27, 2020. URL: https://www2.deloitte.com/us/en/insights/focus/industry-4-0.html, Access Date: March 16, 2020.

12. Germany - The Federal Government. Artificial Intelligence Strategy. November 2018. https://www.ki-strategie-deutschland.de/home.html?file=files/downloads/Nationale_KI-Strategie_engl.pdf, Access Date: November 22, 2019.

13. UK Department for Business. Energy & Industrial Strategy and Department for Digital, Culture, Media & Sport. Policy Paper: AI Sector Deal. 21 May 2019, https://www.gov.uk/government/publications/artificial-intelligence-sector-deal/ai-sector-deal#:~:text=A%20key%20ambition%20of%20the,of%20AI%20across%20the%20country,   Access   Date: September 18, 2019.

14. UK Department for Business, Energy & Industrial Strategy. Policy Paper - The Grand Challenges. 13 September 2019. https://www.gov.uk/government/publications/industrial-strategy-the-grand-challenges/industrial-strategy-the-grand-challenges, Access Date: November 22, 2019.
15. Professor Dame Wendy Hall and Jérôme Pesenti. Growing The Artificial Intelligence Industry in the UK. 2017. URL: https://assets.publishing.service.gov.uk/government/uploads/system/uploads/attachment_data/file/652097/Growing_the_artificial_intelligence_industry_in_the_UK.pdf, Access Date: December 25, 2019.
16. UN Global Pulse. The Big Data & SDGs. 2017. URL: https://www.un.org/en/pdfs/Bigdata_SDGs_single_spread_2017.pdf, Access Date: August 6, 2019.

CHAPTER 9

Future

Democratization of Artificial Intelligence for the Future of Humanity

The agriculture industry has employed the first food harvesting citizen data scientists from the beginning of human civilization. Yet the conundrum is that today's agriculture industry lacks the data that is globally representative to address some of the most important challenges of our time: the global warming, infest of pesticides on humans, food security for 7+ billion population, food vulnerabilities in the face of rapid decrease of natural resources, agroeconomics issues, suicide of farmers and rapid depletion of natural species. One of the chief reasons for the lack of globally well represented data in agriculture is due to unavailability of compute technologies that's affordable, contextual, localized, connected and easy to plug-in and consume by small-scale farmers.

With the emergence of Cloud technologies and with the development of Internet of Things (IoT), the call for action is to express connect small-scale farmers to Cloud and IoT technologies and to globally collect agriculture datasets for solving humanity issues. That's our future!

APPENDIX

Appendix A

Windows AI Platform—AI Platform for Windows Developers

At Microsoft, we are making huge investments in AI and ML across the company. AI capabilities in Office 365 help subscribers with productivity at work, intelligent features in the Photos app for Windows 10 make it easier for people to create videos and search through massive photo collections, and Windows Hello uses AI to recognize your face and get you quickly logged in to your Windows devices.[342] We also use AI to answer your questions when using Bing Search and Cortana, to enable advertisers build deeper connections with customers, and to help security professionals safeguard businesses against modern threats. Microsoft Research continues to drive the state of the art with speech recognition, machine reading and comprehension, computer vision.[343] We've scaled AI to tiny devices, and even built AI that can draw! All this compression work will only make existing machine-learning models 10 to 100 times smaller. To deploy machine learning onto Cortex M0s, the smallest of the ARM processors—they are physically about the size of a red-pepper flake and the models need to be made 1,000 to 10,000 times smaller.[344]

nVidia Jetson TX2

Harness AI at the Edge with the Jetson TX2 Developer Kit[345]

The Jetson TX2 Developer Kit gives you a fast, easy way to develop hardware and software for the Jetson TX2 AI supercomputer on a module. It exposes the hardware capabilities and interfaces of the developer board, comes with design guides and other documentation, and is pre-flashed with a Linux development environment.

It also supports NVIDIA Jetpack—a complete SDK that includes the BSP, libraries for deep learning, computer vision, GPU computing, multimedia processing, and much more.

Detailed specs—https://www.nvidia.com/en-us/autonomous-machines/embedded-systems/jetson-nano/

---

[342] Windows AI Platform - https://blogs.windows.com/windowsdeveloper/2018/03/07/ai-platform-windows-developers/
[343] https://blogs.windows.com/windowsdeveloper/2018/03/07/ai-platform-windows-developers/#qU389aiSTD8WyYRh.97
[344] AI's big leap to tiny devices opens world of possibilities - https://blogs.microsoft.com/ai/ais-big-leap-tiny-devices-opens-world-possibilities/
[345] Jetson TX2 - https://developer.nvidia.com/embedded/jetson-tx2-developer-kit

Google Edge TPU

AI at the Edge[346]

AI is pervasive today, from consumer to enterprise applications. With the explosive growth of connected devices, combined with a demand for privacy/confidentiality, low latency and bandwidth constraints, AI models trained in Cloud increasingly need to be run at the edge. Edge TPU is Google's purpose-built ASIC designed to run AI at the edge. It delivers high performance in a small physical and power footprint, enabling the deployment of high-accuracy AI at the edge.

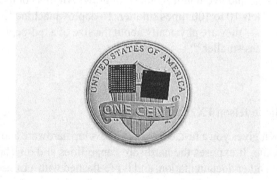

Intel Low Power VPU

Intel's Myriad™X VPU[347] is the third generation and most advanced VPU from Movidius™, an Intel company. Intel's Myriad™ X VPU is the first of it's class to feature the Neural Compute Engine—a dedicated hardware accelerator for deep neural network inferences. The Neural Compute Engine in conjunction with the 16 powerful SHAVE cores and an ultra-high throughput intelligent memory fabric makes Myriad X the industry leader for on-device deep neural networks and computer vision applications. Intel's Myriad™ X VPU has received additional upgrades to imaging and vision engines including additional programmable SHAVE cores, upgraded and expanded vision accelerators, and a new native 4K ISP pipeline with support for up to 8 HD sensors connecting directly to the VPU.

---

[346] Google Edge TPU - https://cloud.google.com/edge-tpu/
[347] https://www.movidius.com/applications

Neural Compute Engine[348]

Hardware Based Acceleration for Deep Neural Networks

Myriad™X features the all-new Neural Compute Engine—a purpose-built hardware accelerator designed to dramatically increase performance of deep neural networks without compromising the low power characteristics of the Myriad VPU product line. Featuring an array of MAC blocks and directly interfacing with the intelligent memory fabric, the Neural Compute Engine is able to rapidly perform the calculations necessary for deep inference without hitting the so-called "data wall" bottleneck encountered by other processor designs. Combining the neural network performance of the 16 proprietary SHAVE cores with the neural compute engine, Myriad X delivers 10X the performance compared to previous generations*.

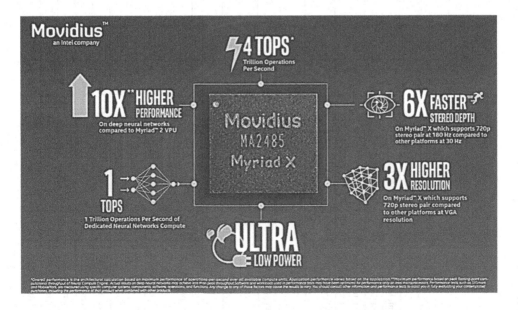

[348] Intel VPUs - https://www.movidius.com/myriadx

Arduino Board Specification[349]

Table 1: Arduino Board specification

Name	Processor	Operating/Input Voltage	CPU Speed	Analog In/Out	Digital IO/PWM	EEPROM [kB]	SRAM [kB]	Flash [kB]	UART
101	Intel® Curie	3.3 V / 7-12V	32MHz	6/0	14/4	-	24	196	-
Gemma	ATtiny85	3.3 V / 4-16 V	8 MHz	1/0	3/2	0.5	0.5	8	0
LilyPad	ATmega168V / ATmega328P	2.7-5.5 V / 2.7-5.5 V	8MHz	6/0	14/6	0.512	1	16	-
LilyPad SimpleSnap	ATmega328P	2.7-5.5 V / 2.7-5.5 V	8 MHz	4/0	9/4	1	2	32	-
LilyPad USB	ATmega32U4	3.3 V / 3.8-5 V	8 MHz	4/0	9/4	1	2.5	32	-
Mega 2560	ATmega2560	5 V / 7-12 V	16 MHz	16/0	54/15	4	8	256	4
Micro	ATmega32U4	5 V / 7-12 V	16 MHz	12/0	20/7	1	2.5	32	1
MKR1000	SAMD21 Cortex-M0+	3.3 V/5V	48MHz	7/1	8/4	-	32	256	1
Pro	ATmega168 / ATmega328P	3.3 V / 3.35-12 V / 5 V / 5-12 V	8 MHz / 16 MHz	6/0	14/6	0.512 / 1	1 / 2	16 / 32	1
Pro Mini	ATmega328P	3.3 V / 3.35-12 V / 5 V / 5-12 V	8 MHz / 16 MHz	6/0	14/6	1	2	32	1
Uno	ATmega328P	5 V / 7-12 V	16 MHz	6/0	14/6	1	2	32	1
Zero	ATSAMD21G18	3.3 V / 7-12 V	48 MHz	6/1	14/10	-	32	256	2
Due	ATSAM3X8E	3.3 V / 7-12 V	84 MHz	12/2	54/12	-	96	512	4
Esplora	ATmega32U4	5 V / 7-12 V	16 MHz	-	-	1	2.5	32	-
Ethernet	ATmega328P	5 V / 7-12 V	16 MHz	6/0	14/4	1	2	32	-
Leonardo	ATmega32U4	5 V / 7-12 V	16 MHz	12/0	20/7	1	2.5	32	1
Mega ADK	ATmega2560	5 V / 7-12 V	16 MHz	16/0	54/15	4	8	256	4
Mini	ATmega328P	5 V / 7-9 V	16 MHz	8/0	14/6	1	2	32	-

[349] Arduino Boards specification - https://www.arduino.cc/en/products/compare

Nano	ATmega168 ATmega328P	5 V / 7-9 V	16 MHz	8/0	14/6	0.512 1	1 2	16 32	1
Yún	ATmega32U4 AR9331 Linux	5 V	16 MHz 400MHz	12/0	20/7	1	2.5 16MB	32 64MB	1
Arduino Robot	ATmega32u4	5 V	16 MHz	6/0	20/6	1 KB (ATmega32u4)/ 512 Kbit (I2C)	2.5 KB (ATmega32u4)	32 KB (ATmega32u4) of which 4 KB used by bootloader	1
MKRZero	SAMD21 Cortex-M0+ 32bit low power ARM MCU	3.3 V	48 MHz	7 (ADC 8/10/12 bit)/1 (DAC 10 bit)	22/12	No	32 KB	256 KB	1

Milk Producing Data Center using AI

Dairy Cows & Peak of Health

Connected Dairy Analytics (CDA) monitor dairy cattle electronically to make sure the cattle are always in the peak of health.

CDA will help to identify dairy cattle-related health issues such as Heat Stress and Bovine Respiratory Disease (BRD). Heat Stress (HS) causes cattle to produce less milk with the same nutritional input, which effectively increases farmers' production costs. The economic toll due to higher temperature heat stress is a $1 billion annual problem. Not only in the United States, but also around the globe heat stress causes an adverse impact on dairy productivity.

Benefits

Connected Dairy will yield huge operational efficiencies, cost savings, and actionable insights to address Dairy cattle-related critical issues. Connected dairy, importantly, is a data enabled insightful tool that facilitates better management of dairy activities. Finally, connected dairy provides forecasting insights that provides window of time opportunity to dairy operational management so that they can better plan to handle any unexpected weather related abnormalities, dairy cattle health and emergencies.

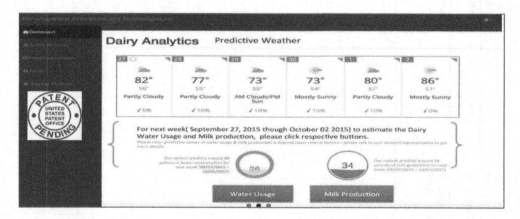

Electronic Monitoring

The opportunity for the dairy industry is to electronically monitor cattle temperature and implement appropriate measures so that the impact of HS can be minimized. The U.S. Department of Agriculture estimates nearly $2.4 billion a year in losses from animal illnesses that lead to death can be prevented by electronically checking on cattle's' vital signs.

Our Approach

The 'Smart Connected Objects", aka, 'the Internet of Things (IoT)', that enables dairies to minimize the economic impact of HS and, at the same, capture the higher Return on Assets (ROA) & Return on Investment (ROI) by improving operational efficiencies.

Happy Cow, more importantly, means happier, more profitable, dairy industry and richer and creamer dairy products.

Reduce medication costs by about 15% per animal and save more sick cattle from death. WSJ: High-Tech Tagging Comes to the Ranch

Contact details:

Hanumayamma Innovations and Technologies Inc.,
628 Crescent Terrace,
Fremont, CA 94536
Phone: (510) 791 – 5759

Index

Printed in the United States
by Baker & Taylor Publisher Services